Staying with the Trouble

EXPERIMENTAL FUTURES:

TECHNOLOGICAL LIVES, SCIENTIFIC ARTS, ANTHROPOLOGICAL VOICES

A series edited by Michael M. J. Fischer and Joseph Dumit

Staying with the Trouble

Making Kin in the Chthulucene

DONNA J. HARAWAY

DUKE UNIVERSITY PRESS DURHAM AND LONDON 2016

Printed in the United States of America
on acid-free paper ∞
Designed by Amy Ruth Buchanan
Typeset in Chaparral Pro by Copperline

Library of Congress Cataloging-in-Publication Data. Names: Haraway, Donna Jeanne, author. Title: Staying with the trouble : making kin in the Chthulucene / Donna J. Haraway. Description: Durham : Duke University Press, 2016. | Series: Experimental futures: technological lives, scientific arts, anthropological voices | Includes bibliographical references and index. | Description based on print version record and CIP data provided by publisher; resource not viewed. Identifiers: LCCN 2016019477 (print) | LCCN 2016018395 (ebook) | ISBN 9780822362142 (hardcover : alk. paper) | ISBN 9780822362241 (pbk. : alk. paper) | ISBN 9780822373780 (e-book) | Subjects: LCSH: Human-animal relationships. | Human-plant relationships. | Human ecology. | Nature—Effect of human beings on. Classification: LCC QL85 (print) | LCC QL85 .H369 2016 (ebook) | DDC 599.9/5—dc23. LC record available at https://lccn.loc.gov/2016019477

Chapter 1 appeared as "Jeux de ficelles avecs des espèces compagnes: Rester avec le trouble," in *Les animaux: deux ou trois choses que nous savons d'eux*, ed. Vinciane Despret and Raphaël Larrère (Paris: Hermann, 2014), 23–59. © Éditions Hermann. Translated by Vinciane Despret.

Chapter 4 is lightly revised from "Anthropocene, Capitalocene, Plantationocene, Chthulucene: Making Kin," which was originally published in *Environmental Humanities,* vol. 6, under Creative Commons license CC BY-NC-ND 3.0. © Donna Haraway.

Chapter 5 is reprinted from *WSQ: Women's Studies Quarterly* 40, nos. 3/4 (spring/summer 2012): 301–16. Copyright © 2012 by the Feminist Press at the City University of New York. Used by permission of The Permissions Company, Inc., on behalf of the publishers, feministpress.org. All Rights reserved.

Chapter 6 is reprinted from *Beyond the Cyborg: Adventures with Donna Haraway*, ed. Margaret Grebowicz and Helen Merrick, 137–46, 173–75. Copyright © Columbia University Press, 2013.

Chapter 7 is reprinted from *Angelaki* 20, no. 2 (2015): 5–14. Reprinted by permission of the publisher. © Taylor and Francis Ltd., tandfonline.com.

Figure 3.1, "Endosymbiosis: Homage to Lynn Margulis." © Shoshanah Dubiner, www.cybermuse.com.

FOR KIN MAKERS OF ALL THE ODDKIN

Contents

List of Illustrations ix
Acknowledgments xi

Introduction 1

1. Playing String Figures with Companion Species 9

2. Tentacular Thinking 30
Anthropocene, Capitalocene, Chthulucene

3. Sympoiesis 58
Symbiogenesis and the Lively Arts of Staying with the Trouble

4. Making Kin 99
Anthropocene, Capitalocene, Plantationocene, Chthulucene

5. Awash in Urine 104
DES and Premarin in Multispecies Response-ability

6. Sowing Worlds 117
A Seed Bag for Terraforming with Earth Others

7. A Curious Practice 126

8. The Camille Stories 134
Children of Compost

Notes 169

Bibliography 229

Index 265

Illustrations

Figure 1.1 *Multispecies Cat's Cradle*. Drawing by Nasser Mufti, 2011 9

Figure 1.2. *Ma'ii Ats'áá' Yílwoí* (Coyotes running opposite ways) 14

Figure 1.3. *Bird Man of the Mission*, mural by Daniel Doherty, 2006 17

Figure 1.4. The PigeonBlog Team of human beings, pigeons, and electronic technologies 23

Figure 1.5. *Capsule*, designed by Matali Crasset, 2003 26

Figure 1.6. Pigeon loft in Batman Park, Melbourne 28

Figure 2.1. *Pimoa cthulhu* 32

Figure 2.2. *Cat's Cradle / String Theory*, Baila Goldenthal, 2008 35

Figure 2.3. Icon for the Anthropocene: Flaming Forests 45

Figure 2.4. Icon for the Capitalocene: Sea Ice Clearing from the Northwest Passage 48

Figure 2.5. *Octopi Wall Street*. Art by Marley Jarvis, Laurel Hiebert, Ḳira Treibergs, 2011 51

Figure 2.6. Icon for the Chthulucene: Potnia Theron with a Gorgon Face 53

Figure 2.7. Day octopus, *Octopus cyanea* 57

Figure 3.1. *Endosymbiosis: Homage to Lynn Margulis*, Shoshanah Dubiner, 2012 59

Figure 3.2. Bee orchid 70

Figure 3.3. Beaded jellyfish made by Vonda N. McIntyre 77

Figure 3.4. Green turtles (*Chelonia mydas*) 80

Figure 3.5. Page from *Tik-Tik the Ringtailed Lemur / Tikitiki Ilay Maky* 81

Figure 3.6. Painting for *Tsambiki Ilamba Fotsy / Bounce the White Sifaka* 84

Figure 3.7. Cover image for *Never Alone (Kisima Ingitchuna)* 86

Figure 3.8. Navajo rug, Two Gray Hills 90

Figure 6.1. An ant of the species *Rhytidoponera metallica* in western Australia, holding a seed of *Acacia neurophylla* by the elaiosome during seed transport 121

Figure 8.1. *Mariposa* mask, Guerrero, Mexico, UBC Museum of Anthropology 135

Figure 8.2. Make Kin Not Babies sticker 139

Figure 8.3. Monarch butterfly caterpillar *Danaus plexippus* on a milkweed pod 145

Figure 8.4. Monarch butterfly resting on fennel in the Pismo Butterfly Grove, 2008 153

Figure 8.5. Mural in La Hormiga, Putumayo, Colombia, depicting landscapes before and after aerial fumigation during the U.S.-Colombia "War on Drugs" 158

Figure 8.6. Kenojuak Ashevak, *Animals of Land and Sea*, 1991 161

Figure 8.7. Monarch butterfly infected with the protozoan parasite *Ophryocystis elektroscirrha*, stuck to the chrysalis, with paper wasp 163

Figure 8.8. Make Kin Not Babies 164

Figure 8.9. Monarch butterfly caterpillar sharing milkweed food plant with oleander aphids (*Aphis nerii*) 167

Acknowledgments

Cooking over many years, the compost pile of colleagues, students, and friends who have made this book possible is promiscuous, layered, and hot. While the holobiome that makes up this book is full of human and nonhuman critters to think and feel with, I especially need to thank Rusten Hogness, Susan Harding, Anna Tsing, Scott Gilbert, Vinciane Despret, Isabelle Stengers, Bruno Latour, Marilyn Strathern, John Law, Jim Clifford, Katie King, Chris Connery, Lisa Rofel, Dai Jinhua, Carla Freccero, Marisol de la Cadena, Jenny Reardon, Beth Stephens, Annie Sprinkle, Helene Moglen, Sheila Namir, Gildas Hamel, Martha Kenney, Karen DeVries, Natasha Myers, Maria Puig de la Bellacasa, Megan Moodie, Margaret Wertheim, Christine Wertheim, Val Hartouni, Michael Hadfield, Margaret McFall-Ngai, Deborah Gordon, Carolyn Hadfield, Thelma Rowell, Sarah Franklin, Marc Bekoff, Rosi Braidotti, Allison Jolly, Adele Clarke, Colin Dayan, Cary Wolfe, Joanne Barker, Kim TallBear, Thom van Dooren, Hugh Raffles, Michael Fischer, Emily Martin, Rayna Rapp, Shelly Errington, Jennifer Gonzalez, Warren Sack, Jason Moore, Faye Ginsberg, Holly Hughes, Thyrza Goodeve, Eduardo Kohn, Beatriz da Costa, Eva Hayward, Harlan Weaver, Sandra Azeredo, Eric Stanley, Eben Kirksey, Lindsay Kelley, Scout Calvert, Kris

Weller, Ron Eglash, Deborah Rose, Karen Barad, Marcia Ochoa, Lisbeth Haas, Eileen Crist, Stefan Helmreich, Carolyn Christov-Bakargiev, Sharon Ghamari, Allison Athens, Bettina Stoetzer, Juno Parreñas, Danny Solomon, Raissa DeSmet, Mark Diekhans, Andrew Matthews, Jake Metcalf, Lisette Olivares, Kami Chisholm, and Lucien Gomoll. Every one of these companions has given me something special for this book; there are so many more I should name.

My home at the University of California at Santa Cruz nurtures vital research groups and centers that are stem cells in the marrow of my bones. Both visitors and UCSC people of the Center for Cultural Studies, the Science and Justice Research Network, the Center for Emerging Worlds, the Research Cluster on Crisis in the Cultures of Capitalism, the Institute of Arts and Sciences, and the History of Consciousness Department shape *Staying with the Trouble* profoundly.

Many of the chapters began as lectures and workshops, and the people who participated infuse my thinking in obvious and subtle ways. I especially want to thank Kavita Philip, Gabriele Schwab, the Critical Theory Institute at UC Irvine, and Jennifer Crewe of Columbia University Press for the opportunity to deliver the Wellek Lectures in 2011.

Over four years, I participated in a writing workshop on Worlding, in which both the writing and the generous critical comments on my own scribblings by Susan Harding, Anna Tsing, Katie Stewart, Lesley Stern, Allen Shelton, Stephen Muecke, and Lauren Berlant remolded the figures, voice, stories, and textures of this book.

Vinciane Despret invited me to Cerisy in Normandy in 2010 to take part in a week-long colloquium asking how we know with other animals. When meals were announced, the staff called our gaggle at the chateau "les animaux" to distinguish us from the more strictly humanistic scholars that summer, and we felt proud. Isabelle Stengers invited me back to Cerisy in the summer of 2013 for her weeklong colloquium called "Gestes spéculatifs," an extraordinary affair marked for me by afternoons in the speculative narration workshop. People I worked and played with at Cerisy inhabit every chapter of *Staying with the Trouble*. I can't name everybody, but want especially to thank Jocelyn Porcher, Benedikte Zitouni, Fabrizio Terranova, Raphaël Larrère, Didier Debaise, Lucienne Strivay, Émelie Hache, and Marcelle Stroobants.

Growing partly out of discussions at Cerisy, the Thousand Names of Gaia / Os Mil Nomes de Gaia in Rio de Janeiro in 2014 refocused my thinking about the geographies, temporalities, and human and nonhuman peoples of our epoch. Thanks especially to Eduardo Viveiros de Castro, Déborah Danowski, and Juliana Fausto.

Marisol de la Cadena invited me to participate twice in her amazing Indigenous Cosmopolitics Sawyer Seminars at UC Davis in 2012. I am grateful for the chance to make string figures with her and her colleagues and students, and with Marilyn Strathern and Isabelle Stengers. I have special debts to Joe Dumit, Kim Stanley Robinson, James Griesemer, and Kristina Lyons from these events.

Both at UCSC and in Denmark, my work has been shaped by the ferment of AURA (Aarhus University Research on the Anthropocene), organized by Anna Tsing with a core group of researchers in biology and anthropology. My thanks go especially to Nils Bubandt and Peter Funch, along with Elaine Gan, Heather Swanson, Rachel Cypher, and Katy Overstreet.

The graduate students and faculty in science studies at UC San Diego entered my book at a special time in 2013, and I want particularly to thank Monica Hoffman and Val Hartouni.

Multispecies studies have surged in many forms around the world, and I owe a special debt to the people of British, Australian, New Zealand, South African, and U.S. animal studies and environmental humanities. Perhaps the fact that all of us inherit the trouble of colonialism and imperialism in densely related, mostly white, Anglophone webs makes us need each other even more as we learn to rethink and refeel with situated earth critters and their people. Invited twice to address the gatherings of the British Animal Studies Network, including the gang assembling for Cosmopolitical Animals, I want to thank especially Erica Fudge, Donna Landry, Garry Marvin, Kaori Nagai, John Lock, and Lynda Birke. Annie Potts, Thom van Dooren, Deborah Bird Rose, Lesley Green, Anthony Collins, and others make me remember that thinking about these matters from the "global South" can help undo some of the arrogance of the "global North." And then I remember too that this problematic "North" is the "South" for the decolonial struggles of humans and nonhumans of the indigenous circumpolar North, a perspective I owe especially to Susan Harding.

SF people are crucial to this book, both as writers and as colleagues, especially Ursula K. Le Guin, Kim Stanley Robinson, Octavia Butler, Vonda McIntyre, Gweneth Jones, Julie Czerneda, Sheryl Vint, Marleen Barr, Sha La Bare, Istvan Csicsery-Ronay, Helen Merrick, Margaret Grebowicz, and, always, Samuel R. Delany.

Colleagues in Sweden, Norway, and the Netherlands contributed richly to this book with their generous responses to my lectures and seminars, as well as by their own research. Thanks especially to Rosi Braidotti, Piet van de Kar, Iris van der Tuin, Tora Holmberg, Cecelia Åsberg, Ulrike Dahl, Marianne Lien, Britta Brena, Kristin Asdal, and Ingunn Moser.

The think tank on Methodologies and Ecologies on Research-Creation in 2014 at the University of Alberta in Edmonton helped me rethink a chapter at a critical time. I am in debt to Natalie Loveless and her extraordinary colleagues and students. I also want to thank the people at the Institute for Humanities Research at Arizona State University in 2013, as well as Laura Hobgood-Oster and her colleagues at the 2011 meetings of the American Academy of Religion for their innovative thinking about humans and other animals.

The Children of Compost in this book owe a great deal to the gathering of the American Association for Literature and the Environment in June 2015, with the theme Notes from Underground: The Depths of Environmental Arts, Culture and Justice. Thanks especially to Anna Tsing, my partner in tunneling, and Cate Sandilands, Giovanna Di Chiro, T. V. Reed, Noël Sturgeon, and Sandra Koelle.

I heartily thank the smart, skilled, and generous people of Duke University Press, especially Ken Wissoker and Elizabeth Ault. Their warmth as well as their intelligence sustained me in making this book. Saving me from some real bloopers, the astute blind reviewers made me less blinkered. Without the extensive and mostly invisible work of such reviewers, scholarship would come undone.

Publication Histories

Chapter 1, "Playing String Figures with Companion Species," is lightly revised from "Jeux de ficelles avecs des espèces compagnes: Rester avec le trouble," in *Les animaux: Deux ou trois choses que nous savons d'eux*, edited by Vinciane Despret and Raphaël Larrère (Paris: Hermann, 2014), 23–59, translated by Vinciane Despret. Chapter 2, "Tentacular Thinking: Anthropocene, Capitalocene, Chthulucene," is significantly revised from "Staying with the Trouble: Sympoièse, figures de ficelle, embrouilles multispécifiques," in *Gestes spéculatifs*, edited and translated by Isabelle Stengers (Paris: Les presses du réel, 2015). A greatly abbreviated version of chapter 3, "Sympoiesis: Symbiogenesis and the Lively Arts of Staying with the Trouble," will appear in *Arts of Living on a Damaged Planet:*

Stories from the Anthropocene, edited by Anna Lowenhaupt Tsing, Nils Bubandt, Elaine Gan, and Heather Swanson, forthcoming from University of Minnesota Press. Chapter 4, "Making Kin: Anthropocene, Capitalocene, Plantationocene, Chthulucene," is lightly revised from *Environmental Humanities* 6 (2015). Chapter 5, "Awash in Urine: DES and Premarin® in Multispecies Response-ability," is lightly revised from *WSQ: Women's Studies Quarterly* 40, nos. 3/4 (spring/summer 2012): 301–16. Copyright © 2012 by the Feminist Press at the City University of New York. Used by permission of The Permissions Company, Inc., on behalf of the publishers, feministpress.org, all rights reserved. Chapter 6, "Sowing Worlds: A Seed Bag for Terraforming with Earth Others," is lightly revised from *Beyond the Cyborg: Adventures with Donna Haraway*, edited by Margaret Grebowicz and Helen Merrick, 137–46, 173–75, copyright © Columbia University Press, 2013. Chapter 7, "A Curious Practice," is lightly revised from *Angelaki* 20, no. 2 (2015): 5–14, reprinted by permission of the publisher (Taylor and Francis Ltd., tandfonline.com). Chapter 8, "The Camille Stories," is published for the first time in this volume.

Introduction

Trouble is an interesting word. It derives from a thirteenth-century French verb meaning "to stir up," "to make cloudy," "to disturb." We—all of us on Terra—live in disturbing times, mixed-up times, troubling and turbid times. The task is to become capable, with each other in all of our bumptious kinds, of response. Mixed-up times are overflowing with both pain and joy—with vastly unjust patterns of pain and joy, with unnecessary killing of ongoingness but also with necessary resurgence. The task is to make kin in lines of inventive connection as a practice of learning to live and die well with each other in a thick present. Our task is to make trouble, to stir up potent response to devastating events, as well as to settle troubled waters and rebuild quiet places. In urgent times, many of us are tempted to address trouble in terms of making an imagined future safe, of stopping something from happening that looms in the future, of clearing away the present and the past in order to make futures for coming generations. Staying with the trouble does not require such a relationship to times called the future. In fact, staying with the trouble requires learning to be truly present, not as a vanishing pivot between awful or edenic pasts and apocalyptic or salvific futures, but as mortal critters entwined in myriad unfinished configurations of places, times, matters, meanings.[1]

Chthulucene is a simple word.[2] It is a compound of two Greek roots (*khthôn* and *kainos*) that together name a kind of timeplace for learning to stay with the trouble of living and dying in response-ability on a damaged earth. *Kainos* means now, a time of beginnings, a time for ongoing, for freshness. Nothing in *kainos* must mean conventional pasts, presents, or futures. There is nothing in times of beginnings that insists on wiping out what has come before, or, indeed, wiping out what comes after. *Kainos* can be full of inheritances, of remembering, and full of comings, of nurturing what might still be. I hear *kainos* in the sense of thick, ongoing presence, with hyphae infusing all sorts of temporalities and materialities.

Chthonic ones are beings of the earth, both ancient and up-to-the-minute. I imagine chthonic ones as replete with tentacles, feelers, digits, cords, whiptails, spider legs, and very unruly hair. Chthonic ones romp in multicritter humus but have no truck with sky-gazing Homo. Chthonic ones are monsters in the best sense; they demonstrate and perform the material meaningfulness of earth processes and critters. They also demonstrate and perform consequences. Chthonic ones are not safe; they have no truck with ideologues; they belong to no one; they writhe and luxuriate in manifold forms and manifold names in all the airs, waters, and places of earth. They make and unmake; they are made and unmade. They are who are. No wonder the world's great monotheisms in both religious and secular guises have tried again and again to exterminate the chthonic ones. The scandals of times called the Anthropocene and the Capitalocene are the latest and most dangerous of these exterminating forces. Living-with and dying-with each other potently in the Chthulucene can be a fierce reply to the dictates of both Anthropos and Capital.

Kin is a wild category that all sorts of people do their best to domesticate. Making kin as oddkin rather than, or at least in addition to, godkin and genealogical and biogenetic family troubles important matters, like to whom one is actually responsible. Who lives and who dies, and how, in this kinship rather than that one? What shape is this kinship, where and whom do its lines connect and disconnect, and so what? What must be cut and what must be tied if multispecies flourishing on earth, including human and other-than-human beings in kinship, are to have a chance?

An ubiquitous figure in this book is SF: science fiction, speculative fabulation, string figures, speculative feminism, science fact, so far. This reiterated list whirls and loops throughout the coming pages, in words and in visual pictures, braiding me and my readers into beings and patterns at stake. Science fact and speculative fabulation need each other, and both need speculative feminism. I think of SF and string figures in a triple sense of figuring. First, promiscuously plucking out fibers in clotted and dense events and practices, I try to follow the threads where they lead in order to track them and find their tangles and patterns crucial for staying with the trouble in real and particular places and times. In that sense, SF is a method of tracing, of following a thread in the dark, in a dangerous true tale of adventure, where who lives and who dies and how might become clearer for the cultivating of multispecies justice. Second, the string figure is not the tracking, but rather the actual thing, the pattern and assembly that solicits response, the thing that is not oneself but with which one must go on. Third, string figuring is passing on and receiving, making and unmaking, picking up threads and dropping them. SF is practice and process; it is becoming-with each other in surprising relays; it is a figure for ongoingness in the Chthulucene.

The book and the idea of "staying with the trouble" are especially impatient with two responses that I hear all too frequently to the horrors of the Anthropocene and the Capitalocene. The first is easy to describe and, I think, dismiss, namely, a comic faith in technofixes, whether secular or religious: technology will somehow come to the rescue of its naughty but very clever children, or what amounts to the same thing, God will come to the rescue of his disobedient but ever hopeful children. In the face of such touching silliness about technofixes (or techno-apocalypses), sometimes it is hard to remember that it remains important to embrace situated technical projects and their people. They are not the enemy; they can do many important things for staying with the trouble and for making generative oddkin.

The second response, harder to dismiss, is probably even more destructive: namely, a position that the game is over, it's too late, there's no sense trying to make anything any better, or at least no sense having any active trust in each other in working and playing for a resurgent world. Some scientists I know express this kind of bitter cynicism, even as they actually work very hard to make a positive difference for both people and other critters. Some people who describe themselves as critical cultural theorists or political progressives express these ideas too. I think the odd coupling of actually working and playing for multispecies flourishing with tenacious energy and skill, while expressing an explicit "game over" attitude that can and does discourage others, including students, is facilitated by various kinds of futurisms. One kind seems to imagine that only if things work do they matter—or, worse, only if what I and my fellow experts do works to fix things does anything matter. More generously, sometimes scientists and others who think, read, study, agitate, and care know too much, and it is too heavy. Or, at least we think we know enough to reach the conclusion that life on earth that includes human people in any tolerable way really is over, that the apocalypse really is nigh.

That attitude makes a great deal of sense in the midst of the earth's sixth great extinction event and in the midst of engulfing wars, extractions, and immiserations of billions of people and other critters for something called "profit" or "power"—or, for that matter, called "God." A game-over attitude imposes itself in the gale-force winds of feeling, not just knowing, that human numbers are almost certain to reach more than 11 billion people by 2100. This figure represents a 9-billion-person increase over 150 years from 1950 to 2100, with vastly unequal consequences for the poor and the rich—not to mention vastly unequal burdens imposed on the earth by the rich compared to the poor—and even worse consequences for nonhumans almost everywhere. There are many other examples of dire realities; the Great Accelerations of the post–World War II era gouge their marks in earth's rocks, waters, airs, and critters. There is a fine line between acknowledging the extent and seriousness of the troubles and succumbing to abstract futurism and its affects of sublime despair and its politics of sublime indifference.

This book argues and tries to perform that, eschewing futurism, staying with the trouble is both more serious and more lively. Staying with the trouble requires making oddkin; that is, we require each other in unexpected collaborations and combinations, in hot compost piles. We become-with each other or not at all. That kind of material semiotics is always situated, someplace and not noplace, entangled and worldly. Alone, in our separate kinds of expertise and experience, we know both too much and too little, and so we succumb to despair or to hope, and neither is a sensible attitude. Neither despair nor hope is tuned to the senses, to mindful matter, to material semiotics, to mortal earthlings in thick copresence. Neither hope nor despair knows how to teach us to "play string figures with companion species," the title of the first chapter of this book.

Three long chapters open *Staying with the Trouble*. Each chapter tracks stories and figures for making kin in the Chthulucene in order to cut the bonds of the Anthropocene and Capitalocene. Pigeons in all their worldly diversity—from creatures of empire, to working men's racing birds, to spies in war, to scientific research partners, to collaborators in art activisms on three continents, to urban companions and pests—are the guides in chapter 1.

In their homely histories, pigeons lead into a practice of "tentacular thinking," the title of the second chapter. Here, I expand the argument that bounded individualism in its many flavors in science, politics, and philosophy has finally become unavailable to think with, truly no longer thinkable, technically or any other way. Sympoiesis—making-with—is a keyword throughout the chapter, as I explore the gifts for needed thinking offered by theorists and storytellers. My partners in science studies, anthropology, and storytelling—Isabelle Stengers, Bruno Latour, Thom van Dooren, Anna Tsing, Marilyn Strathern, Hannah Arendt, Ursula Le Guin, and others—are my companions throughout tentacular thinking. With their help, I introduce the three timescapes of the book: the Anthropocene, the Capitalocene, and the Chthulucene. Allied with the Pacific day octopus, Medusa, the only mortal Gorgon, figured as the Mistress of the Animals, saves the day and ends the chapter.

"Symbiogenesis and the Lively Arts of Staying with the Trouble," chapter 3, spins out the threads of sympoiesis in ecological evolutionary developmental biology and in art/science activisms committed to four iconic troubled places: coral reef holobiomes, Black Mesa coal country in Navajo and Hopi lands and other fossil fuel extraction zones impacting especially ferociously on indigenous peoples, complex lemur forest habitats in Madagascar, and North American circumpolar lands and seas subject to new and old colonialisms in the grip of rapidly melting ice. This chapter makes string figures with the threads of reciprocating energies of biologies, arts, and activisms for multispecies resurgence. Navajo-Churro sheep, orchids, extinct bees, lemurs, jellyfish, coral polyps, seals, and microbes play leading roles with their artists, biologists, and activists throughout the chapter. Here and throughout the book, the sustaining creativity of people who care and act animates the action. Not surprisingly, contemporary indigenous people and peoples, in conflict and collaboration with many sorts of partners, make a sensible difference. Biologists, beginning with the incomparable Lynn Margulis, infuse the thinking and playing of this chapter.

"Making Kin," chapter 4, is both a reprise of the timescapes of Anthropocene, Capitalocene, and Chthulucene, and a plea to "Make Kin

Not Babies." Antiracist, anticolonial, anticapitalist, proqueer feminists of every color and from every people have long been leaders in the movement for sexual and reproductive freedom and rights, with particular attention to the violence of reproductive and sexual orders for poor and marginalized people. Feminists have been leaders in arguing that sexual and reproductive freedom means being able to bring children, whether one's own or those of others, to robust adulthood in health and safety in intact communities. Feminists have also been historically unique in insisting on the power and right of every woman, young or old, to choose *not* to have a child. Cognizant of how easily such a position repeats the arrogances of imperialism, feminists of my persuasion insist that motherhood is not the telos of women and that a woman's reproductive freedom trumps the demands of patriarchy or any other system. Food, jobs, housing, education, the possibility of travel, community, peace, control of one's body and one's intimacies, health care, usable and woman-friendly contraception, the last word on whether or not a child will be born, joy: these and more are sexual and reproductive rights. Their absence around the world is stunning. For excellent reasons, the feminists I know have resisted the languages and policies of population control because they demonstrably often have the interests of biopolitical states more in view than the well-being of women and their people, old and young. Resulting scandals in population control practices are not hard to find. But, in my experience, feminists, including science studies and anthropological feminists, have not been willing seriously to address the Great Acceleration of human numbers, fearing that to do so would be to slide once again into the muck of racism, classism, nationalism, modernism, and imperialism.

But that fear is not good enough. Avoidance of the urgency of almost incomprehensible increases in human numbers since 1950 can slip into something akin to the way some Christians avoid the urgency of climate change because it touches too closely on the marrow of one's faith. *How* to address the urgency is the question that must burn for staying with the trouble. What is decolonial feminist reproductive freedom in a dangerously troubled multispecies world? It cannot be just a humanist affair, no matter how anti-imperialist, antiracist, anticlassist, and pro-woman. It also cannot be a "futurist" affair, attending mainly to abstract numbers and big data, but not to the differentiated and layered lives and deaths of actual people. Still, a 9 billion increase of human beings over 150 years, to a level of 11 billion by 2100 if we are lucky, is not just a number; and it cannot be explained away by blaming Capitalism or any other word starting with a capital letter. The need is stark to think together anew across differences of historical position and of kinds of knowledge and expertise.

"Awash in Urine," chapter 5, begins with personal and intimate relations, luxuriating in the consequences of following estrogens that connect an aging woman and her elder dog, specifically, me and my companion and research associate Cayenne. Before the threads of the string figure have been tracked far, remembering their cyborg littermates, woman and dog find themselves in histories of veterinary research, Big Pharma, horse farming for estrogen, zoos, DES feminist activism, interrelated animal rights and women's health actions, and much more. Intensely inhabiting specific bodies and places as the means to cultivate the capacity to respond to worldly urgencies with each other is the core theme.

Ursula K. Le Guin, Octavia Butler, and ants and acacia seeds populate chapter 6, "Sowing Worlds." The task is to tell an SF adventure story with acacias and their associates as the protagonists. It turns out that Le Guin's carrier bag theory of narrative comes to the rescue, along with biologist Deborah Gordon's theories about ant interactions and colony behavior, to elaborate the possibilities of ecological evolutionary developmental biology and nonhierarchical systems theories for shaping the best stories. Science fiction and science fact cohabit happily in this tale. With Le Guin as their scribe, the prose of acacia seeds and the lyrics of lichens give way to the mute poetics of rocks in the final passages.

"A Curious Practice," chapter 7, draws close to the philosopher, psychologist, animal-human student, and cultural theorist Vinciane Despret because of her incomparable ability to think-with other beings, human or not. Despret's work on attunement and on critters rendering each other capable of unexpected feats in actual encounters is necessary to staying with the trouble. She attends not to what critters are supposed to be able to do, by nature or education, but to what beings evoke from and with each other that was truly not there before, in nature or culture. Her kind of thinking enlarges the capacities of all the players; that is her worlding practice. The urgencies of the Anthropocene, Capitalocene, and Chthulucene demand that kind of thinking beyond inherited categories and capacities, in homely and concrete ways, like the sorts of things Arabian babblers and their scientists get up to in the Negev desert. Despret teaches how to be curious, as well as how to mourn by bringing the dead into active presence; and I needed her touch before writing the concluding stories of *Staying with the Trouble*. Her curious practice prepared me to write about the Communities of Compost and the tasks of speakers for the dead, as they work for earthly multispecies recuperation and resurgence.

"The Camille Stories: Children of Compost" closes this book. This invitation to a collective speculative fabulation follows five generations of a symbiogenetic join of a human child and monarch butterflies along the many lines and nodes of these insects' migrations between Mexico and the United States and Canada. These lines trace socialities and materialities crucial to living and dying with critters on the edge of disappearance so that they might go on. Committed to nurturing capacities to respond, cultivating ways to render each other capable, the Communities of Compost appeared all over the world in the early twenty-first century on ruined lands and waters. These communities committed to help radically reduce human numbers over a few hundred years while developing practices of multispecies environmental justice of myriad kinds. Every new child had at least three human parents; and the pregnant parent exercised reproductive freedom in the choice of an animal symbiont for the child, a choice that ramified across the generations of all the species. The relations of symbiogenetic people and unjoined humans brought many surprises, some of them deadly, but perhaps the deepest surprises emerged from the relations of the living and the dead, in symanimagenic complexity, across the holobiomes of earth.

Lots of trouble, lots of kin to be going on with.

CHAPTER 1

Playing String Figures with Companion Species

In honor of G. Evelyn Hutchinson (1903–91)
and Beatriz da Costa (1974–2012).
Hutchinson, my PhD adviser, wrote a biographical
memoir called *The Kindly Fruits of the Earth*, a title that
enfolds all the "reliable voyageurs" of this chapter.

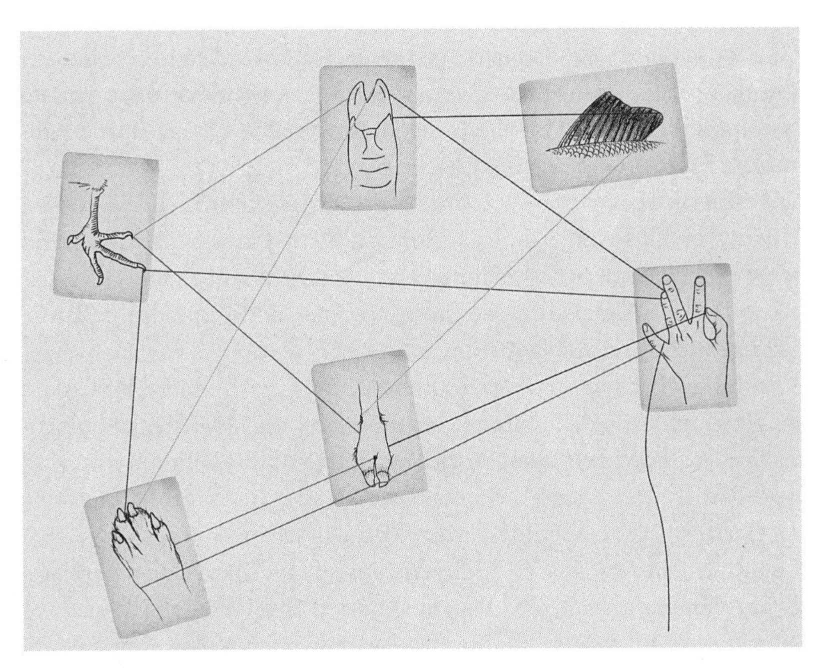

1.1. *Multispecies Cat's Cradle*. Drawing by Nasser Mufti, 2011.

Multispecies Storytelling and the Practices of Companions

String figures are like stories; they propose and enact patterns for participants to inhabit, somehow, on a vulnerable and wounded earth.[1] My multispecies storytelling is about recuperation in complex histories that are as full of dying as living, as full of endings, even genocides, as beginnings. In the face of unrelenting historically specific surplus suffering in companion species knottings, I am not interested in reconciliation or restoration, but I am deeply committed to the more modest possibilities of partial recuperation and getting on together. Call that staying with the trouble. And so I look for real stories that are also speculative fabulations and speculative realisms. These are stories in which multispecies players, who are enmeshed in partial and flawed translations across difference, redo ways of living and dying attuned to still possible finite flourishing, still possible recuperation.

SF is a sign for science fiction, speculative feminism, science fantasy, speculative fabulation, science fact, and also, string figures. Playing games of string figures is about giving and receiving patterns, dropping threads and failing but sometimes finding something that works, something consequential and maybe even beautiful, that wasn't there before, of relaying connections that matter, of telling stories in hand upon hand, digit upon digit, attachment site upon attachment site, to craft conditions for finite flourishing on terra, on earth. String figures require holding still in order to receive and pass on. String figures can be played by many, on all sorts of limbs, as long as the rhythm of accepting and giving is sustained. Scholarship and politics are like that too—passing on in twists and skeins that require passion and action, holding still and moving, anchoring and launching.

Racing pigeons in Southern California, along with their diverse people, geographies, other critters, technologies, and knowledges, shape practices of living and dying in rich worldings that I think of as string figure games. This chapter, enabled by diverse actual pigeons and their rich tracings, is the opening pattern of a cluster of knots. The critters of all my stories inhabit an *n*-dimensional niche space called Terrapolis. My fabulated multiple integral equation for Terrapolis is at once a story, a speculative fabulation, and a string figure for multispecies worlding.

$$\int_{\alpha}^{\Omega} \text{Terra}[x]_n = \iiiint \ldots \iint \text{Terra}(x_1, x_2, x_3, x_4, \ldots, x_n, t)\, dx_1\, dx_2\, dx_3\, dx_4 \ldots dx_n dt = \text{Terrapolis}$$

x_1 = stuff/physis, x_2 = capacity, x_3 = sociality, x_4 = materiality, x_n = dimensions-yet-to-come

α (alpha) = EcologicalEvolutionaryDevelopmental Biology's multispecies epigenesis

Ω (omega) = recuperating terra's pluriverse

t = worlding time, not container time, entangled times of past/present/yet to come

Terrapolis is a fictional integral equation, a speculative fabulation.

Terrapolis is *n*-dimensional niche space for multispecies becoming-with.

Terrapolis is open, worldly, indeterminate, and polytemporal.

Terrapolis is a chimera of materials, languages, histories.

Terrapolis is for companion species, *cum panis*, with bread, at table together—not "posthuman" but "com-post."

Terrapolis is in place; Terrapolis makes space for unexpected companions.

Terrapolis is an equation for guman, for humus, for soil, for ongoing risky infection, for epidemics of promising trouble, for permaculture.

Terrapolis is the SF game of response-ability.[2]

Companion species are engaged in the old art of terraforming; they are the players in the SF equation that describes Terrapolis. Finished once and for all with Kantian globalizing cosmopolitics and grumpy human-exceptionalist Heideggerian worlding, *Terrapolis* is a mongrel word composted with a mycorrhiza of Greek and Latin rootlets and their symbionts. Never poor in world, Terrapolis exists in the SF web of always-too-much connection, where response-ability must be cobbled together, not in the existentialist and bond-less, lonely, Man-making gap theorized by Heidegger and his followers. Terrapolis is rich in world, inoculated against posthumanism but rich in com-post, inoculated against human exceptionalism but rich in humus, ripe for multispecies storytelling. This Terrapolis is not the home world for the human as *Homo*, that ever parabolic, re- and de-tumescing, phallic self-image of the same; but for the human that is transmogrified in etymological Indo-European sleight of tongue into guman, that worker of and in the soil.[3] My SF critters are beings of the mud more than the sky, but the stars too shine in

Terrapolis. In Terrapolis, shed of masculinist universals and their politics of inclusion, guman are full of indeterminate genders and genres, full of kinds-in-the-making, full of significant otherness. My scholar-friends in linguistics and ancient civilizations tell me that this guman is adama/adam, composted from all available genders and genres and competent to make a home world for staying with the trouble. This Terrapolis has kin-making, string figure, SF relations with Isabelle Stengers's kind of fleshy cosmopolitics and SF writers' practices of worlding.

The British social anthropologist Marilyn Strathern, who wrote *The Gender of the Gift* based on her ethnographic work in highland Papua New Guinea (Mt. Hagen), taught me that "it matters what ideas we use to think other ideas (with)."[4] Strathern is an ethnographer of thinking practices. She embodies for me the arts of feminist speculative fabulation in the scholarly mode. It matters what matters we use to think other matters with; it matters what stories we tell to tell other stories with; it matters what knots knot knots, what thoughts think thoughts, what descriptions describe descriptions, what ties tie ties. It matters what stories make worlds, what worlds make stories. Strathern wrote about accepting the risk of relentless contingency; she thinks about anthropology as the knowledge practice that studies relations with relations, that puts relations at risk with other relations, from unexpected other worlds. In 1933, Alfred North Whitehead, the American mathematician and process philosopher who infuses my sense of worlding, wrote *The Adventures of Ideas*.[5] SF is precisely full of such adventures. Isabelle Stengers, a chemist, scholar of Whitehead and Gilles Deleuze, radical thinker about materiality in sciences, and an unruly feminist philosopher, gives me "speculative thinking" in abundance. With Isabelle Stengers we cannot denounce the world in the name of an ideal world. In the spirit of feminist communitarian anarchism and the idiom of Whitehead's philosophy, she maintains that decisions must take place somehow in the presence of those who will bear their consequences. That is what she means by cosmopolitics.[6]

In relay and return, SF morphs in my writing and research into speculative fabulation and string figures. Relays, string figures, passing patterns back and forth, giving and receiving, patterning, holding the unasked-for pattern in one's hands, response-ability; that is core to what I mean by staying with the trouble in serious multispecies worlds. Becoming-with, not becoming, is the name of the game; becoming-with is how partners are, in Vinciane Despret's terms, rendered capable.[7] Ontologically heterogeneous partners become who and what they are in relational material-semiotic worlding. Natures, cultures, subjects, and objects do not preexist their intertwined worldings.

Companion species are relentlessly becoming-with. The category companion species helps me refuse human exceptionalism without invoking posthumanism. Companion species play string figure games where who is/are to be in/of the world is constituted in intra-and interaction.[8] The partners do not precede the knotting; species of all kinds are consequent upon worldly subject- and object-shaping entanglements. In human-animal worlds, companion species are ordinary beings-in-encounter in the house, lab, field, zoo, park, truck, office, prison, ranch, arena, village, human hospital, forest, slaughterhouse, estuary, vet clinic, lake, stadium, barn, wildlife preserve, farm, ocean canyon, city streets, factory, and more.

Although they are among humanity's oldest games, string figures are not everywhere the same game. Like all offspring of colonizing and imperial histories, I—we—have to relearn how to conjugate worlds with partial connections and not universals and particulars. In the late nineteenth and early twentieth centuries, European and Euro-American ethnologists collected string figure games from all over the world; these discipline-making travelers were surprised that when they showed the string figure games they had learned as children at home, their hosts already knew such games and often in greater variety. String figure games came late to Europe, probably from Asian trade routes. All of the epistemological desires and fables of this period of the history of comparative anthropology were ignited by the similarities and differences, with their undecidably independent inventions or cultural diffusions, tied together by the threads of hand and brain, making and thinking, in the relays of patterning in "Native" and "Western" string figure games.[9] In comparative tension, the figures were both the same and not the same at all; SF is still a risky game of worlding and storying; it is staying with the trouble.

Figure 1.2 shows the hands of the science writer and natural history radio producer Rusten Hogness[10] learning a Navajo string figure called Ma'ii Ats'áá' Yílwoí (in English "Coyotes Running Opposite Ways"). Coyote is the trickster who constantly scatters the dust of disorder into the orderly star patterns made by the Fire God, setting up the noninnocent world-making performances of disorder and order that shape the lives of terran critters. In the Navajo language, string games are called *na'atł'o'*. Navajo string games will reappear in my multispecies storytelling about

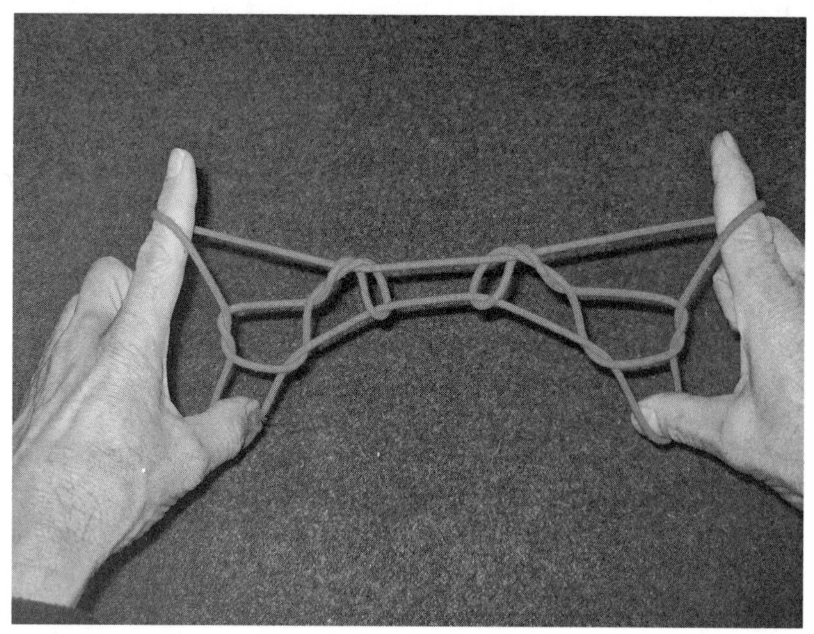

1.2. *Ma'ii Ats'áá' Yílwoí* (Coyotes Running Opposite Ways). Photograph by Donna Haraway.

Navajo-Churro sheep and the women and men who wove and weave lives with and from them, but these games are needed in this chapter too, for thinking with pigeons in Los Angeles and beyond. Cat's cradle and *jeux de ficelle* are not enough; the knots must ramify and double back in many attachment sites in Terrapolis. Navajo string games are one form of "continuous weaving," practices for telling the stories of the constellations, of the emergence of the People, of the Diné.[11]

These string figures are *thinking* as well as *making* practices, pedagogical practices and cosmological performances. Some Navajo thinkers describe string games as one kind of patterning for restoring *hózhó*, a term imperfectly translated into English as "harmony," "beauty," "order," and "right relations of the world," including right relations of humans and nonhumans. Not *in* the world, but *of* the world; that crucial difference in English prepositions is what leads me to weave Navajo string figures, *na'atł'o'*, into the web of SF worlding. The worlds of SF are not containers; they are patternings, risky comakings, speculative fabulations. In SF on Terrapolis, recuperation is in partial connection to *hózhó*. It matters which ideas we think other ideas with; my thinking or making cat's cradle with *na'atł'o'* is not an innocent universal gesture, but a risky proposition in relentless historical relational contingency. And these contingencies include abundant histories of conquest, resistance, recuperation, and resurgence. Telling stories together with historically situated critters is fraught with the risks and joys of composing a more livable cosmopolitics.

Pigeons will be my first guides. Citizens of Terrapolis, pigeons are members of opportunistic social species who can and do live in a myriad of times and places. Highly diverse, they occupy many categories in many languages, sorted in English terms into wild and domestic worlds, but those particular oppositions are not general or universal, even in the so-called West. The varied and proliferating specificities of pigeons are astonishing. Codomesticated with their people, these other-than-human critters nurture the kind of trouble important to me. Pigeons have very old histories of becoming-with human beings. These birds tie their people into knots of class, gender, race, nation, colony, postcolony, and—just maybe—recuperating terra-yet-to-come.

Pigeons are also "creatures of empire"—that is, animals who went with European colonists and conquerors all over the world, including places where other varieties of their kind were already well established, transforming ecologies and politics for everybody in ways that still ramify through multispecies flesh and contested landscapes.[12] Hardly always colonists, pigeons belong to kinds and breeds indigenous to many places, in uncounted configurations of living and dying. Building naturalcultural economies and lives for thousands of years, these critters are also infamous for ecological damage and biosocial upheaval. They are treasured kin and despised pests, subjects of rescue and of invective, bearers of rights and components of the animal-machine, food and neighbor, targets of extermination and of biotechnological breeding and multiplication, companions in work and play and carriers of disease, contested subjects and objects of "modern progress" and "backward tradition." Besides all that, *kinds* of pigeons vary, and vary, and then vary some more, with kinds for nearly every spot on terra.

Becoming-with people for several thousand years, domestic pigeons (*Columba livia domestica*) emerged from birds native to western and southern Europe, North Africa, and western and southern Asia. Rock doves came with Europeans to the Americas, entering North America through Port Royal in Nova Scotia in 1606. Everywhere they have gone, these cosmopolitical pigeons occupy cities with gusto, where they incite human love and hatred in extravagant measure. Called "rats with wings," feral pigeons are subjects of vituperation and extermination, but they also become cherished opportunistic companions who are fed and watched avidly all over the world. Domestic rock doves have worked as spies carrying messages, racing birds, fancy pigeons at fairs and bird markets, food for working families, psychological test subjects, Darwin's interlocutors on the power of artificial selection, and more. Feral pigeons are a favorite food for urban raptors, like peregrine falcons, who, after recovering from near extermination from DDT-thinned eggshells, have taken up life on bridges and ledges of city skyscrapers.

Pigeons are competent agents—in the double sense of both delegates and actors—who render each other and human beings capable of situated social, ecological, behavioral, and cognitive practices. Their worlding is expansive, and the SF games in this chapter do not touch very many, much less all, of the threads tied with and by these birds.[13] My SF game tracks modest, daring, contemporary, risk-filled projects for recuperation, in which people and animals tangle together in innovative ways that might, just barely possibly, render each other capable of a finite flourishing—now and yet to come. The collaborations among differently situated people—and peoples—are as crucial as, and enabled by, those between the humans and animals. Pigeons fly us not into collaborations in general, but into specific crossings from familiar worlds into uncomfortable and unfamiliar ones to weave something that might come unraveled, but might also nurture living and dying in beauty in the *n*-dimensional niche space of Terrapolis. My hope is that these knots propose promising patterns for multispecies response-ability inside ongoing trouble.

California Racing Pigeons and Their People: Collaborating Arts for Worldly Flourishing

Becoming-With; Rendering-Capable

The capabilities of pigeons surprise and impress human beings, who often forget how they themselves are rendered capable by and with both things and living beings. Shaping response-abilities, things and living beings can be inside and outside human and nonhuman bodies, at different scales of time and space. All together the players evoke, trigger, and call forth what—and who—exists. Together, becoming-with and rendering-capable invent *n*-dimensional niche space and its inhabitants.

1.3. *Bird Man of the Mission*, mural of a homeless mentally ill man called Lone Star Swan and some of the urban pigeons who have been his friends and companions on the street in San Francisco's Mission District. Painted by Daniel Doherty in 2006 within the Clarion Alley Mural Project, this work was heavily tagged and finally painted over in 2013. Written for the Street Art SF team by Jane Bregman and posted on October 7, 2014, the story of *The Bird Man of the Mission* is on the website of Street Art SF. Photograph by James Clifford, ©2009. Courtesy of Daniel Doherty and the Clarion Alley Mural Project.

What results is often called nature. Pigeon natures in these coproduced senses matter to my SF story.

Pigeons released in unfamiliar places find their way back to their home lofts from thousands of kilometers away even on cloudy days.[14] Pigeons have the map sense and compass sense that have endeared them to pigeon fanciers who race them for sport, scientists who study them for the behavioral neurobiology of orientation and navigation, spies who wish to send messages across enemy territory, and writers of mystery novels who call on a good pigeon to carry secrets.[15] Almost always men and boys, racing enthusiasts around the world—with perhaps the hottest spots of the sport on the rooftops of cities like Cairo and Istanbul and of immigrant Muslim neighborhoods in European cities like Berlin—selectively breed and elaborately nurture their talented birds to specialize in fast and accurate homing from release points. Ordinary feral pigeons are no slouches at getting home either.

Pigeons will use familiar landmarks to find their way, and they are very good at recognizing and discriminating objects and masses below them during flight. In Project Sea Hunt in the 1970s and '80s, the U.S. Coast Guard worked with pigeons, who were better at spotting men and equipment in open water than human beings.[16] Indeed, pigeons were accurate 93 percent of the time, compared to human accuracy in similar problems of 38 percent. The pigeons perched in an observation bubble on the underside of a helicopter, where they pecked keys to indicate their finds. When they worked with their people instead of in isolation, pigeons were nearly 100 percent accurate. Clearly, the pigeons and Coast Guard personnel had to learn how to communicate with each other, and the pigeons had to learn what their humans were interested in seeing. In nonmimetic ways, people and birds had to invent pedagogical and technological ways to render each other capable in problems novel to all of them. The pigeons never graduated to jobs to save real shipwreck victims, however, because in 1983, after two helicopters crashed and federal money was cut for the research, the project was ended.

Not very many kinds of other-than-human critters have convinced human skeptics that the animals recognize themselves in a mirror—a talent made known to scientists by such actions as picking at paint spots or other marks on one's body that are visible only in a mirror. Pigeons share this capacity with, at least, human children over two years old, rhesus macaques, chimpanzees, magpies, dolphins, and elephants.[17] So-called self-recognition carries great weight in Western-influenced psychology and philosophy, besotted by individualism in theory and method, as these fields have been. Devising tests to show who can and can't do it is something of a competitive epistemological sport. Pigeons passed their first mirror tests in the laboratories of B. F. Skinner in 1981.[18] In 2008, *Science News* reported that Keio University researchers showed that, even with five- to seven-second time delays, pigeons did better at self-recognition tests with both mirrors and live video images of themselves than three-year-old human children.[19] Pigeons pick out different people in photographs very well too, and in Professor Shigeru Watanabe's Laboratory of Comparative Cognitive Neuroscience at Keio University, pigeons could tell the difference between paintings by Monet or Picasso, and even generalize to discriminate unfamiliar paintings from different styles and schools by various painters.[20] It would be a mistake to start building the predictable arguments along the lines of "my bird-brain cognition is better than or equal to your ape-brain cognition." What is happening seems to me to be more interesting than that, and more pregnant with consequences for getting on well with each other, for caring in both emergent similarity and difference. Pigeons, people, and apparatus have teamed up to make each other capable of something new in the world of multispecies relationships.

It is all very well to offer proof of becoming a self-recognizing self in certain kinds of setups, but it is surely as critical to be able to recognize one another and other beings in ways that make sense to the sorts of lives the critters will lead, whether in racing-pigeon lofts or urban squares. Scientists do very interesting research on these topics, but here I want instead to tune in to the online *Racing Pigeon Post* essays by Tanya Berokoff. A teacher in speech communication and lifelong companion with other animals, she is a member of the Palomar Racing Pigeon Club in California with her husband, John Berokoff, who races the birds with mostly other men. Drawing on her social science knowledge and on American popular culture, Tanya Berokoff explicitly uses psychologist John Bowlby's attachment theory and Tina Turner's lyrics for "What's Love Got to Do with It?" to talk about how fanciers assist pigeon parents to raise their youngsters and help them feel competent and safe as they mature into calm, confident, reliable, socially competent, home-seeking racers.[21] She describes pigeon people's obligation to put themselves in the place of the pigeons to understand their ways of knowing and their social practices, and the idiom Berokoff uses for knowledge is love, including but not only instrumental love. The actors are both pigeons and people, in inter- and intraspecies relatings. She describes the details of the gestures and postures of pigeons with each other, the time they spend with each other, and what they do to fill that time. She concludes, "It would seem that our pigeons do quite a good job of exhibiting an agape type of love toward each other . . . Our pigeons are actually doing the work of real love." For her, the "work of real love" is "not about an emotional need to fall in love but to be genuinely loved by another."[22] Meeting that need for their columbine social partners, she says, is what the pigeons seem to do, and that is also what their people owe the pigeons. Berokoff uses Bowlby's attachment theory in detail to describe the needs of young pigeons as they mature, and their partners are both other pigeons and human beings response-able with them. The scene she describes is not all rosy. Pigeon bullying, the taxing labor of racing for birds and people, competition for attention and love—and recipes for cooking some of the pigeons—are all in these posts. My point is not that this discourse or this sport is innocent, but that here is a scene of great relational complexity, a vigorous multispecies SF practice.

PigeonBlog

Recuperation and staying with the trouble are the themes of my SF practice. It is all too possible to address these questions through human brutality toward pigeons, or indeed through pigeon damage to other species or to human-built structures. Instead, I want to turn to the differential burdens of urban air pollution that contribute to different rates of human (and other-than-human, but that is not rated) mortality and illness, often distributed by race and class. Working pigeons will be our companions in projects of environmental justice in California that seek to repair both blighted neighborhoods and social relations. We will stay with the trouble in the tissues of an art activism project called PigeonBlog. This was a project by artist-researcher Beatriz da Costa with her students Cina Hazegh and Kevin Ponto; they tied SF patterns with many human, animal, and cyborg coshapers.

In August 2006, racing pigeons flew as participants in three public social experiments that intimately joined communication technologies with city people and urban sporting birds. The pigeons flew once as part of a Seminar in Experimental Critical Theory at the University of California at Irvine and twice for the festival called Seven Days of Art and Interconnectivity of the Inter-Society for Electronic Arts in

San Jose, California.[23] PigeonBlog required extensive collaboration between "homing pigeons, artists, engineers, and pigeon fanciers engaged in a grass-roots scientific data gathering initiative designed to collect and distribute information about air quality conditions to the general public."[24] Worldwide, racing pigeons are no strangers to alliances with working-class people in relations of competitive masculine sport and profound cross-species affection, and their historical capabilities in surveillance and communication technologies and networks are very old and very important. These pigeons have been workers and subjects in ornithology and psychology research labs for many decades. But sporting homing pigeons had not, before PigeonBlog, been invited to join all of that heritage together with another set of players, namely, art activists. The project sought to join savvy, inexpensive, do-it-yourself electronics with citizen science and interspecies coproduced art and knowledge "in the pursuit of resistant action."[25] The data were intended to provoke, motivate, amplify, inspire, and illustrate, not to substitute for or surpass professional air pollution science and monitoring. These were data produced to generate further imaginative and knowing action in many domains of practice. Da Costa set out not to become an air pollution scientist, but to spark collaboration in something quite different: multispecies art in action for mundane worlds in need of—and capable of—recuperation across consequential differences.

Air pollution is legendary in Southern California, especially Los Angeles County, and it impacts the health of people and other critters especially fiercely near highways, power plants, and refineries. These sites often cluster in and near the neighborhoods of working-class people, people of color, and immigrants—hardly mutually exclusive categories. Official government air pollution monitoring devices in Southern California are placed at fixed points away from high-traffic areas and known pollution sources and at altitudes higher than the zones in which people and lots of other plants and animals breathe. Each monitoring device costs many thousands of dollars and can only measure gases in its immediate vicinity, relying on various models to extrapolate to the volume of the air basin. Properly equipped racing pigeons can gather continuous real-time air pollution data while moving through the air at key heights not accessible to the official instruments, as well as from the ground where they are released for their homing flights. These data could also be streamed in real time to the public via the Internet. What would it take to enlist the cooperation of such birds and their people, and what kind of caring and response-ability could such a collaboration evoke? Who would render whom capable of what?

Da Costa explained the equipment: "The pigeon 'backpack' developed for this project consisted of a combined GPS (latitude, longitude, altitude)/GSM (cell phone tower communication) unit and corresponding antennas, a dual automotive CO/NO_x pollution sensor, a temperature sensor, a Subscriber Identity Module (SIM) card interface, a microcontroller and standard supporting electronic components. Designed in this manner, we essentially ended up developing an open-platform Short Message Service (SMS) enabled cell phone, ready to be rebuilt and repurposed by anyone who is interested in doing so."[26] The researcher-artist-engineers took about three months to design the basic technology, but making the pack small, comfortable, and safe enough for the pigeons took almost a year of building hands-on multispecies trust and knowledge essential to joining the birds, technology, and people. No one wanted an overloaded homing pigeon plucked from the air by an opportunistic falcon that was not a member of the project! Nobody, least of all the men who bred, raised, handled, and loved their racing pigeons, would tolerate anxious and unhappy birds lumbering home under duress. The artist-researchers and the pigeon fanciers had to render each other capable of mutual trust so that they could ask the birds for their confidence and skill. That meant lots of fitting sessions and pigeon balance training in lofts and lots of learning to learn with a generous and knowledgeable pigeon fancier, Bob Matsuyama, who was also a middle school shop and science teacher, and his talented and educated fliers. The pigeons were not SIM cards; they were living coproducers, and the artist-researchers and pigeons had to learn to interact and to train together with the mentoring of the men of the pigeon fancy. All the players rendered each other capable; they "became-with" each other in speculative fabulation. Many trials and test flights later, the multispecies team was ready to trace the air in string figure patterns of electronic tracks.[27]

There were many press reports and reactions to the 2006 performances and to the PigeonBlog website. Da Costa reported that an engineer from Texas contacted her about coauthoring a grant proposal to the U.S. Defense Advanced Research Projects Agency to collaborate in the development of small autonomous aerial surveillance vehicles designed around the aerodynamics of birds. If only that had been a joke! But the long military use of other-than-human animals as weapons and spy systems has only become fancier and more "techy" in the twenty-

1.4. The PigeonBlog team of human beings, pigeons, and electronic technologies. Photograph by Deborah Forster for PigeonBlog. Courtesy of Robert Niediffer, artistic executor for Beatriz da Costa.

first century.[28] In another vein, People for the Ethical Treatment of Animals (PETA) tried to shut PigeonBlog down as abuse of animals. PETA issued a public statement calling on the administration of the University of California at Irvine, where da Costa was a faculty member, to take action. The rationale was fascinating: PigeonBlog was not justified in its use of nonhuman animals because it was not even conducting scientifically grounded experiments, to which PETA might also object, but less so because that at least would have teleological, functional reason (cure disease, map genomes, etc.) on its side. Art was trivial, mere play compared to the serious work of expanding subjects of rights or advancing science. Da Costa took seriously questions about the cosmopolitics and material-semiotics of collaboration for animals in art, politics, or science. Who renders whom capable of what, and at what price, borne by whom? But, she asked, "Is human-animal work as part of political [and art] action less legitimate than the same type of activity when framed under the umbrella of science?"[29] Perhaps it is precisely in the realm of play, outside the dictates of teleology, settled categories, and function, that serious worldliness and recuperation become possible. That is surely the premise of SF.

Well before PETA noticed da Costa's art research, racing-pigeon men's fear of the kind of controversy and attack that some (not all) parts of the animal rights movement bring to many organized working/playing human-animal relationships, including becoming-with pigeons in competitive sport, almost stopped PigeonBlog before it got started.[30] In the early stages of her project, da Costa contacted the American Racing Pigeon Union in an effort to meet pigeon fanciers to see if they and their pigeons would participate. The first contact person was interested but frankly afraid of animal rights people and tactics. He referred da Costa to Bob Matsuyama, who worked with the project extensively and also helped the art researchers meet pigeon fanciers in San Jose, relaying earned trust. When PigeonBlog was finished, the American Racing Pigeon Union gave da Costa a formal "Certificate of Appreciation" for the work she did for the birds and their people by showing a wider public the accomplishments and capabilities of racing pigeons.

There are lots of fans of PigeonBlog, including green and environmental activists, but one response in particular made da Costa feel that the racing pigeons of California had flown well, opening up something promising in the world across species. The Cornell University Laboratory of Ornithology asked da Costa to serve on its board for "Urban Bird Gardens" as part of the lab's citizen science initiative. Data collected by ordinary people, from elderly walkers to schoolchildren, could and did become part of databases bringing together university research and the affections and questions of citizens. Consider a closely related Cornell-linked citizen science initiative, Project PigeonWatch, which surveys regional differences in color types in different populations of common feral pigeons. One PigeonWatch project is in Washington, DC, and enlists city school groups to observe and record urban pigeons. Many things happen in this work in Terrapolis. City kids, overwhelmingly from "minority" groups, learn to see despised birds as valuable and interesting city residents, as worth notice. Neither the kids nor the pigeons are urban "wildlife"; both sets of beings are civic subjects and objects in intra-action. But I cannot and will not forget that these pigeons and black kids in DC both carry the marks of U.S. racist iconography as unruly, dirty, out of place, feral. The actual kids move from seeing pigeons as "rats with wings" to sociable birds with lives and deaths. The kids transmute from bird hecklers and sometimes physical abusers to astute observers and advocates of beings whom they had not known how to see or respect. The schoolchildren became response-able. Perhaps, because pigeons have long histories of affective and cognitive relations with people, the pigeons looked back at the kids too, and at least the birds were not heckled. I know this account is a story, an invitation as much as an accomplishment, but the space for recuperation across despised cross-species categories of city dwellers deserves to be widened, not shut down.[31]

Writing of another art project joining homing pigeons and their people in collaboration in the face of the danger of the loss of the very community of pigeon fanciers (colombophiles) that nurtures them both, Vinciane Despret asked what the pigeon loft (pigeonnier) designed by the artist Matali Crasset at Chaudry, France, in 2003 commemorates:

> But without the lover of pigeons (pigeon fancier), without the knowledge and know-how of men and birds, without selection, apprenticeship, without transmission of practices, what then would remain would be pigeons, but not homing pigeons, not voyageurs. What is commemorated, then, is not the animal alone, nor the practice alone, but the activation of two "becomings-with" that are written explicitly into the origin of the project. Otherwise said, what is brought into existence are the relations by which pigeons transform men into talented pigeon fanciers and by which the fanciers transform the pigeons into reliable racing pigeons. This is how the work commemorates. It tasks itself with crafting a memory in the sense of prolonging the achievement into the present. This is a kind of "reprise."[32]

To re-member, to com-memorate, is actively to reprise, revive, retake, recuperate. Committed to the multispecies, SF, string figure worlding of becoming-with, da Costa and Despret are companion species. They remember; they entice and prolong into the fleshly present what would disappear without the active reciprocity of partners. Homing or racing pigeons and feral pigeons call both their emergent and traditional peoples to response-ability, and vice versa. City dwellers and rural people of different species and modes of living and dying make each other *colombophiles talentueux* in company with *voyageurs fiables*.

Despret and da Costa are playing string figure games with Matali Crasset, relaying knotted patterns and possibilities in Terrapolis. Crasset is an industrial designer, a profession that requires listening to and collaborating with partners in ways fine artists need not engage, but

1.5. *Capsule*, designed by Matali Crasset, 2003, for the project of La Fondation de France. *Les nouveaux commanditaires*. Médiation-Production: artconnexion. Lille, France. © André Morin.

which da Costa also practices in her work and play as artist researcher and multispecies art activist. The pigeon loft Crasset proposed was commissioned by La Défense, the association of pigeon fanciers in Beauvois en Cambresis, and by La Base de Loisirs de Caudry (the leisure park of Caudry). The interior space of the capsule is functionally organized like a tree, a kind of axis of the world, and the exterior shape echoes old Egyptian designs for pigeon lofts. Historical, mythical, and material worlds are in play here, in this home for birds commissioned by those who breed, raise, fly, and become-with them.

Another pigeon loft in the shape of a tower imposes itself on my memory; another proposal for multispecies recuperation for creatures of empire is held out to those of whatever species who might grasp it. This time we are in Melbourne in Australia, in Batman Park along the Yarra River, part of the Wurundjeri people's territory prior to European settlement. This colonized area along the Yarra became a wasteland, sewage dump, and site for cargo and rail transport, destroying the wetlands (Anglo scientific term) and destroying country (Anglo-Aboriginal term for multidimensional and storied place). Wetlands and country are as alike and as different as cat's cradle, *jeux de ficelle*, *na'atł'o'*, and *matjka-wuma*;

for staying with the trouble, the names and patterns are necessary to each other, but they are not isomorphic.[33] They inhabit linked, split, and tangled histories.

The small Batman Park was established in 1982 along a disused freight train rail yard, and the pigeon loft was built in the 1990s to encourage pigeons to roost away from city buildings and streets. The loft is a tower structure built as part of the city's management plan for feral pigeons. These are not the beloved sporting pigeons of fanciers or colombophiles, but the urban "rats of the sky" we met a few paragraphs ago in a Washington, DC, city parks program tied to the internationally eminent Cornell University Laboratory of Ornithology. Melbourne's pigeons came with Europeans and thrived in the ecosystems and worlds that replaced the Yarra River wetlands and dispossessed most of the Aboriginal traditional owners of the land responsible for taking care of country. In 1985, the Wurundjeri Tribe Land Compensation and Cultural Heritage Council was established partly to develop awareness of Wurundjeri culture and history within contemporary Australia. I do not know if this council played any role in the partial recuperation of the land of Batman Park; I do know that sites along the Yarra River were places of significance to the Wurundjeri. In 1835, businessman and explorer John Batman signed a document with a group of Wurundjeri elders for the purchase of land in the first and only documented time that Europeans "negotiated their presence and occupation of Aboriginal lands directly with the traditional owners . . . For 600,000 acres of Melbourne, including most of the land now within the suburban area, John Batman paid 40 pairs of blankets, 42 tomahawks, 130 knives, 62 pairs scissors, 40 looking glasses, 250 handkerchiefs, 18 shirts, 4 flannel jackets, 4 suits of clothes and 150 lb. of flour."[34] The British governor of New South Wales repudiated this impudent treaty for its trespass on the rights of the Crown. Somehow, this fraught history must be inherited, must be re-membered, in that little park strip of reclaimed urban land with its striking pigeon tower.

Batman Park's pigeon loft is not art research for citizen science or industrial design commissioned by the racing-pigeon community, but a birth control—or, better, hatching control—technology crucial to multispecies urban flourishing. Feral pigeon fecundity is itself a material urban force, and also a potent signifier of the overfilling of the land with settlers and immigrants and depriving the land of endemic wetland birds and Aboriginal peoples. Staying with the trouble, the task is multispecies recuperation and somehow, in that suggestive Australian

1.6. Pigeon loft in Batman Park, Melbourne. Photograph by Nick Carson, 2008.

idiom, "getting on together" with less denial and more experimental justice. I want to see the pigeon loft as a small, practical enactment and a reminder to further opening to the response-ability of staying with the trouble. Response-ability is about both absence and presence, killing and nurturing, living and dying—and remembering who lives and who dies and how in the string figures of naturalcultural history. The loft has two hundred nesting boxes for pigeons, inviting them to lay their eggs. People come from below and replace their eggs with artificial ones to brood. People are allowed—encouraged—to feed pigeons near the loft but not elsewhere. *Pitchfork*, a blog dedicated to writing about "projects to do with permaculture, education, and growing food," took note of the Batman Park loft not just for its efforts to deal with pigeon-human conflict in innovative ways, but also for a rich product of concentrated roosting birds—compostable droppings. The blogger noted suggestively, "The easiest way to get pigeon manure into your food system is to get the pigeons to fly it in for you."[35] In a park that was a sewage dump not so long ago, this suggestion from the permaculture world has a definite charm. This pigeon loft is not a prolife project; in my view, no serious animal-human becoming-with can be a prolife project in the chilling American sense of that term. And the municipal pigeon tower certainly cannot undo unequal treaties, conquest, and wetlands destruction; but it is nonetheless a possible thread in a pattern for ongoing, noninnocent, interrogative, multispecies getting on together.

Reliable Voyageurs

Companion species infect each other all the time. Pigeons are world travelers, and such beings are vectors and carry many more, for good and for ill. Bodily ethical and political obligations are infectious, or they should be. *Cum panis*, companion species, at table together. Why tell stories like my pigeon tales, when there are only more and more openings and no bottom lines? Because there are quite definite response-abilities that are strengthened in such stories.

The details matter. The details link actual beings to actual response-abilities. As spies, racers, messengers, urban neighbors, iridescent sexual exhibitionists, avian parents, gender assistants for people, scientific subjects and objects, art-engineering environmental reporters, search-and-rescue workers at sea, imperialist invaders, discriminators of painting styles, native species, pets, and more, around the earth pigeons and their partners of many kinds, including people, make history. Each time a story helps me remember what I thought I knew, or introduces me to new knowledge, a muscle critical for caring about flourishing gets some aerobic exercise. Such exercise enhances collective thinking and movement in complexity. Each time I trace a tangle and add a few threads that at first seemed whimsical but turned out to be essential to the fabric, I get a bit straighter that staying with the trouble of complex worlding is the name of the game of living and dying well together on terra, in Terrapolis. We are all responsible to and for shaping conditions for multispecies flourishing in the face of terrible histories, and sometimes joyful histories too, but we are not all response-able in the same ways. The differences matter—in ecologies, economies, species, lives.

If only we could all be so lucky as to have a savvy artist design our lofts, our homes, our messaging packs! If only we all had the map sense to navigate in the troubled times and places!

CHAPTER 2

Tentacular Thinking

Anthropocene, Capitalocene, Chthulucene

We are all lichens.
—Scott Gilbert, "We Are All Lichens Now"

Think we must. We must think.
—Stengers and Despret, *Women Who Make a Fuss*

What happens when human exceptionalism and bounded individualism, those old saws of Western philosophy and political economics, become unthinkable in the best sciences, whether natural or social? Seriously unthinkable: not available to think with. Biological sciences have been especially potent in fermenting notions about all the mortal inhabitants of the earth since the imperializing eighteenth century. *Homo sapiens*—the Human as species, the Anthropos as the human species, Modern Man—was a chief product of these knowledge practices. What happens when the best biologies of the twenty-first century cannot do their job with bounded individuals plus contexts, when organisms plus environments, or genes plus whatever they need, no longer sustain the overflowing richness of biological knowledges, if they ever did? What happens when organisms plus environments can hardly be remembered for the same reasons that even Western-indebted people can no longer figure themselves as individuals and societies of individuals in human-only histories? Surely such a transformative time on earth must not be named the Anthropocene!

In this chapter, with all the unfaithful offspring of the sky gods, with my littermates who find a rich wallow in multispecies muddles, I want to make a critical and joyful fuss about these matters. I want to stay with the trouble, and the only way I know to do that is in generative joy, terror, and collective thinking.

My first demon familiar in this task will be a spider, *Pimoa cthulhu*, who lives under stumps in the redwood forests of Sonoma and Mendocino Counties, near where I live in North Central California.[1] Nobody lives everywhere; everybody lives somewhere. Nothing is connected to everything; everything is connected to something.[2] This spider is in place, has a place, and yet is named for intriguing travels elsewhere. This spider will help me with returns, and with roots and routes.[3] The eight-legged tentacular arachnid that I appeal to gets her generic name from the language of the Goshute people of Utah and her specific name from denizens of the depths, from the abyssal and elemental entities, called chthonic.[4] The chthonic powers of Terra infuse its tissues everywhere, despite the civilizing efforts of the agents of sky gods to astralize them and set up chief Singletons and their tame committees of multiples or subgods, the One and the Many. Making a small change in the biologist's taxonomic spelling, from cthulhu to chthulu, with renamed *Pimoa chthulu* I propose a name for an elsewhere and elsewhen that was, still is, and might yet be: the Chthulucene. I remember that *tentacle* comes from the Latin *tentaculum*, meaning "feeler," and *tentare*, meaning "to feel" and "to try"; and I know that my leggy spider has many-armed allies. Myriad tentacles will be needed to tell the story of the Chthulucene.[5]

The tentacular ones tangle me in SF. Their many appendages make string figures; they entwine me in the poiesis—the making—of speculative fabulation, science fiction, science fact, speculative feminism, *soin de ficelle*, so far. The tentacular ones make attachments and detachments; they ake cuts and knots; they make a difference; they weave paths and consequences but not determinisms; they are both open and knotted in some ways and not others.[6] SF is storytelling and fact telling; it is the patterning of possible worlds and possible times, material-semiotic worlds, gone, here, and yet to come. I work with string figures as a theoretical trope, a way to think-with a host of companions in sympoietic threading, felting, tangling, tracking, and sorting. I work with and in SF as material-semiotic composting, as theory in the mud, as muddle.[7]

2.1. *Pimoa cthulhu*. Photograph by Gustavo Hormiga.

The tentacular are not disembodied figures; they are cnidarians, spiders, fingery beings like humans and raccoons, squid, jellyfish, neural extravaganzas, fibrous entities, flagellated beings, myofibril braids, matted and felted microbial and fungal tangles, probing creepers, swelling roots, reaching and climbing tendrilled ones. The tentacular are also nets and networks, IT critters, in and out of clouds. Tentacularity is about life lived along lines—and such a wealth of lines—not at points, not in spheres. "The inhabitants of the world, creatures of all kinds, human and non-human, are wayfarers"; generations are like "a series of interlaced trails."[8] String figures all.

All the tentacular stringy ones have made me unhappy with posthumanism, even as I am nourished by much generative work done under that sign. My partner Rusten Hogness suggested compost instead of posthuman(ism), as well as humusities instead of humanities, and I jumped into that wormy pile.[9] Human as humus has potential, if we could chop and shred human as Homo, the detumescing project of a self-making and planet-destroying CEO. Imagine a conference not on the Future of the Humanities in the Capitalist Restructuring University, but instead on the Power of the Humusities for a Habitable Multispecies Muddle! Ecosexual artists Beth Stephens and Annie Sprinkle made a bumper sticker for me, for us, for SF: "Composting is so hot!"[10]

The earth of the ongoing Chthulucene is sympoietic, not autopoietic. Mortal Worlds (Terra, Earth, Gaia, Chthulu, the myriad names and powers that are not Greek, Latin, or Indo-European at all)[11] do not make themselves, no matter how complex and multileveled the systems, no matter how much order out of disorder might be produced in generative autopoietic system breakdowns and relaunchings at higher levels of order. Autopoietic systems are hugely interesting—witness the history of cybernetics and information sciences; but they are not good models for living and dying worlds and their critters. Autopoietic systems are not closed, spherical, deterministic, or teleological; but they are not quite good enough models for the mortal SF world. Poiesis is symchthonic, sympoietic, always partnered all the way down, with no starting and subsequently interacting "units."[12] The Chthulucene does not close in on itself; it does not round off; its contact zones are ubiquitous and continuously spin out loopy tendrils. Spider is a much better figure for sympoiesis than any inadequately leggy vertebrate of whatever pantheon. Tentacularity is symchthonic, wound with abyssal and dreadful graspings, frayings, and weavings, passing relays again and again, in the generative recursions that make up living and dying.

After I used the term *sympoiesis* in a grasp for something other than the lures of autopoiesis, Katie King told me about M. Beth Dempster's Master of Environmental Studies thesis written in 1998, in which she suggested the term *sympoiesis* for "collectively-producing systems that do not have self-defined spatial or temporal boundaries. Information and control are distributed among components. The systems are evolutionary and have the potential for surprising change." By contrast, autopoietic systems are "self-producing" autonomous units "with self defined spatial or temporal boundaries that tend to be centrally controlled, homeostatic, and predictable."[13] Dempster argued that many systems are mistaken for autopoietic that are really sympoietic. I think this point is important for thinking about rehabilitation (making livable again) and sustainability amid the porous tissues and open edges of damaged but still ongoing living worlds, like the planet earth and its denizens in current times being called the Anthropocene. If it is true that neither biology nor philosophy any longer supports the notion of independent organisms in environments, that is, interacting units plus contexts/rules, then sympoiesis is the name of the game in spades. Bounded (or neoliberal) individualism amended by autopoiesis is not good enough figurally or scientifically; it misleads us down deadly paths.

Barad's agential realism and intra-action become common sense, and perhaps a lifeline for Terran wayfarers.

SF, string figuring, is sympoietic. Thinking-with my work on cat's cradle, as well as with the work of another of her companions in thinking, Félix Guattari, Isabelle Stengers relayed back to me how players pass back and forth to each other the patterns-at-stake, sometimes conserving, sometimes proposing and inventing.

> More precisely, com-menting, if it means thinking-with, that is becoming-with, is in itself a way of relaying . . . But knowing that what you take has been held out entails a particular thinking "between." It does not demand fidelity, still less fealty, rather a particular kind of loyalty, the answer to the trust of the held out hand. Even if this trust is not in "you" but in "creative uncertainty," even if the consequences and meaning of what has been done, thought or written, do not belong to you anymore than they belonged to the one you take the relay from, one way or another the relay is now in your hands, together with the demand that you do not proceed with "mechanical confidence." [In cat's cradling, at least] two pairs of hands are needed, and in each successive step, one is "passive," offering the result of its previous operation, a string entanglement, for the other to operate, only to become active again at the next step, when the other presents the new entanglement. But it can also be said that each time the "passive" pair is the one that holds, and is held by the entanglement, only to "let it go" when the other one takes the relay.[14]

In passion and action, detachment and attachment, this is what I call cultivating response-ability; that is also collective knowing and doing, an ecology of practices. Whether we asked for it or not, the pattern is in our hands. The answer to the trust of the held-out hand: think we must.

Marilyn Strathern is an ethnographer of thinking practices. She defines anthropology as studying relations with relations—a hugely consequential, mind- and body-altering sort of commitment.[15] Nourished by her lifelong work in highland Papua New Guinea (Mt. Hagen), Strathern writes about accepting the risk of relentless contingency, of putting relations at risk with other relations, from unexpected worlds. Embodying the practice of feminist speculative fabulation in the scholarly mode, Strathern taught me—taught us—a simple but game-changing thing: "It matters what ideas we use to think other ideas."[16] I compost my soul in this hot pile. The worms are not human; their undulating bodies in-

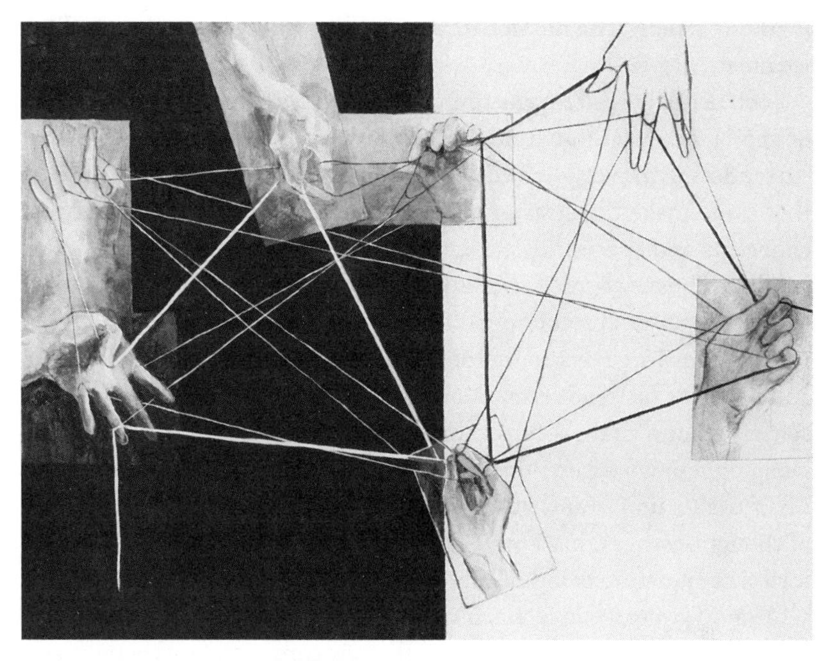

2.2. *Cat's Cradle/String Theory*, Baila Goldenthal, 2008. Oil on canvas, 36 × 48 in. Courtesy of Maurya Simon and Tamara Ambroson.

gest and reach, and their feces fertilize worlds. Their tentacles make string figures.

It matters what thoughts think thoughts. It matters what knowledges know knowledges. It matters what relations relate relations. It matters what worlds world worlds. It matters what stories tell stories. Paintings by Baila Goldenthal are eloquent testimony to this mattering.[17]

What is it to surrender the capacity to think? These times called the Anthropocene are times of multispecies, including human, urgency: of great mass death and extinction; of onrushing disasters, whose unpredictable specificities are foolishly taken as unknowability itself; of refusing to know and to cultivate the capacity of response-ability; of refusing to be present in and to onrushing catastrophe in time; of unprecedented looking away. Surely, to say "unprecedented" in view of the realities of the last centuries is to say something almost unimaginable. How can we think in times of urgencies *without* the self-indulgent and self-fulfilling myths of apocalypse, when every fiber of our being is interlaced, even complicit, in the webs of processes that must somehow be engaged and repatterned? Recursively, whether we asked for it or not, the pattern is in our hands. The answer to the trust of the held-out hand: think we must.

Instructed by Valerie Hartouni, I turn to Hannah Arendt's analysis of the Nazi war criminal Adolf Eichmann's inability to think. In that surrender of thinking lay the "banality of evil" of the particular sort that could make the disaster of the Anthropocene, with its ramped-up genocides and speciescides, come true.[18] This outcome is still at stake; think we must; we must think! In Hartouni's reading, Arendt insisted that thought was profoundly different from what we might call disciplinary knowledge or science rooted in evidence, or the sorting of truth and belief or fact and opinion or good and bad. Thinking, in Arendt's sense, is not a process for evaluating information and argument, for being right or wrong, for judging oneself or others to be in truth or error. All of that is important, but not what Arendt had to say about the evil of thoughtlessness that I want to bring into the question of the geohistorical conjuncture being called the Anthropocene.

Arendt witnessed in Eichmann not an incomprehensible monster, but something much more terrifying—she saw commonplace thoughtlessness. That is, here was a human being unable to make present to himself what was absent, what was not himself, what the world in its sheer not-one-selfness is and what claims-to-be inhere in not-oneself. Here was someone who could not be a wayfarer, could not entangle, could not track the lines of living and dying, could not cultivate response-ability, could not make present to itself what it is doing, could not live in consequences or with consequence, could not compost. Function mattered, duty mattered, but the world did not matter for Eichmann. The world does not matter in ordinary thoughtlessness. The hollowed-out spaces are all filled with assessing information, determining friends and enemies, and doing busy jobs; negativity, the hollowing out of such positivity, is missed, an astonishing abandonment of thinking.[19] This quality was not an emotional lack, a lack of compassion, although surely that was true of Eichmann, but a deeper surrender to what I would call immateriality, inconsequentiality, or, in Arendt's and also my idiom, thoughtlessness. Eichmann was astralized right out of the muddle of thinking into the practice of business as usual no matter what. There was no way the world could become for Eichmann and his heirs—us?—a "matter of care."[20] The result was active participation in genocide.

The anthropologist, feminist, cultural theorist, storyteller, and connoisseur of the tissues of heterogeneous capitalism, globalism, traveling worlds, and local places Anna Tsing examines the "arts of living on a damaged planet,"[21] or, in the subtitle of her book, "the possibility of life in Capitalist ruins." She performs thinking of a kind that must be cultivated in the all-too-ordinary urgencies of onrushing multispecies extinctions, genocides, immiserations, and exterminations. I name these things urgencies rather than emergencies because the latter word connotes something approaching apocalypse and its mythologies. Urgencies have other temporalities, and these times are ours. These are the times we must think; these are the times of urgencies that need stories.

Following matsutake mushrooms in their fulminating assemblages of Japanese, Americans, Chinese, Koreans, Hmong, Lao, Mexicans, fungal spores and mats, oak and pine trees, mycorrhizal symbioses, pickers, buyers, shippers, restaurateurs, diners, businessmen, scientists, foresters, DNA sequencers and their changing species, and much more, Tsing practices sympoietics in edgy times. Refusing either to look away or to reduce the earth's urgency to an abstract system of causative destruction, such as a Human Species Act or undifferentiated Capitalism, Tsing argues that precarity—failure of the lying promises of Modern Progress—characterizes the lives and deaths of all terran critters in these times. She looks for the eruptions of unexpected liveliness and the contaminated and nondeterministic, unfinished, ongoing practices of living in the ruins. She performs the force of stories; she shows in the flesh how it matters which stories tell stories as a practice of caring and thinking. "If a rush of troubled stories is the best way to tell contaminated diversity, then it's time to make that rush part of our knowledge practices . . . Matsutake's willingness to emerge in blasted landscapes allows us to explore the ruins that have become our collective home. To follow matsutake guides us to possibilities of coexistence within environmental disturbance. This is not an excuse for further human damage. Still, matsutake show one kind of collaborative survival."

Driven by radical curiosity, Tsing does the ethnography of "salvage accumulation" and "patchy capitalism," the kind that can no longer promise progress but can and does extend devastation and make precarity the name of our systematicity. There is no simple ethical, political, or theoretical point to take from Tsing's work; there is instead the force of engaging the world in the kind of thinking practices impossible for Eichmann's heirs. "Matsutake tell us about surviving collaboratively in disturbance and contamination. We need this skill for living in ruins."[22] This is not a longing for salvation or some other sort of optimistic politics; neither is it a cynical quietism in the face of the depth of the trouble. Rather, Tsing proposes a commitment to living and dying with response-ability in unexpected company. Such living and dying have the best chance of cultivating conditions for ongoingness.

The ecological philosopher and multispecies ethnographer Thom van Dooren also inhabits the layered complexities of living in times of extinction, extermination, and partial recuperation; he deepens our consideration of what thinking means, of what not becoming thoughtless exacts from all of us. In his extraordinary book *Flight Ways*, van Dooren accompanies situated bird species living on the extended edge of extinction, asking what it means to hold open space for another.[23] Such holding open is far from an innocent or obvious material or ethical practice; even when successful, it exacts tolls of suffering as well as surviving as individuals and as kinds. In his examination of the practices of the North American whooping crane species survival plan, for example, van Dooren details multiple kinds of hard multispecies captivities and labors, forced life, surrogate reproductive labor, and substitute dying—none of which should be forgotten, especially in successful projects. Holding open space might—or might not—delay extinction in ways that make possible composing or recomposing flourishing naturalcultural assemblages. *Flight Ways* shows how extinction is not a point, not a single event, but more like an extended edge or a widened ledge. Extinction is a protracted slow death that unravels great tissues of ways of going on in the world for many species, including historically situated people.[24]

Van Dooren proposes that mourning is intrinsic to cultivating response-ability. In his chapter on conservation efforts for Hawaiian crows (ʻAlalā for Hawaiians, *Corvus hawaiiensis* for Linneans), whose forest homes and foods as well as friends, chicks, and mates have largely disappeared, van Dooren argues that it is not just human people who mourn the loss of loved ones, of place, of lifeways; other beings mourn as well. Corvids grieve loss. The point rests on biobehavioral studies as well as intimate natural history; neither the capacity nor the practice of mourning is a human specialty. Outside the dubious privileges of human exceptionalism, thinking people must learn to grieve-with.

> Mourning is about dwelling with a loss and so coming to appreciate what it means, how the world has changed, and how we must *ourselves* change and renew our relationships if we are to move forward from here. In this context, genuine mourning should open us into an aware-

> ness of our dependence on and relationships with those countless others being driven over the edge of extinction . . . The reality, however, is that there *is* no avoiding the necessity of the difficult cultural work of reflection and mourning. This work is not opposed to practical action, rather it is the foundation of any sustainable and informed response.

Grief is a path to understanding entangled shared living and dying; human beings must grieve *with*, because we are in and of this fabric of undoing. Without sustained remembrance, we cannot learn to live with ghosts and so cannot think. Like the crows and with the crows, living and dead "we are at stake in each other's company."[25]

At least one more SF thread is crucial to the practice of thinking, which must be thinking-with: storytelling. It matters what thoughts think thoughts; it matters what stories tell stories. "Urban Penguins: Stories for Lost Places," van Dooren's chapter on Sydney Harbor's Little Penguins (*Eudyptula minor*), succeeds in crafting a nonanthropomorphic, nonanthropocentric sense of storied place. In their resolutely "philopatric" (home loving) nesting and other life practices, these urban penguins—real, particular birds—story place, *this* place, not just any place. Establishing the reality and vivid specificity of penguin-storied place is a major material-semiotic accomplishment. Storying cannot any longer be put into the box of human exceptionalism. Without deserting the terrain of behavioral ecology and natural history, this writing achieves powerful attunement to storying in penguin multimodal semiotics.[26]

Ursula Le Guin taught me the carrier bag theory of storytelling and of naturalcultural history. Her theories, her stories, are capacious bags for collecting, carrying, and telling the stuff of living. "A leaf a gourd a shell a net a bag a sling a sack a bottle a pot a box a container. A holder. A recipient."[27] So much of earth history has been told in the thrall of the fantasy of the first beautiful words and weapons, of the first beautiful weapons *as* words and vice versa. Tool, weapon, word: that is the word made flesh in the image of the sky god; that is the Anthropos. In a tragic story with only one real actor, one real world-maker, the hero, this is the Man-making tale of the hunter on a quest to kill and bring back the terrible bounty. This is the cutting, sharp, combative tale of action that defers the suffering of glutinous, earth-rotted passivity beyond bearing. All others in the prick tale are props, ground, plot space, or prey. They don't matter; their job is to be in the way, to be overcome, to be the road, the conduit, but not the traveler, not the begetter. The last thing the hero wants to know is that his beautiful words and weapons will be worthless without a bag, a container, a net.

Nonetheless, no adventurer should leave home without a sack. How did a sling, a pot, a bottle suddenly get in the story? How do such lowly things keep the story going? Or maybe even worse for the hero, how do those concave, hollowed-out things, those holes in Being, from the get-go generate richer, quirkier, fuller, unfitting, ongoing stories, stories with room for the hunter but which weren't and aren't about him, the self-making human, the human-making machine of history? The slight curve of the shell that holds just a little water, just a few seeds to give away and to receive, suggests stories of becoming-with, of reciprocal induction, of companion species whose job in living and dying is not to end the storying, the worlding. With a shell and a net, becoming human, becoming humus, becoming terran, has another shape—that is, the side-winding, snaky shape of becoming-with. To think-with is to stay with the naturalcultural multispecies trouble on earth. There are no guarantees, no arrow of time, no Law of History or Science or Nature in such struggles. There is only the relentlessly contingent SF worlding of living and dying, of becoming-with and unbecoming-with, of sympoiesis, and so, just possibly, of multispecies flourishing on earth.

Like Le Guin, Bruno Latour passionately understands the need to change the story, to learn somehow to narrate—to think—outside the prick tale of Humans in History, when the knowledge of how to murder each other—and along with each other, uncountable multitudes of the living earth—is not scarce. Think we must; we must think. That means, simply, we *must* change the story; the story *must* change. Le Guin writes, "Hence it is with a certain feeling of urgency that I seek the nature, subject, words of the other story, the untold one, the life story."[28] In this terrible time called the Anthropocene, Latour argues that the fundamentals of geopolitics have been blasted open. None of the parties in crisis can call on Providence, History, Science, Progress, or any other god trick outside the common fray to resolve the troubles.[29] A common livable world must be composed, bit by bit, or not at all. What used to be called nature has erupted into ordinary human affairs, and vice versa, in such a way and with such permanence as to change fundamentally means and prospects for going on, including going on at all. Searching for compositionist practices capable of building effective new collectives, Latour argues that we must learn to tell "Gaïa stories." If that word is too hard, then we can call our narrations "geostories," in which "all the former props and passive agents have become active without, for that, being part of a giant plot written by some overseeing entity."[30] Those who tell Gaia stories or geostories are the "Earthbound," those who eschew the dubious pleasures of transcendent plots of modernity and the purifying division of society and nature. Latour argues that we face a stark divide: "Some are readying themselves to live as Earthbound in the Anthropocene; others decided to remain as Humans in the Holocene."[31]

In much of his writing, Latour develops the language and imagery of trials of strength; and in thinking about the Anthropocene and the Earthbound, he extends that metaphor to develop the difference between a police action, where peace is restored by an already existing order, and war or politics, where real enemies must be overcome to establish what will be. Latour is determined to avoid the idols of a ready-to-hand fix, such as Laws of History, Modernity, the State, God, Progress, Reason, Decadence, Nature, Technology, or Science, as well as the debilitating disrespect for difference and shared finitude inherent in those who already know the answers toward those who only need to learn them—by force, faith, or self-certain pedagogy. Those who "believe" they have the answers to the present urgencies are terribly dangerous. Those who refuse to be *for* some ways of living and dying and not others are equally dangerous. Matters of fact, matters of concern,[32] and matters of care are knotted in string figures, in SF.

Latour embraces sciences, not Science. In geopolitics, "the important point here is to realize that the facts of the matter cannot be delegated to a higher unified authority that would have done the choice *in our stead*. Controversies—no matter how spurious they might be—are no excuse to delay the *decision* about which side represents our world *better*."[33] Latour *aligns* himself with the reports of the Intergovernmental Panel on Climate Change (IPCC); he does not *believe* its assessments and reports; he *decides* what is strong and trustworthy and what is not. He casts his lot with some worlds and worldings and not others. One need not hear Latour's "decision" discourse with an individualist ear; he is a compositionist intent on understanding how a common world, how collectives, are built-with each other, where all the builders are not human beings. This is neither relativism nor rationalism; it is SF, which Latour would call both sciences and scientifiction and I would call both sciences and speculative fabulation—all of which are political sciences, in our aligned approaches.

"Alignment" is a rich metaphor for wayfarers, for the Earthbound, and does not as easily as "decision" carry the tones of modernist liberal choice discourse, at least in the United States. Further, the refusal of the modernist category of belief is also crucial to my effort to persuade us to take up the Chthulucene and its tentacular tasks.[34] Like Stengers and like myself, Latour is a thoroughgoing materialist committed to an ecology of practices, to the mundane articulating of assemblages through situated work and play in the muddle of messy living and dying. Actual players, articulating with varied allies of all ontological sorts (molecules, colleagues, and much more), must compose and sustain what is and will be. Alignment in tentacular worlding must be a seriously tangled affair!

Intent on the crucial refusal of self-certainty and preexisting god tricks, which I passionately share, Latour turns to a resource—relentless reliance on the material-semiotic trope of trials of strength—that, I think, makes it unnecessarily hard to tell his and our needed new story. He defines war as the absence of a referee so that trials of strength must determine the legitimate authority. Humans in History and the Earthbound in the Anthropocene are engaged in trials of strength where there is no Referee who/which can establish what is/was/will be. History versus Gaia stories are at stake. Those trials—the war of the Earthbound with the Humans—would not be conducted with rockets and bombs; they would be conducted with every other imaginable resource and with no god trick from above to decide life and death, truth and error. But still, we are in the story of the hero and the first beautiful words and weapons, not in the story of the carrier bag. Anything not decided in the presence of the Authority is war; Science (singular and capitalized) is the Authority; the Authority conducts police actions. In contrast, sciences (always rooted in practices) are war. Therefore, in Latour's passionate speculative fabulation, such war is our only hope for real politics. The past is as much the contested zone as the present or future.

Latour's thinking and stories need a specific kind of enemies. He draws on Carl Schmitt's "political theology," which is a theory of peace through war, with the enemy as *hostis*, with all its tones of host, hostage, guest, and worthy enemy. Only with such an enemy, Schmitt and Latour hold, is there respect and a chance to be less, not more, deadly in conflict. Those who operate within the categories of Authority and of belief are notoriously prone to exterminationist and genocidal combat (it's hard to deny that!). They are lost without a pre-established Referee. The *hostis* demands much better. But all the action remains within the narrative vise of trials of strength, of mortal combat, within which the knowledge of how to murder each other remains well entrenched. Latour makes clear that he does not *want* this story, but he does not propose another. The only real possibility for peace lies in the tale of the respected enemy, the *hostis*, and trials of strength. "But when you are at war, it is only through the throes of the encounters that the authority you have or don't have will be decided *depending whether you win or lose*."[35]

Schmitt's enemies do not allow the story to change in its marrow; the Earthbound need a more tentacular, less binary life story. Latour's Gaia stories deserve better companions in storytelling than Schmitt. The question of whom to think-with is immensely material. I do not think Latour's dilemma can be resolved in the terms of the Anthropocene. His Earthbound will have to trek into the Chthulucene to entangle with the ongoing, snaky, unheroic, tentacular, dreadful ones, the ones which/who craft material-semiotic netbags of little use in trials of strength but of great use in bringing home and sharing the means of living and dying well, perhaps even the means of ecological recuperation for human and more-than-human critters alike.

Shaping her thinking about the times called Anthropocene and "multi-faced Gaïa" (Stengers's term) in companionable friction with Latour, Isabelle Stengers does not ask that we recompose ourselves to become able, perhaps, to "face Gaïa." But like Latour and even more like Le Guin, one of her most generative SF writers, Stengers is adamant about changing the story. Focusing on intrusion rather than composition, Stengers calls Gaia a fearful and devastating power that intrudes on our categories of thought, that intrudes on thinking itself.[36] Earth/Gaia is maker and destroyer, not resource to be exploited or ward to be protected or nursing mother promising nourishment. Gaia is not a person but complex systemic phenomena that compose a living planet. Gaia's intrusion into our affairs is a radically materialist event that collects up multitudes. This intrusion threatens not life on earth itself—microbes will adapt, to put it mildly—but threatens the livability of earth for vast kinds, species, assemblages, and individuals in an "event" already under way called the Sixth Great Extinction.[37]

Stengers, like Latour, evokes the name of Gaia in the way James Lovelock and Lynn Margulis did, to name complex nonlinear couplings between processes that compose and sustain entwined but nonadditive subsystems as a partially cohering systemic whole.[38] In this hypothesis, Gaia is autopoietic—self-forming, boundary maintaining, contingent, dynamic, and stable under some conditions but not others. Gaia is not reducible to the sum of its parts, but achieves finite systemic coherence in the face of perturbations within parameters that are themselves responsive to dynamic systemic processes. Gaia does not and could not care about human or other biological beings' intentions or desires or needs, but Gaia puts into question our very existence, we who have provoked its brutal mutation that threatens both human and nonhuman livable presents and futures. Gaia is not about a list of questions waiting for rational policies;[39] Gaia is an intrusive event that undoes thinking as usual. "She is what specifically questions the tales and refrains of modern history. There is only one real mystery at stake, here: it is the answer we, meaning those who belong to this history, may be able to create as we face the consequences of what we have provoked."[40]

Anthropocene

So, what have we provoked? Writing in the midst of California's historic multiyear drought and the explosive fire season of 2015, I need the photograph of a fire set deliberately in June 2009 by Sustainable Resource Alberta near the Saskatchewan River Crossing on the Icefields Parkway in order to stem the spread of mountain pine beetles, to create a fire barrier to future fires, and to enhance biodiversity. The hope is that this fire acts as an ally for resurgence. The devastating spread of the pine beetle across the North American West is a major chapter of climate change in the Anthropocene. So too are the predicted megadroughts and the extreme and extended fire seasons. Fire in the North American West has a complicated multispecies history; fire is an essential element for ongoing, as well as an agent of double death, the killing of ongoingness. The material semiotics of fire in our times are at stake.

Thus it is past time to turn directly to the time-space-global thing called Anthropocene.[41] The term seems to have been coined in the early 1980s by University of Michigan ecologist Eugene Stoermer (d. 2012), an expert in freshwater diatoms. He introduced the term to refer to growing evidence for the transformative effects of human activities on the earth. The name Anthropocene made a dramatic star appearance in globalizing discourses in 2000 when the Dutch Nobel Prize–winning atmospheric chemist Paul Crutzen joined Stoermer to propose that human activities had been of such a kind and magnitude as to merit the use of a new geological term for a new epoch, superseding the Holocene,

2.3. Icon for the Anthropocene: Flaming Forests. From Rocky Mountain House, Alberta, Canada, June 2, 2009. Photograph by Cameron Strandberg.

which dated from the end of the last ice age, or the end of the Pleistocene, about twelve thousand years ago. Anthropogenic changes signaled by the mid-eighteenth-century steam engine and the planet-changing exploding use of coal were evident in the airs, waters, and rocks.[42] Evidence was mounting that the acidification and warming of the oceans are rapidly decomposing coral reef ecosystems, resulting in huge ghostly white skeletons of bleached and dead or dying coral. That a symbiotic system—coral, with its watery world-making associations of cnidarians and zooanthellae with many other critters too—indicated such a global transformation will come back into our story.

But for now, notice that the Anthropocene obtained purchase in popular and scientific discourse in the context of ubiquitous urgent efforts to find ways of talking about, theorizing, modeling, and managing a Big Thing called Globalization. Climate-change modeling is a powerful positive feedback loop provoking change-of-state in systems of political and ecological discourses.[43] That Paul Crutzen was both a Nobel laureate and an atmospheric chemist mattered. By 2008, many scientists around the world had adopted the not-yet-official but increasingly indispensable term;[44] and myriad research projects, performances, installations, and conferences in the arts, social sciences, and humanities found the term mandatory in their naming and thinking, not least for facing both accelerating extinctions across all biological taxa and also multispecies, including human, immiseration across the expanse of Terra. Fossil-burning human beings seem intent on making as many new fossils as possible as fast as possible. They will be read in the strata of the rocks on the land and under the waters by the geologists of the very near future, if not already. Perhaps, instead of the fiery forest, the icon for the Anthropocene should be Burning Man![45]

The scale of burning ambitions of fossil-making man—of this Anthropos whose hot projects for accelerating extinctions merits a name for a geological epoch—is hard to comprehend. Leaving aside all the other accelerating extractions of minerals, plant and animal flesh, human homelands, and so on, surely, we want to say, the pace of development of renewable energy technologies and of political and technical carbon pollution-abatement measures, in the face of palpable and costly ecosystem collapses and spreading political disorders, will mitigate, if not eliminate, the burden of planet-warming excess carbon from burning still more fossil fuels. Or, maybe the financial troubles of the global coal and oil industries by 2015 would stop the madness. Not so. Even casual acquaintance with the daily news erodes such hopes, but the trouble is worse than what even a close reader of IPCC documents and the press will find. In "The Third Carbon Age," Michael Klare, a professor of Peace and World Security Studies at Hampshire College, lays out strong evidence against the idea that the old age of coal, replaced by the recent age of oil, will be replaced by the age of renewables.[46] He details the large and growing global national and corporate investments in renewables; clearly, there are big profit and power advantages to be had in this sector. And at the same time, every imaginable, and many unimaginable, technologies and strategic measures are being pursued by all the big global players to extract every last calorie of fossil carbon, at whatever depth and in whatever formations of sand, mud, or rock, and with whatever horrors of travel to distribution and use points, to burn before someone else gets at that calorie and burns it first in the great prick story of the first and the last beautiful words and weapons.[47] In what he calls the Age of Unconventional Oil and Gas, hydro-fracking is the tip of the (melting) iceberg. Melting of the polar seas, terrible for polar bears and for coastal peoples, is very good for big competitive military, exploration, drilling, and tanker shipping across the northern passages. Who needs an ice-breaker when you can count on melting ice?[48]

A complex systems engineer named Brad Werner addressed a session at the meetings of the American Geophysical Union in San Francisco in 2012. His point was quite simple: scientifically speaking, global capitalism "has made the depletion of resources so rapid, convenient and barrier-free that 'earth-human systems' are becoming dangerously unstable in response." Therefore, he argued, the only scientific thing to do is revolt! Movements, not just individuals, are critical. What is required is action and thinking that do not fit within the dominant capitalist culture; and, said Werner, this is a matter not of opinion, but of geophysical dynamics. The reporter who covered this session summed up Werner's address: "He is saying that his research shows that our entire economic paradigm is a threat to ecological stability."[49] Werner is not the first or the last researcher and maker of matters of concern to argue this point, but his clarity at a scientific meeting is bracing. Revolt! Think we must; we must think. Actually think, not like Eichmann the Thoughtless. Of course, the devil is in the details—how to revolt? How to matter and not just want to matter?

Capitalocene

But at least one thing is crystal clear. No matter how much he might be caught in the generic masculine universal and how much he only looks up, the Anthropos did not do this fracking thing and he should not name this double-death-loving epoch. The Anthropos is not Burning Man after all. But because the word is already well entrenched and seems less controversial to many important players compared to the Capitalocene, I know that we will continue to need the term *Anthropocene*. I will use it too, sparingly; what and whom the Anthropocene collects in its refurbished netbag might prove potent for living in the ruins and even for modest terran recuperation.

Still, if we could only have one word for these SF times, surely it must be the Capitalocene.[50] Species Man did not shape the conditions for the Third Carbon Age or the Nuclear Age. The story of Species Man as the agent of the Anthropocene is an almost laughable rerun of the great phallic humanizing and modernizing Adventure, where man, made in the image of a vanished god, takes on superpowers in his secular-sacred ascent, only to end in tragic detumescence, once again. Autopoietic, self-making man came down once again, this time in tragic system failure, turning biodiverse ecosystems into flipped-out deserts of slimy mats

2.4. Icon for the Capitalocene: Sea Ice Clearing from the Northwest Passage, Data 2012. NASA Visible Earth image by Jesse Allen, 2015, using data from the Land Atmosphere Near Real-Time Capability for EOS (LANCE). National Snow and Ice Data Center.

and stinging jellyfish. Neither did technological determinism produce the Third Carbon Age. Coal and the steam engine did not determine the story, and besides the dates are all wrong, not because one has to go back to the last ice age, but because one has to at least include the great market and commodity reworldings of the long sixteenth and seventeenth centuries of the current era, even if we think (wrongly) that we can remain Euro-centered in thinking about "globalizing" transformations shaping the Capitalocene.[51] One must surely tell of the networks of sugar, precious metals, plantations, indigenous genocides, and slavery, with their labor innovations and relocations and recompositions of critters and things sweeping up both human and nonhuman workers of all kinds. The infectious industrial revolution of England mattered hugely, but it is only one player in planet-transforming, historically situated, new enough, worlding relations. The relocation of peoples, plants, and animals; the leveling of vast forests; and the violent mining of metals preceded the steam engine; but that is not a warrant for wringing one's hands about the perfidy of the Anthropos, or of Species Man, or of Man the Hunter.

The systemic stories of the linked metabolisms, articulations, or coproductions (pick your metaphor) of economies and ecologies, of histories and human and nonhuman critters, must be relentlessly opportunistic and contingent. They must also be relentlessly relational, sympoietic, and consequential.[52] They are terran, not cosmic or blissed or cursed into outer space. The Capitalocene is terran; it does not have to be the last biodiverse geological epoch that includes our species too. There are so many good stories yet to tell, so many netbags yet to string, and not just by human beings.

As a provocation, let me summarize my objections to the Anthropocene as a tool, story, or epoch to think with: (1) The myth system associated with the Anthropos is a setup, and the stories end badly. More to the point, they end in double death; they are not about ongoingness. It is hard to tell a good story with such a bad actor. Bad actors need a story, but not the whole story. (2) Species Man does not make history. (3) Man plus Tool does not make history. That is the story of History human exceptionalists tell. (4) That History must give way to geostories, to Gaia stories, to symchthonic stories; terrans do webbed, braided, and tentacular living and dying in sympoietic multispecies string figures; they do not do History. (5) The human social apparatus of the Anthropocene tends to be top-heavy and bureaucracy prone. Revolt needs other forms of action and other stories for solace, inspiration, and effectiveness. (6) Despite its reliance on agile computer modeling and autopoietic systems theories, the Anthropocene relies too much on what should be an "unthinkable" theory of relations, namely the old one of bounded utilitarian individualism—preexisting units in competition relations that take up all the air in the atmosphere (except, apparently, carbon dioxide). (7) The sciences of the Anthropocene are too much contained within restrictive systems theories and within evolutionary theories called the Modern Synthesis, which for all their extraordinary importance have proven unable to think well about sympoiesis, symbiosis, symbiogenesis, development, webbed ecologies, and microbes. That's a lot of trouble for adequate evolutionary theory. (8) Anthropocene is a term most easily meaningful and usable by intellectuals in wealthy classes and regions; it is not an idiomatic term for climate, weather, land, care of country, or much else in great swathes of the world, especially but not only among indigenous peoples.

I am aligned with feminist environmentalist Eileen Crist when she writes against the managerial, technocratic, market-and-profit besotted, modernizing, and human-exceptionalist business-as-usual commitments of so much Anthropocene discourse. This discourse is not simply wrong-headed and wrong-hearted in itself; it also saps our capacity for imagining and caring for other worlds, both those that exist precariously now (including those called wilderness, for all the contaminated history of that term in racist settler colonialism) and those we need to bring into being in alliance with other critters, for still possible recuperating pasts, presents, and futures. "Scarcity's deepening persistence, and the suffering it is auguring for all life, is an artifact of human exceptionalism at every level." Instead, a humanity with more earthly integrity "invites the *priority* of our pulling back and scaling down, of welcoming limitations of our numbers, economies, and habitats for the sake of a higher, more inclusive freedom and quality of life."[53]

If Humans live in History and the Earthbound take up their task within the Anthropocene, too many Posthumans (and posthumanists, another gathering altogether) seem to have emigrated to the Anthropocene for my taste. Perhaps my human and nonhuman people are the dreadful Chthonic ones who snake within the tissues of Terrapolis.

Note that insofar as the Capitalocene is told in the idiom of fundamentalist Marxism, with all its trappings of Modernity, Progress, and History, that term is subject to the same or fiercer criticisms. The stories of both the Anthropocene and the Capitalocene teeter constantly on the brink of becoming much Too Big. Marx did better than that, as did Darwin. We can inherit their bravery and capacity to tell big-enough stories without determinism, teleology, and plan.[54]

Historically situated relational worldings make a mockery both of the binary division of nature and society and of our enslavement to Progress and its evil twin, Modernization. The Capitalocene was relationally made, and not by a secular godlike anthropos, a law of history, the machine itself, or a demon called Modernity. The Capitalocene must be relationally unmade in order to compose in material-semiotic SF patterns and stories something more livable, something Ursula K. Le Guin could be proud of. Shocked anew by our—billions of earth habitants', including your and my—ongoing daily assent in practice to this thing called capitalism, Philippe Pignarre and Isabelle Stengers note that denunciation has been singularly ineffective, or capitalism would have long ago vanished from the earth. A dark bewitched commitment to the lure of Progress (and its polar opposite) lashes us to endless infernal alternatives, as if we had no other ways to reworld, reimagine, relive, and

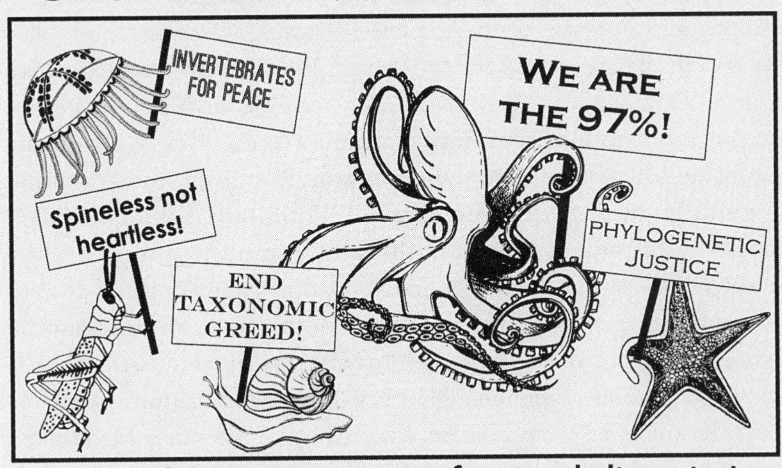

2.5. *Octopi Wall Street*: Symchthonic revolt. Art by Marley Jarvis, Laurel Hiebert, Kira Treibergs, 2011. Oregon Institute of Marine Biology.

reconnect with each other, in multispecies well-being. This explication does not excuse us from doing many important things better; quite the opposite. Pignarre and Stengers affirm on-the-ground collectives capable of inventing new practices of imagination, resistance, revolt, repair, and mourning, and of living and dying well. They remind us that the established disorder is not necessary; another world is not only urgently needed, it is possible, but not if we are ensorcelled in despair, cynicism, or optimism, and the belief/disbelief discourse of Progress.[55] Many Marxist critical and cultural theorists, at their best, would agree.[56] So would the tentacular ones.[57]

Chthulucene

Reaching back to generative complex systems approaches by Lovelock and Margulis, Gaia figures the Anthropocene for many contemporary Western thinkers. But an unfurling Gaia is better situated in the Chthulucene, an ongoing temporality that resists figuration and dating and demands myriad names. Arising from Chaos,[58] Gaia was and is a powerful intrusive force, in no one's pocket, no one's hope for salvation, capable of provoking the late twentieth century's best autopoietic complex systems thinking that led to recognizing the devastation caused by anthropogenic processes of the last few centuries, a necessary counter to the Euclidean figures and stories of Man.[59] Brazilian anthropologists and philosophers Eduardo Viveiros de Castro and Déborah Danowski exorcise lingering notions that Gaia is confined to the ancient Greeks and subsequent Eurocultures in their refiguring the urgencies of our times in the post-Eurocentric conference "The Thousand Names of Gaia."[60] Names, not faces, not morphs of the same, something else, a thousand somethings else, still telling of linked ongoing generative and destructive worlding and reworlding in this age of the earth. We need another figure, a thousand names of something else, to erupt out of the Anthropocene into another, big-enough story. Bitten in a California redwood forest by spidery *Pimoa chthulhu*, I want to propose snaky Medusa and the many unfinished worldings of her antecedents, affiliates, and descendants. Perhaps Medusa, the only mortal Gorgon, can bring us into the holobiomes of Terrapolis and heighten our chances for dashing the twenty-first-century ships of the Heroes on a living coral reef instead of allowing them to suck the last drop of fossil flesh out of dead rock.

The terra-cotta figure of Potnia Theron, the Mistress of the Animals, depicts a winged goddess wearing a split skirt and touching a bird with each hand.[61] She is a vivid reminder of the breadth, width, and temporal reach into pasts and futures of chthonic powers in Mediterranean and Near Eastern worlds and beyond.[62] Potnia Theron is rooted in Minoan and then Mycenean cultures and infuses Greek stories of the Gorgons (especially the only mortal Gorgon, Medusa) and of Artemis. A kind of far-traveling Ur-Medusa, the Lady of the Beasts is a potent link between Crete and India. The winged figure is also called Potnia Melissa, Mistress of the Bees, draped with all their buzzing-stinging-honeyed gifts. Note the acoustic, tactile, and gustatory senses elicited by the Mistress and her sympoietic, more-than-human flesh. The snakes and bees are more like stinging tentacular feelers than like binocular eyes, although these critters see too, in compound-eyed insectile and many-armed optics.

In many incarnations around the world, the winged bee goddesses are very old, and they are much needed now.[63] Potnia Theron/Melissa's snaky locks and Gorgon face tangle her with a diverse kinship of chthonic earthly forces that travel richly in space and time. The Greek word Gorgon translates as dreadful, but perhaps that is an astralized, patriarchal hear-

2.6. Icon for the Chthulucene. Potnia Theron with a Gorgon Face. Type of Potnia Theron, Kameiros, Rhodes, circa 600 BCE, terracotta, 13 in. diameter, British Museum, excavated by Auguste Salzmann and Sir Alfred Bilotti; purchased 1860. Photograph by Marie-Lan Nguyen, © 2007.

ing of much more awe-ful stories and enactments of generation, destruction, and tenacious, ongoing terran finitude. Potnia Theron/Melissa/Medusa give faciality a profound makeover, and that is a blow to modern humanist (including technohumanist) figurations of the forward-looking, sky-gazing Anthropos. Recall that the Greek *chthonios* means "of, in, or under the earth and the seas"—a rich terran muddle for SF, science fact, science fiction, speculative feminism, and speculative fabulation. The chthonic ones are precisely not sky gods, not a foundation for the Olympiad, not friends to the Anthropocene or Capitalocene, and definitely not finished. The Earthbound can take heart—as well as action.

The Gorgons are powerful winged chthonic entities without a proper genealogy; their reach is lateral and tentacular; they have no settled lineage and no reliable kind (genre, gender), although they are figured and storied as female. In old versions, the Gorgons twine with the Erinyes (Furies), chthonic underworld powers who avenge crimes against the natural order. In the winged domains, the bird-bodied Harpies carry out these vital functions.[64] Now, look again at the birds of Potnia Theron and ask what they do. Are the Harpies their cousins? Around 700 BCE Hesiod imagined the Gorgons as sea demons and gave them sea deities for parents. I read Hesiod's *Theogony* as laboring to stabilize a very bumptious queer family. The Gorgons erupt more than emerge; they are intrusive in a sense akin to what Stengers understands by Gaia.

The Gorgons turned men who looked into their living, venomous, snake-encrusted faces into stone. I wonder what might have happened if those men had known how to politely greet the dreadful chthonic ones. I wonder if such manners can still be learned, if there is time to learn now, or if the stratigraphy of the rocks will only register the ends and end of a stony Anthropos.[65]

Because the deities of the Olympiad identified her as a particularly dangerous enemy to the sky gods' succession and authority, mortal Medusa is especially interesting for my efforts to propose the Chthulucene as one of the big-enough stories in the netbag for staying with the trouble of our ongoing epoch. I resignify and twist the stories, but no more than the Greeks themselves constantly did.[66] The hero Perseus was dispatched to kill Medusa; and with the help of Athena, head-born favorite daughter of Zeus, he cut off the Gorgon's head and gave it to his accomplice, this virgin goddess of wisdom and war. Putting Medusa's severed head face-forward on her shield, the Aegis, Athena, as usual, played traitor to the Earthbound; we expect no better from motherless mind children. But great good came of this murder-for-hire, for from Medusa's dead body came the winged horse Pegasus. Feminists have a special friendship with horses. Who says these stories do not still move us materially?[67] And from the blood dripping from Medusa's severed head came the rocky corals of the western seas, remembered today in the taxonomic names of the Gorgonians, the coral-like sea fans and sea whips, composed in symbioses of tentacular animal cnidarians and photosynthetic algal-like beings called zooanthellae.[68]

With the corals, we turn definitively away from heady facial representations, no matter how snaky. Even Potnia Theron, Potnia Melissa, and Medusa cannot alone spin out the needed tentacularities. In the tasks of thinking, figuring, and storytelling, the spider of my first pages, *Pimoa chthulhu*, allies with the decidedly nonvertebrate critters of the seas. Corals align with octopuses, squids, and cuttlefish. Octopuses are called spiders of the seas, not only for their tentacularity, but also for their predatory habits. The tentacular chthonic ones have to eat; they are at table, *cum panis*, companion species of terra. They are good figures for the luring, beckoning, gorgeous, finite, dangerous precarities of the Chthulucene. This Chthulucene is neither sacred nor secular; this earthly worlding is thoroughly terran, muddled, and mortal—and at stake now.

Mobile, many-armed predators, pulsating through and over the coral reefs, octopuses are called spiders of the sea. And so *Pimoa chthulhu* and *Octopus cyanea* meet in the webbed tales of the Chthulucene.[69]

All of these stories are a lure to proposing the Chthulucene as a needed third story, a third netbag for collecting up what is crucial for ongoing, for staying with the trouble.[70] The chthonic ones are not confined to a vanished past. They are a buzzing, stinging, sucking swarm now, and human beings are not in a separate compost pile. We are humus, not Homo, not anthropos; we are compost, not posthuman. As a suffix, the word *kainos*, "-cene," signals new, recently made, fresh epochs of the thick present. To renew the biodiverse powers of terra is the sympoietic work and play of the Chthulucene. Specifically, unlike either the Anthropocene or the Capitalocene, the Chthulucene is made up of ongoing multispecies stories and practices of becoming-with in times that remain at stake, in precarious times, in which the world is not finished and the sky has not fallen—yet. We are at stake to each other. Unlike the dominant dramas of Anthropocene and Capitalocene discourse, human beings are not the only important actors in the Chthulucene, with all other beings able simply to react. The order is reknitted: human beings are with and of the earth, and the biotic and abiotic powers of this earth are the main story.

However, the doings of situated, actual human beings matter. It matters with which ways of living and dying we cast our lot rather than others. It matters not just to human beings, but also to those many critters across taxa which and whom we have subjected to exterminations, extinctions, genocides, and prospects of futurelessness. Like it or not, we are in the string figure game of caring for and with precarious worldings made terribly more precarious by fossil-burning man making new fossils as rapidly as possible in orgies of the Anthropocene and Capitalocene. Diverse human and nonhuman players are necessary in every fiber of the tissues of the urgently needed Chthulucene story. The chief actors are not restricted to the too-big players in the too-big stories of Capitalism and the Anthropos, both of which invite odd apocalyptic panics and even odder disengaged denunciations rather than attentive practices of thought, love, rage, and care.

Both the Anthropocene and the Capitalocene lend themselves too readily to cynicism, defeatism, and self-certain and self-fulfilling predictions, like the "game over, too late" discourse I hear all around me these days, in both expert and popular discourses, in which both technotheocratic geoengineering fixes and wallowing in despair seem to coinfect any possible common imagination. Encountering the sheer not-us, more-than-human worlding of the coral reefs, with their requirements for ongoing living and dying of their myriad critters, is also to encounter the knowledge that at least 250 million human beings today depend directly on the ongoing integrity of these holobiomes for their own ongoing living and dying well. Diverse corals and diverse people and peoples are at stake to and with each other. Flourishing will be cultivated as a multispecies response-ability without the arrogance of the sky gods and their minions, or else biodiverse terra will flip out into something very slimy, like any overstressed complex adaptive system at the end of its abilities to absorb insult after insult.

Corals helped bring the Earthbound into consciousness of the Anthropocene in the first place. From the start, uses of the term *Anthropocene* emphasized human-induced warming and acidification of the oceans from fossil-fuel-generated CO_2 emissions. Warming and acidification are known stressors that sicken and bleach coral reefs, killing the photosynthesizing zooanthellae and so ultimately their cnidarian symbionts and all of the other critters belonging to myriad taxa whose worlding depends on intact reef systems. Corals of the seas and lichens of the land also bring us into consciousness of the Capitalocene, in which deep-sea mining and drilling in oceans and fracking and pipeline construction across delicate lichen-covered northern landscapes are fundamental to accelerating nationalist, transnationalist, and corporate unworlding.

But coral and lichen symbionts also bring us richly into the storied tissues of the thickly present Chthulucene, where it remains possible—just barely—to play a much better SF game, in nonarrogant collaboration with all those in the muddle. We are all lichens; so we can be scraped off the rocks by the Furies, who still erupt to avenge crimes against the earth. Alternatively, we can join in the metabolic transformations between and among rocks and critters for living and dying well. "'Do you realize,' the phytolinguist will say to the aesthetic critic, 'that [once upon a time] they couldn't even read Eggplant?' And they will smile at our

2.7. Day octopus, *Octopus cyanea*, in the water near Lanai, Hawaii. Photograph by David Fleethham. © OceanwideImages.com.

ignorance, as they pick up their rucksacks and hike on up to read the newly deciphered lyrics of the lichen on the north face of Pike's Peak.'"[71]

Attending to these ongoing matters returns me to the question that began this chapter. What happens when human exceptionalism and the utilitarian individualism of classical political economics become unthinkable in the best sciences across the disciplines and interdisciplines? Seriously unthinkable: not available to think with. Why is it that the epochal name of the Anthropos imposed itself at just the time when understandings and knowledge practices about and within symbiogenesis and sympoietics are wildly and wonderfully available and generative in all the humusities, including noncolonizing arts, sciences, and politics? What if the doleful doings of the Anthropocene and the unworldings of the Capitalocene are the last gasps of the sky gods, not guarantors of the finished future, game over? It matters which thoughts think thoughts. We must think!

The unfinished Chthulucene must collect up the trash of the Anthropocene, the exterminism of the Capitalocene, and chipping and shredding and layering like a mad gardener, make a much hotter compost pile for still possible pasts, presents, and futures.

CHAPTER 3

Sympoiesis

Symbiogenesis and the Lively Arts of Staying with the Trouble

Symbiogenesis

Sympoiesis is a simple word; it means "making-with." Nothing makes itself; nothing is really autopoietic or self-organizing. In the words of the Inupiat computer "world game," earthlings are *never alone*.[1] That is the radical implication of sympoiesis. *Sympoiesis* is a word proper to complex, dynamic, responsive, situated, historical systems. It is a word for worlding-with, in company. Sympoiesis enfolds autopoiesis and generatively unfurls and extends it.

The vivid four-by-six-foot painting called *Endosymbiosis* hangs in the hallway joining the Departments of Geosciences and Biology at UMass Amherst, near the Life and Earth Café, surely a spatial clue to how critters become-with each other.[2] Perhaps as sensual molecular curiosity and definitely as insatiable hunger, irresistible attraction toward enfolding each other is the vital motor of living and dying on earth. Critters interpenetrate one another, loop around and through one another, eat each another, get indigestion, and partially digest and partially assimilate one another, and thereby establish sympoietic arrangements that are otherwise known as cells, organisms, and ecological assemblages.

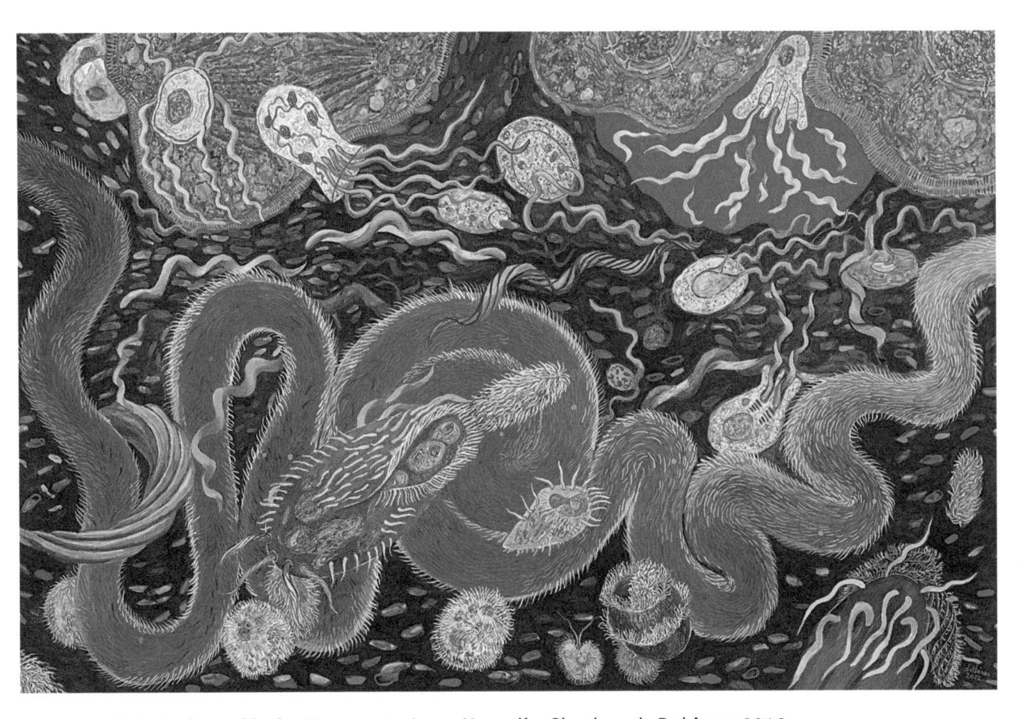

3.1. *Endosymbiosis: Homage to Lynn Margulis*, Shoshanah Dubiner, 2012. www.cybermuse.com.

Another word for these sympoietic entities is *holobionts*, or, etymologically, "entire beings" or "safe and sound beings."[3]

That is decidedly not the same thing as One and Individual. Rather, in polytemporal, polyspatial knottings, holobionts hold together contingently and dynamically, engaging other holobionts in complex patternings. Critters do not precede their relatings; they make each other through semiotic material involution, out of the beings of previous such entanglements. Lynn Margulis knew a great deal about "the intimacy of strangers," a phrase she proposed to describe the most fundamental practices of critters becoming-with each other at every node of intra-action in earth history. I propose *holoents* as a general term to replace "units" or "beings."

Like Margulis, I use *holobiont* to mean symbiotic assemblages, at whatever scale of space or time, which are more like knots of diverse intra-active relatings in dynamic complex systems, than like the entities of a biology made up of preexisting bounded units (genes, cells, organisms, etc.) in interactions that can only be conceived as competitive or cooperative. Like hers, my use of *holobiont* does not designate host + symbionts because all of the players are symbionts to each other, in diverse kinds of relationalities and with varying degrees of openness to attachments and assemblages with other holobionts. *Symbiosis* is not a synonym for "mutually beneficial." The array of names needed to designate the heterogeneous webbed patterns and processes of situated and dynamic dilemmas and advantages for the symbionts/holobionts is only beginning to surface as biologists let go of the dictates of possessive individualism and zero-sum games as the template for explanation.

An adept in the study of microbes, cell biology, chemistry, geology, and paleogeography, as well as a lover of languages, arts, stories, systems theories, and alarmingly generative critters, including human beings, Margulis was a radical evolutionary theorist. Her first and most intense loves were the bacteria and archaea of Terra and all their bumptious doings. The core of Margulis's view of life was that new *kinds* of cells, tissues, organs, and species evolve primarily through the long-lasting intimacy of strangers. The fusion of genomes in symbioses, followed by natural selection—with a very modest role for mutation as a motor of system level change—leads to increasingly complex levels of good-enough quasi-individuality to get through the day, or the aeon. Margulis called this basic and mortal life-making process symbiogenesis.

Bacteria and archaea did it first. My sense is that in her heart of hearts

Margulis felt that bacteria and archaea did it all, and there wasn't much left for so-called higher-order biological entities to do or invent. Eventually, however, by fusing with each other in stabilized, ongoing ways, archaea and bacteria invented the modern complex cell, with its nucleus full of ropy chromosomes made of DNA and proteins, and diverse other sorts of extranuclear organelles, from undulating whips and spinning blades for locomotion to specialized vesicles and tubules for a zillion functions that work better kept a bit separate from each other.[4] Because she was a founder of Gaia theory with James Lovelock and a student of interlocked and multileveled systemic processes of nonreductionist organization and maintenance that make earth itself and earth's living beings unique, Margulis called these processes autopoietic.[5] Perhaps she would have chosen the term *sympoietic*, but the word and concept had not yet surfaced.[6] As long as autopoiesis does not mean self-sufficient "self making," autopoiesis and sympoiesis, foregrounding and backgrounding different aspects of systemic complexity, are in generative friction, or generative enfolding, rather than opposition.

In 1998, a Canadian environmental studies graduate student named M. Beth Dempster suggested the term *sympoiesis* for "collectively-producing systems that do not have self-defined spatial or temporal boundaries. Information and control are distributed among components. The systems are evolutionary and have the potential for surprising change." By contrast, autopoietic systems are "self-producing" autonomous units "with self defined spatial or temporal boundaries that tend to be centrally controlled, homeostatic, and predictable."[7] Symbiosis makes trouble for autopoiesis, and symbiogenesis is an even bigger troublemaker for self-organizing individual units. The more ubiquitous symbiogenesis seems to be in living beings' dynamic organizing processes, the more looped, braided, outreaching, involuted, and sympoietic is terran worlding.

Mixotricha paradoxa is everyone's favorite critter for explaining complex "individuality," symbiogenesis, and symbiosis. Margulis described this critter that is/are made up of at least five different taxonomic *kinds* of cells with their genomes this way:

> Under low magnification, *M. paradoxa* looks like a single-celled swimming ciliate. With the electron microscope, however, it is seen to consist of five distinct kinds of creatures. Externally, it is most obviously the kind of one-celled organism that is classified as a protist. But in-

> side each nucleated cell, where one would expect to find mitochondria, are many spherical bacteria. On the surface, where cilia should be, are some 250,000 hairlike *Treponema spirochetes* (resembling the type that causes syphilis), as well as a contingent of large rod bacteria that is also 250,000 strong. In addition, we have redescribed 200 spirochetes of a larger type and named them *Canaleparolina darwiniensis*.[8]

Leaving out viruses, each *M. paradoxa* is not one, not five, not several hundred thousand, but a poster critter for holobionts. This holobiont lives in the gut of an Australian termite, *Mastotermes darwiniensis*, which has its own SF stories to tell about ones and manys, or holoents. Termite symbioses, including their doings with people, not to mention mushrooms, are the stuff of legends—and cuisine. Check out the holobiomes of *Macrotermes natalensis* and its cultivated fungus *Termitomyces*, recently in the science news.[9] *M. paradoxa* and their ilk have been my companions in writing and thinking for decades.

Since Darwin's *On the Origin of Species* in 1859, biological evolutionary theory has become more and more essential to our ability to think, feel, and act well; and the interlinked Darwinian sciences that came together roughly between the 1930s and 1950s into "the Modern Synthesis" or "New Synthesis" remain astonishing. How could one be a serious person and not honor such works as Theodosius Dobzhansky's *Genetics and the Origin of Species* (1937), Ernst Mayr's *Systematics and the Origin of Species* (1942), George Gaylord Simpson's *Tempo and Mode in Evolution* (1944), and even Richard Dawkins's later sociobiological formulations within the Modern Synthesis, *The Selfish Gene* (1976)? However, bounded units (code fragments, genes, cells, organisms, populations, species, ecosystems) and relations described mathematically in competition equations are virtually the only actors and story formats of the Modern Synthesis. Evolutionary momentum, always verging on modernist notions of progress, is a constant theme, although teleology in the strict sense is not. Even as these sciences lay the groundwork for scientific conceptualization of the Anthropocene, they are undone in the very thinking of Anthropocene systems that require enfolded autopoietic and sympoietic analysis.

Rooted in units and relations, especially competitive relations, the sciences of the Modern Synthesis, for example, population genetics, have a hard time with four key biological domains: embryology and development, symbiosis and collaborative entanglements of holobionts and holobiomes, the vast worldings of microbes, and exuberant critter biobehavioral inter- and intra-actions.[10] Approaches tuned to "multispecies becoming-with" better sustain us in staying with the trouble on terra. An emerging "New New Synthesis"—an extended synthesis—in transdisciplinary biologies and arts proposes string figures tying together human and nonhuman ecologies, evolution, development, history, affects, performances, technologies, and more.

Indebted first to Margulis, I can only sketch a few aspects of the "Extended Evolutionary Synthesis" unfolding in the early twenty-first century.[11] Forming part of her cosmopolitan heritage, formulations of symbiogenesis predate Margulis in the early twentieth-century work of the Russian Konstantin Mereschkowsky and others.[12] However, Margulis, her successors, and her colleagues bring together symbiogenetic imaginations and materialities with all of the powerful cyborg tools of the late twentieth-century molecular and ultrastructural biological revolutions, including electron microscopes, nucleic acid sequencers, immunoassay techniques, immense and comparative genomic and proteomic databases, and more. The strength of the Extended Synthesis is precisely in the intellectual, cultural, and technical convergence that makes it possible to develop new model systems, concrete experimental practices, research collaborations, and both verbal and mathematical explanatory instruments. Such a convergence was materially impossible before the 1970s and after.

A model is a work object; a model is not the same *kind* of thing as a metaphor or analogy. A model is worked, and it does work. A model is like a miniature cosmos, in which a biologically curious Alice in Wonderland can have tea with the Red Queen and ask how this world works, even as she is worked by the complex-enough, simple-enough world. Models in biological research are stabilized systems that can be shared among colleagues to investigate questions experimentally and theoretically. Traditionally, biology has had a small set of hard-working living models, each shaped in knots and layers of practice to be apt for some kinds of questions and not others. Listing seven basic model systems of developmental biology (namely, fruit flies *Drosophila melanogaster*; a nematode, *Caenorhabditis elegans*; the mouse *Mus musculis*; a frog, *Xenopus laevis*; the zebrafish *Danio rerio*; the chicken *Gallus gallus*; and the mustard *Arabidopsis thaliana*), Scott Gilbert wrote,

The recognition that one's organism is a model system provides a platform upon which one can apply for funds, and it assures one of a community of like-minded researchers who have identified problems that the community thinks are important. There has been much lobbying for the status of a model system and the fear is that if your organism is not a recognized model, you will be relegated to the backwaters of research. Thus, "model organisms" have become the center for both scientific and political discussions in contemporary developmental biology.[13]

Excellent for studying how parts (genes, cells, tissues, etc.) of well-defined entities fit together into cooperating and/or competing units, all seven of these individuated systems fail the researcher studying webbed inter- and intra-actions of symbiosis and sympoiesis, in heterogeneous temporalities and spatialities. Holobionts require models tuned to an expandable number of quasi-collective/quasi-individual partners in constitutive relatings; these relationalities *are* the objects of study. The partners do not precede the relatings. Such models are emerging for the transformative processes of EcologicalEvolutionaryDevelopmental biology.

Margulis gave us dynamic multipartnered entities like *Mixotricha paradoxa* to study the evolutionary invention of complex cells from the intra- and interactions of bacteria and archaea. I will briefly introduce two more models, each proposed and elaborated in the laboratory to study a transformation of organizational patterning in the living world: (1) a choanoflagellate-bacteria model for the invention of animal multicellularity, and (2) a squid-bacteria model for the elaboration of developmental symbioses between and among critters necessary to each other's becoming. A third symbiogenetic model for the formation of complex ecosystems immediately suggests itself in the holobiomes of coral reefs, and I will approach this model through science art worldings rather than the experimental laboratory.

Although multicellular plants appeared on earth half a million years earlier, because of its robustness and sympoietic richness, I focus on a proposed model system for the emergence of animal multicellularity. Every living thing has emerged and persevered (or not) bathed and swaddled in bacteria and archaea. Truly nothing is sterile; and that reality is a terrific danger, basic fact of life, and critter-making opportunity. Using molecular and comparative genomic approaches and proposing infectious—symbiogenetic—processes, Nicole King's laboratory at the

University of California, Berkeley, works to reconstruct possible origins and development of animal multicellularity.[14] These scientists show that interspecies—really, interkingdom—meetings and enfoldings can produce entities that hold together, develop, communicate, and form layered tissues like animals do.

As Alegado and King put it,

> Comparisons among modern animals and their closest living relatives, the choanoflagellates, suggest that the first animals used flagellated collar cells to capture bacterial prey. The cell biology of prey capture, such as cell adhesion between predator and prey, involves mechanisms that may have been co-opted to mediate intercellular interactions during the evolution of animal multicellularity. Moreover, a history of bacterivory may have influenced the evolution of animal genomes by driving the evolution of genetic pathways for immunity and facilitating lateral gene transfer. Understanding the interactions between bacteria and the progenitors of animals may help to explain the myriad ways in which bacteria shape the biology of modern animals, including ourselves.[15]

In Marilyn Strathern's sense, partial connections abound. Getting hungry, eating, and partially digesting, partially assimilating, and partially transforming: these are the actions of companion species.

King's ambitious program is crafting a stabilized and genomically well-characterized model system of cultures of choanoflagellates (*Salpingoeca rosetta*) and bacteria from the genus *Algoriphagus* to investigate critical aspects of the formation of multicellular animals. Choanoflagellates can live as either single cells or multicellular colonies; what determines the transitions? The close evolutionary relationship between choanoflagellates and animals lends strength to the model.[16] The symbiogenetic theory of origins of multicellularity is contested; there are attractive alternate explanations. What distinguishes King's lab is its production of a model system that is experimentally tractable, transferable in principle to other sites, and generative of testable questions at the heart of being animal. To be animal is to become-with bacteria (and, no doubt, viruses and many other sorts of critters; a basic aspect of sympoiesis is its expandable set of players). No wonder the best science writers bring Nicole King's lab into my dinner conversations on a regular basis.[17]

Next, I hold out a tasty model system for studying developmental symbioses. The question here is not how animals hold themselves together at all, but rather, how they craft developmental patternings that take them through time in astonishing morphogeneses. My favorite model is the diminutive Hawaiian bobtail squid, *Euprymna scolopes*, and its bacterial symbionts, *Vibrio fischeri*, which are essential for the squid's constructing its ventral pouch that houses luminescing bacteria, so that the hunting squid can look like a starry sky to its prey below on dark nights, or appear not to cast a shadow on moonlit nights. The squid-bacterial symbiosis has proven remarkably generative for many kinds of studies, "from ecology and evolution of a symbiotic system to the underlying molecular mechanisms of partner interactions that lead to establishment, development, and long-term-persistence of the alliance."[18]

Unless the juvenile squid are infected in the right spot, at the right time, by the right bacteria, they do not develop their own structures for housing bacteria when they are hunting adults. The bacteria are fully part of the squid's developmental biology. In addition, the bacteria produce signals that regulate the adult squids' circadian rhythms. The squid regulate bacterial numbers, exclude unwanted associates, and provide inviting surfaces for setting up vibrio homes. Herself trained in marine invertebrate field biology, biochemistry, and biophysics, McFall-Ngai began work on the naturally occurring squid-bacteria holobiont in 1988, when she started to collaborate with Edward (Ned) Ruby, a microbiologist also interested in symbiosis. Remembering that other vibrio bacteria are responsible for the pathogenic communication that is cholera, I was not surprised to learn what multitalented communicators these sorts of bacteria are. As McFall-Ngai put it, "The Vibrionaceae are a group of bacteria whose members often have broad physiological scope and multiple ecological niches."[19] Material semiotics is exuberantly chemical; the roots of language across taxa, with all its understandings and misunderstandings, lie in such attachments.

The sympoietic collaborations of squid and bacteria are matched by the sympoietic string figures across disciplines and methodologies, including genome sequencing, myriad imaging technologies, functional genomics, and field biology, which make symbiogenesis such a powerful framework for twenty-first-century biology. Working on pea aphid symbiosis with *Buchnera*, Nancy Moran emphasizes this point: "The primary reason that symbiosis research is suddenly active, after decades at the margins of mainstream biology, is that DNA technology and genomics give us enormous new ability to discover symbiont diversity, and more significantly, to reveal how microbial metabolic capabilities contribute to the functioning of hosts and biological communities."[20] I would add the necessity of asking how the multicellular partners in the symbioses affect the microbial symbionts. "Host-symbiont" seems an odd locution for what is happening; at whatever size, all the partners making up holobionts are symbionts to each other.

Two transformative papers embody for me the profound scientific changes afoot.[21] Subtitling their paper "We Have Never Been Individuals," Gilbert, Sapp, and Tauber argue for holobionts and a symbiotic view of life by summarizing the evidence against bounded units from anatomy, physiology, genetics, evolution, immunology, and development. In "Animals in a Bacterial World: A New Imperative for the Life Sciences," the twenty-six coauthors present the growing knowledge of a vast range of animal-bacterial interactions at both ecosystem and intimate symbiosis scales. They argue that this evidence should profoundly alter approaches to five questions: "how have bacteria facilitated the origin and evolution of animals; how do animals and bacteria affect each other's genomes; how does normal animal development depend on bacterial partners; how is homeostasis maintained between animals and their symbionts; and how can ecological approaches deepen our understanding of the multiple levels of animal-bacterial interaction."[22]

Stories about worried colleagues at conferences, uncomprehending reviewers unused to so much evidential and disciplinary boundary crossing in one paper, or initially enthusiastic editors getting cold feet surround these papers. Such stories normally surround risky and generative syntheses and propositions. The critics are a crucial part of the holobiome of making science, and I am not a disinterested observer.[23] Nonetheless, I think it matters that both of these papers were published in prominent places at a critical inflection point in the curve of research on, and explanation of, complex biological systems in the urgent times called the Anthropocene, when the arts for living on a damaged planet demand sympoietic thinking and action.

Interlacing Sciences and Arts with Involutionary Momentum

I am committed to art science worldings as sympoietic practices for living on a damaged planet. Carla Hustak and Natasha Myers gave all of us a beautiful paper titled "Involutionary Momentum" that is a hinge for me between symbiogenesis and the science art worldings I present in the third section of this chapter. These authors reread Darwin's own sensuous writing about his exquisite attention to absurdly sexual orchids and their pollinating insects; Hustak and Myers also themselves attend to the many enfoldings and communications among bees, wasps, orchids, and scientists. The authors suggest that "involution" powers the "evolution" of living and dying on earth. Rolling inward enables rolling outward; the shape of life's motion traces a hyperbolic space, swooping and fluting like the folds of a frilled lettuce, coral reef, or bit of crocheting. Like the biologists of the previous section, Hustak and Myers argue that a zero-sum game based on competing methodological individualists is a caricature of the sensuous, juicy, chemical, biological, material-semiotic, and science-making world. Counting "articulate plants and other loquacious organisms" among their number, living critters love the floridly repetitive mathematics of the pushes and pulls of hyperbolic geometry, not the accountant's hell of a zero-sum game.[24]

> Rather, the orchid and its bee-pollinators are mutually constituted through a reciprocal capture from which neither plant nor insect can be disentangled . . . It is in encounters among orchids, insects, and scientists that we find openings for an ecology of interspecies intimacies and subtle propositions. What is at stake in this involutionary approach is a theory of ecological relationality that takes seriously organisms' practices, their inventions, and experiments crafting interspecies lives and worlds. This is an ecology inspired by a feminist ethic of "response-ability" . . . in which questions of species difference are always conjugated with attentions to affect, entanglement, and rupture; an affective ecology in which creativity and curiosity characterize the experimental forms of life of all kinds of practitioners, not only the humans.[25]

Orchids are famous for their flowers looking like the genitals of the female insects of the particular species needed to pollinate them. The right sort of males seeking females of their own kind are drawn to the color, shape, and alluring insectlike pheromones of a particular species of orchid. These interactions have been explained (away) in neo-Darwinian orthodoxy as nothing but biological deception and exploitation of the insect by the flower—in other words, an excellent example of the selfish gene in action. Hustak and Myers instead read aslant neo-Darwinism, even in this hard case of strong asymmetry of "costs and benefits," to find other necessary models for a science of plant ecology. The stories of mutation, adaptation, and natural selection are not silenced; but they are not turned up so loud as to deafen scientists, as if the evidence demanded it, when increasingly something more complex is audible in research across fields. "This requires reading with our senses attuned to stories told in otherwise muted registers. Working athwart the reductive, mechanistic, and adaptationist logics that ground the ecological sciences, we offer a reading that amplifies accounts of the creative, improvisational, and fleeting practices through which plants and insects *involve* themselves in one another's lives."[26]

But what happens when a partner involved critically in the life of another disappears from the earth? What happens when holobionts break apart? What happens when entire holobiomes crumble into the rubble of broken symbionts? This kind of question has to be asked in the urgencies of the Anthropocene and Capitalocene if we are to nurture arts for living on a damaged planet. In his science fiction novel *The Speaker for the Dead* Orson Scott Card explored how a young boy who had excelled in exterminationist technoscience in a cross-species war with an insectoid hive species later in life took up responsibility for the dead, for collecting up the stories for those left behind when a being, or a way of being, dies. The man had to do what the boy, immersed only in cyber-realities and deadly virtual war, was never allowed to do; the man had to visit, to live with, to face the dead and the living in all of their materialities. The task of the Speaker for the Dead is to bring the dead into the present, so as to make more response-able living and dying possible in times yet to come. My hinge to science art worldings turns on the ongoing performance of memory by an orchid for its extinct bee.

In xkcd's cartoon "Bee Orchid," we know a vanished insect once existed because a living flower still looks like the erotic organs of the avid female bee hungry for copulation. But the cartoon does something very special; it does *not* mistake lures for identity; it does *not* say the flower is exactly like the extinct insect's genitals. Instead, the flower collects up the presence of the bee aslant, in desire and mortality. The shape of the flower is "an idea of what the female bee looked like to the male bee . . . as interpreted by a plant . . . the only memory of the bee is a painting by a dying flower."[27] Once embraced by living buzzing bees, the flower is a speaker for the dead. A stick figure promises to remember the bee flower when it comes time. The practice of the arts of memory enfold all terran critters. That must be part of any possibility for resurgence!

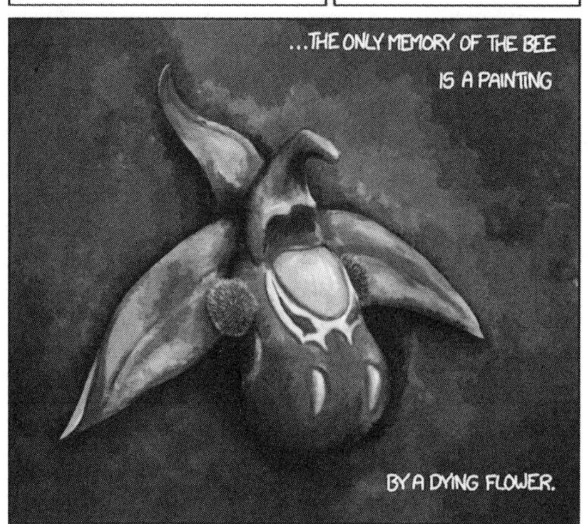

3.2. "Bee Orchid." © xkcd.com (Randall Munroe).

Science Art Worldings for Staying with the Trouble

I end this chapter with four engaged science art activist worldings committed to partial healing, modest rehabilitation, and still possible resurgence in the hard times of the imperial Anthropocene and Capitalocene. I think of these worldings as stinger-endowed, unfurling, grasping tentacles of the ink-spurting, disguise-artist, hunting critters of an ongoing past, present, and future called the Chthulucene.[28] Speaking resurgence to despair, the Chthulucene is the timespace of the symchthonic ones, the symbiogenetic and sympoietic earthly ones, those now submerged and squashed in the tunnels, caves, remnants, edges, and crevices of damaged waters, airs, and lands. The chthonic ones are those indigenous to the earth in myriad languages and stories; and decolonial indigenous peoples and projects are central to my stories of alliance.

Each of the science art worldings cultivates robust response-ability for powerful and threatened places and beings. Each is a model system for sympoietic, multiplayer, multispecies thinking and action located in a particularly sensitive place: (1) the Great Barrier Reef and all the world's coral reefs, with the Crochet Coral Reef project, initiated and coordinated by the Institute for Figuring in Los Angeles; (2) the island Republic of Madagascar, with the Malagasy-English children's natural history book series called the Ako Project, made possible by multinational friendships among scientists and artists; (3) the circumpolar northern lands of the Inupiat in Alaska, site of the *Never Alone* computer game project, centered in story-making practices among the Inupiat[29] and brought into being by the sympoiesis of E-Line Media and the Cook Inlet Tribal Council; and, my most developed case, (4) Black Mesa and the Navajo and Hopi lands enmeshed in Arizona, site of many-threaded coalitional work including Black Mesa Indigenous Support, Black Mesa Trust (Hopi), the scientists and indigenous herding people committed to Navajo-Churro sheep, Black Mesa Weavers for Life and Land, the mostly Diné activists of the Black Mesa Water Coalition, and the people and sheep of Diné be'iiná / The Navajo Lifeway.[30]

Each of these projects is a case of noninnocent, risky, committed "becoming involved in one another's lives."[31] Making-with and tangled-with the tentacular ones, which are gripping and stinging for an ongoing generative Chthulucene, each is a SF string figure of multispecies becoming-with. These science art worldings are holobiomes, or holoents, in which scientists, artists, ordinary members of communities, and nonhuman beings become enfolded in each other's projects, in each other's lives; they come to need each other in diverse, passionate, corporeal, meaningful ways. Each is an animating project in deadly times. They are sympoietic, symbiogenetic, and symanimagenic.

Four Critical Zones

Bathed in hot and acid oceans that are becoming more acidic and hotter by the decade, coral holobiomes everywhere are threatened. Coral reefs have the highest biodiversity of any kind of marine ecosystem. The symbiosis of cnidarian polyps, photosynthesizing dinoflagellates called zooanthellae living in the coral tissue, and a horde of microbes and viruses make up the keystone of the coral holobiome, which is home to multitudes of other critters. Hundreds of millions of human beings, many of them very poor, depend directly on healthy coral ecosystems for their livelihoods.[32] Such sentences hugely understate coral interdependence with human and nonhuman critters. Recognition of dying coral reef ecosystems in warming and acidifying seas was at the heart of advancing the very term *Anthropocene* in 2000. Coral, along with lichens, are also the earliest instances of symbiosis recognized by biologists; these are the critters that taught biologists to understand the parochialism of their own ideas of individuals and collectives. These critters taught people like me that we are all lichens, all coral. In addition, deepwater reefs in some locations seem to be able to function as refugia for replenishing damaged corals in shallower waters.[33] Coral reefs are the forests of the sea, like Anna Tsing's forest refugia of the land. Besides all of this, coral reef worlds are achingly beautiful. I cannot imagine it is only human people who know this beauty in their flesh.

A large island nation off the east coast of Africa, the Republic of Madagascar is home to complex, layered tapestries of historically situated peoples and other critters, including lemurs, close relatives of monkeys and apes. Nine out of ten kinds of Madagascar's nonhuman critters, including all species of lemurs, live nowhere else on earth. The rate of extinction and destruction of the many kinds of Madagascar's forests and watersheds vital for rural people (the large majority of Madagascar's human citizens), urban and town residents, and myriad nonhumans is almost beyond imagination, except that it is well advanced—but not uncontested locally and translocally. Evidence from photography indicates that 40 to 50 percent of the forests of Madagascar that were still thriving in 1950 are gone now, along with their critters, including their people, who for centuries harvested (and cultivated) woodland bounty for their lives. Forest well-being is one of the most urgent priorities for flourishing—indeed, survival—all over the earth. The contestations must matter; it's not a choice, it's a necessity.[34]

The circumpolar North bears the brunt of the Anthropocene and Capitalocene. The Arctic is warming at almost twice the rate of the global average. Sea ice, glaciers, and permafrost melt; people, animals, microbes, and plants can no longer rely on the seasons, nor indeed on the temporally punctuated solid or liquid forms of matter crucial to their perceptions and ways of getting on in life. Eating each other properly requires meeting each other properly, and that requires good-enough synchronicity. Synchronicity is exactly one of the system properties flipping out all over earth. Change on earth is not the problem; rates and distributions of change are very much the problem. In addition, consumption-obsessed imperial circumpolar nations vie with each other in increasingly militarized seas to claim and extract the huge reserves of carbonized fossils encased in the far North, promising a further release of greenhouse gases on a scale that simply cannot be allowed to happen. A geophysical, geopolitical storm of unprecedented proportions is changing practices of living and dying across the North. The coalitions of peoples and critters facing this storm are critical to the possibilities of earth's powers of resurgence.

Located on the four-thousand-square-mile Colorado Plateau, Black Mesa, or Big Mountain, is ancestral land for both Hopi and Diné peoples. Black Mesa is also a contemporary place needed by both Navajo and Hopi families for income, food, water, sociality, and ceremony. The Black Mesa coalfield, once a huge Pleistocene lake, is the largest coal deposit in the United States. Beginning in 1968, this colonizing capitalist extractive nation hosted the largest strip-mining operation in North America, run by the Peabody Western Coal Company, part of Peabody Energy, the world's largest private-sector coal company. For forty years, coal from the Black Mesa strip mine was pulverized, mixed with immense quantities of pristine water from the irreplaceable Navajo aquifer, and carried in a giant slurry pipeline (owned by Southern Pacific) 273 miles to the heavily polluting coal-fired Mohave Generating Station in Nevada, built by the Bechtel Corporation. This plant provided energy for the blooming toxic cities in the desert Southwest, including Los Angeles. People living on Black Mesa to this day have neither assured clean water nor reliable electricity, and many of their wells failed as the Navajo aquifer was depleted. Sheep that drink from sulfate-rich toxic waste ponds die, and groundwater is polluted.

First the slurry pipe, then the Black Mesa mine, and finally the Mohave Generating Station were closed down in 2005 through the concerted work of both indigenous and settler environmentalists.[35] Attempting to combine operations with its nearby Kayenta site under a single renewal permit running to 2026, Peabody currently has plans to reopen and expand the Black Mesa mine, targeting still more land needed by sheep and people, not to mention other critters. The expanded operation would wash coal with water from the Coconino Aquifer.

Coal from the Kayenta strip mine is shipped ninety-seven miles to the Navajo Generating Station (NGS) on the Arizona-Utah border, near Glen Canyon Dam; the NGS is the largest power-generating plant in the U.S. West.[36] The irony of the power station's name should escape no one, since half of Navajo homes do not have electricity and the Navajo Nation does not own the plant. Even setting aside the long-term well-being of people, other critters, land, and water, without a serious share in the profits made from coal and affordable energy for local residents, dependence on coal-related jobs keeps the Navajo Nation, as well as the Hopi, in a vise grip. Unemployment in the Navajo Nation runs around 45 percent, and both Hopi and Diné are among the poorest citizens of the United States. When built by Bechtel in the 1970s on land leased from the Navajo Nation, this plant was the second-largest utility in the United States. The largest owner of the Navajo Generating Station is the federal government's Bureau of Reclamation in the Department of the Interior; the Bureau of Indian Affairs, also in the Department of the Interior, is charged with protecting Native lands and resources. Coyote is well and truly in the sheep corral in that arrangement. In 2010 Peabody's Kayenta mine was listed among the most dangerous in the United States and targeted for increased scrutiny by the federal Mine Safety and Health Administration.[37] This plant powers the pumping stations that transfer the waters of the Colorado River through a 336-mile-long aqueduct to the always fast-growing cities of Tucson and Phoenix. Amid ongoing struggles over both the plant's effect on air quality and access to water in the desert, in 2014 the NGS obtained a permit to continue operation as a conventional coal-fired plant until December 2044.[38]

Hopi ancestors dug coal for their fires out of seams in Black Mesa's sandstone for centuries. Despite a destructive meme to the contrary—a very useful meme for the fossil fuel extraction industry—Diné and Hopi agriculturalists and herders lived adjacent to and among each other in mixed amity and competition until the advent of industrial-scale coal mining on Black Mesa, which engineered intense conflict conveniently misread as timeless tribal feuds. In 1966, transnational corporations obtained leases signed by both tribal councils, without discussion or consent by the great majority of tribal members or collective bodies (kivas, chapters). The terms of bargaining for these leases were both inherently asymmetrical and enabled by ethically compromised legal processes, epitomized by a lawyer and bishop of the Mormon church named John Boyden, who, without the Hopi's knowledge, worked simultaneously for Peabody and selected Hopi leaders. Thousands of Navajo lived on Black Mesa, including some of the most traditional among the Diné. The Navajo tribal council initially refused to work with Boyden, so he cultivated Hopi whose leaders were bitterly factionalized between so-called traditionalists and progressives, beginning at a time when the Hopi had no overall governing council. Boyden worked effectively over a long period to craft legislation to clear the land of Navajo sheep people and to shift legal control to the Hopi, who did not live on the land that would be strip-mined. Traditional Hopi fiercely opposed Boyden, but to no avail. Well connected in Washington, Boyden was crucial to crafting the legal, political, and economic strategy to exploit Black Mesa's coal bounty. A Freedom of Information Act suit filed by the Native American Rights Fund ascertained that out of funds held in a federal trust for the Hopi, over thirty years Boyden was paid $2.7 million for his "pro bono" services to the tribe.[39]

In 1974, the U.S. Congress passed a bill introduced by Arizona senator John McCain, a man with close personal and family ties to mining and energy industries, called the Navajo-Hopi Land Settlement Act. The act has resulted in the forced removal of up to fifteen thousand Diné without serious provision of anywhere for people and animals to go, even if ties to specific places were irrelevant. But both sheep and people know and care a great deal where they come from, where they are, and where they go.[40] In 1980 the federal government purchased a uranium-contaminated site near Chambers, Arizona, as new lands for the evicted Diné. In 1996, McCain, then chairman of the Senate Committee on Indian Affairs, authored a second forced relocation act. The Navajo turned to the United Nations High Commission for Human Rights. The struggle continues, with extraordinary efforts by young activists to heal the coal-scarred wounds dividing Hopi and Navajo. In 2005, 75 percent of overall annual Hopi income and 40 percent of Navajo income derived ultimately from Black Mesa mining operations. The struggle is dauntingly complex.[41]

The stories I will tell about Black Mesa are about resurgence in the face of genocide and extermination, about sheep and weaving, about art science activist worldings, about coalitions in struggle for what the Navajo call *hózhó*—balance, harmony, beauty, right relations of land and people—in this troubled world on the Colorado Plateau.

And so these are four critical zones of the tussle between the Anthropocene and Capitalocene, on the one hand, and the Chthulucene, on the other: coral forests of the oceans, diverse tropical forests of an island nation and ecosystem, rapidly melting arctic lands and seas, and coal seams and aquifers of indigenous lands linked in a global chain of ongoing colonial anthropogenic devastation. It is time to turn to sympoietic worldings, to vital models crafted in SF patterns in each zone, where ordinary stories, ordinary becoming "involved in each other's lives," propose ways to stay with the trouble in order to nurture well-being on a damaged planet. Symchthonic stories are not the tales of heroes; they are the tales of the ongoing.

Resurgence in Four Parts

THE CROCHET CORAL REEF

In 1997, Daina Taimina, a Latvian mathematician at Cornell University, "finally worked out how to make a physical model of hyperbolic space that allows us to feel, and to tacitly explore the properties of this unique geometry. The method she used was crochet."[42] With this tie between math and fiber arts in mind, in 2005, after reading an article on coral bleaching, Christine Wertheim, a crafter and poet, suggested to her twin sister Margaret, a mathematician and artist, "We should crochet a coral reef."[43] We can fight for the coral reefs that way, implied this odd imperative. The sisters were watching an episode of *Xena Warrior Princess*, and Xena's and her sidekick Gabrielle's fabulous fighting action—or maybe just the incomparable Lucy Lawless and Renee O'Connor—inspired them.[44] The consequences have been utterly out of proportion to what the twin sisters in Los Angeles imagined that first night. So far, about eight thousand people, mostly women, in twenty-seven countries—

3.3. Beaded jellyfish made by Vonda N. McIntyre for the Crochet Coral Reef. From the collection of the Institute for Figuring (IFF). Photograph © IFF.

from Ireland, Latvia, the United Arab Emirates, Australia, the United States, the UK, Croatia, and more—have come together to crochet in wool, cotton, plastic bags, discarded reel-to-reel tape, vinyl jelly yarn, Saran wrap, and just about anything else that can be induced to loop and whirl in the codes of crocheting.

The code is so simple: crocheted models of hyperbolic planes achieve their ruffled forms by progressively increasing the number of stitches in each row. The emergent vitalities of this wooly experimental life-form take diverse corporeal shape as crafters increase the numbers from row to row irregularly, oddly, whimsically, or strictly to see what forms they could make—not just any forms, but crenulated beings that take life as marine critters of the vulnerable reefs.[45] "Every woolen form has its fibrous DNA."[46] But wool is hardly the only material. Plastic bottle anemone trees with trash tendrils and anemones made from *New York Times* blue plastic wrappers find their reef habitats. Making fabulated, rarely mimetic, but achingly evocative models of coral reef ecosystems, or maybe of just a few critters, the Crochet Coral Reef has morphed into what is probably the world's largest collaborative art project.

The involutionary momentum of the crochet coral reef powers the sympoietic knotting of mathematics, marine biology, environmental activism, ecological consciousness raising, women's handicrafts, fiber arts, museum display, and community art practices. A kind of hyperbolic embodied knowledge, the crochet reef lives enfolded in the materialities of global warming and toxic pollution; and the makers of the reef practice multispecies becoming-with to cultivate the capacity to respond, response-ability.[47] The crochet reef is the fruit of "algorithmic code, improvisational creativity, and community engagement."[48] The reef works not by mimicry, but by open-ended, exploratory process. "Iterate, deviate, elaborate" are the principles of the process.[49] DNA could not have said it better.

The Crochet Coral Reef has a core set of reefs made for exhibitions, like the first ones at the Warhol Museum in Pittsburgh and the Chicago Cultural Center, both in 2007, to the Coral Forest exhibited in Abu Dhabi in 2014 and beyond. The morphing assemblages are kept at the Los Angeles Institute for Figuring (IFF), and they fill the Wertheims' home. The IFF is the Wertheims' nonprofit organization in LA, founded in 2003 and dedicated to "the aesthetic dimensions of mathematics, science, and engineering."[50] The core concept is material play, and the IFF proposes and enacts not think tanks or work tanks, but play tanks, which I understand as arts for living on a damaged planet. The IFF and the Crochet Coral Reef are art-science-activist worldings, bringing people together to do string figures with math, sciences, and arts in order to make active attachments that might matter to resurgence in the Anthropocene and Capitalocene—that is, to make string figures tangled in the Chthulucene. There are incarnations of a "biodiverse reef," "toxic reef," "bleached reef," "coral forest," "plastic midden," "white spire garden," "bleached bone reef," "beaded coral garden," "coral forest medusa," and more, along with the many satellite reefs made by collectives of crafters that come together all over the world to mount local exhibits. Crafters make fabulated healthy reefs, but my sense is that most of the reefs show the stigmata of plastic trash, bleaching, and toxic pollution. Crocheting with this trash feels to me like the looping of love and rage.

The skills and sensibilities of Margaret and Christine Wertheim, who were born in Brisbane near the Great Barrier Reef, are fundamental, along with the skills and concerns of the thousands of reef crafters. With degrees in mathematics and physics, Margaret Wertheim is a science writer, curator, and artist. She has written extensively on the cultural history of theoretical physics. Her 2009 TED talk "The Beautiful Math of Coral" has been watched by over a million people.[51] With two books written in feminine feminist materialist poetics, Christine Wertheim is a poet, performer, artist, critic, curator, crafter, and teacher. She aptly describes her work as "infesting fertile zones between cunning linguistics, psychoanalysis, poetry and gender studies."[52] Clearly, these twin sisters were primed for sympoietic SF.

Infecting each other and anyone who comes into contact with their fibrous critters, the thousands of crafters crochet psychological, material, and social attachments to biological reefs in the oceans, but not by practicing marine field biology or by diving among the reefs or making some other direct contact. Rather, the crafters stitch "intimacy without proximity," a presence without disturbing the critters that animate the project, but with the potential for being part of work and play for confronting the exterminationist, trashy, greedy practices of global industrial economies and cultures.[53] Intimacy without proximity is not "virtual" presence; it is "real" presence, but in loopy materialities. The abstractions of the mathematics of crocheting are a kind of lure to an affective cognitive ecology stitched in fiber arts. The crochet reef is a practice of caring without the neediness of touching by camera or hand in yet another voyage of discovery. Material play builds caring publics.

3.4. Green turtles (*Chelonia mydas*) crawling out of the ocean onto the beach to lay their eggs. Credit: Mark Sullivan, NOAA, Permit #10137-07.

The result is another strong thread in the holobiome of the reef: we are all corals now.

Returning to the birth tendrils of the Wertheim sisters in coral reef worlds, I close this little section on the Crochet Coral Reef with a gorgeous photo of green sea turtles coming out of the ocean onto the beach to lay their eggs. Laying their eggs in more than eighty countries and endangered or threatened everywhere, green turtles are globally distributed across the tropical and subtropical belt of earth. A portrait of another green turtle flying in the ocean over the Great Barrier Reef in Australia advertises the Regional Chamber of the Rights of Nature Tribunal held in Far North Queensland in 2015.[54] About eighteen thousand female turtles nest each season on Raine Island in the Great Barrier Reef; this population is one of only two large nesting groups on the earth today.[55] The tribunal collected statements from Aboriginal witnesses about proper governance of the reef to present at the International Tribunal for the Rights of Nature during the Climate Change summit in Paris in December 2015. Sea turtles, corals, Aboriginal witnesses on the care of decolonizing Country, the holobiomes of scientists, denizens of

3.5. Page from *Tik-Tik the Ringtailed Lemur/Tikitiki Ilay Maky*. UNICEF Madagascar and the Lemur Conservation Foundation. Text by Alison Jolly and Hanta Rasamimanana. Art by Deborah Ross. Courtesy of Margaretta Jolly.

the Chthulucene, diverse environmental justice activists, and international art science crafters come together in SF, in speculative fabulation for flourishing.

THE MADAGASCAR AKO PROJECT

As a Yale graduate student studying lemur behavior in 1962 in what is now the Berenty Primate Reserve, Alison Jolly fell into noninnocent love and knowledge in her first encounter with female-led, swaggering, opportunistic ring-tailed lemurs in the spiny forest and dry gallery forest of the south of the island. Simply and transformatively, this young six-foot-tall American white woman became a lover and seeker of knowledge and well-being with and for the beings of Madagascar, especially the astonishing species of lemurs, the radically different forest ecosystems the length and breadth of the island, and the land's complex people and peoples. Author of many books and scientific papers and participant in numerous study and conservation teams, Jolly died in 2014. Her contributions to primatology, biodiversity conservation, and historically informed, passionate analyses of conservation conflicts and necessities were legion. But Jolly herself seemed especially to prize the sympoietic gift she helped craft, the Ako Project,[56] which is tuned to practices for resurgence in vulnerable Malagasy worlds. This is the part of her work I most love.[57]

In the marrow of her bones, Jolly understood the terrible contradictions and frictions in her embrace of *both* the rural people, who cut and burn the forests to make small agricultural plots called *tavy*, and her beloved prosimians with all their forest partners.[58] Of course, she knew she was not Malagasy, but at best a guest who might reciprocate appropriately, and at worst another in a long line of colonizers, always taking land and giving advice for the best reasons. Aware of the controversies over whether shifting cultivators destroyed or nurtured and managed the forest, she learned a great deal about what made contemporary, escalating tavy burnings lethal to the future of the forests and of all their critters, including the people who need them not just for their products (including lemurs for food), but to sustain fertility in phosphorous-poor tropical soils. She knew that making tavy had been part of the cycle of forest succession and biodiversity maintenance, with evidence in old stands in Ranomafana Park. But, she argued, not anymore. Nothing has time to regenerate anymore. Jolly knew in detail what the press of rapidly increasing human numbers means to the forests in the situated history of multiple land dispossessions, relocations, violent suppressions, imposition of regimes of private property, insecure markets, a succession of failed national governments, huge solicited and imposed national debt, and broken development promises. She wrote vividly about local people's accurate assessment of the effects of generations of visiting experts, while the experts and visiting research scientists often knew little or nothing about the terrible history of land seizures, colonial and postcolonial search-and-destroy operations, rapacious extraction schemes, and the impact on villagers of the failed projects of usually well intentioned but often ignorant foreign scientists and both local and foreign NGOs. She also knew what sustained committed work of real colleagues and friends could accomplish in Madagascar against the odds and across differences of all sorts. There are many possible examples and many important people, but I want to tell about one little project that might be considered a model system for sympoiesis.

Written in both English and Malagasy, each book in the Ako Project vividly narrates the adventures of a young Malagasy lemur from one of six species, from the tiny mouse lemur or *ny tsididy*, to the queer-fingered aye-aye or *ny aiay*, to singing Indri or *ilay babakoto*. The stories are fleshed-out natural histories, full of the empirical sensuous curiosity of that genre; and they are bumptious adventures of gutsy young lemurs living the joys and dangers of their habitats and of their groups' social arrangements. Surrounding each lemur species with diverse plant and animal critters proper to their habitats, the project provides both teachers' guides in Malagasy and beautifully crafted posters showing the unique regions of Madagascar where the stories take place. The books are *not* textbooks; they are stories, feasts for mind, heart, and body for children (and adults) who have no access to storybooks or to the critters of their own nation or even region. Most Malagasy never see a lemur on the land, on television, or in a book. Those privileged enough to go to a school with books saw pictures of French rabbits, a fact Alison Jolly told me with disgust in the 1980s when I interviewed her for *Primate Visions*. Many villages are still without schools; and the formal curriculum for children, whether modeled on the older French system or newer learner-centered approaches, is irrelevant to most of the population. State financing for rural schools is extremely paltry, and most rural children are taught by community teachers with no teacher training and no income except from fees paid by very poor families. Teaching about local critters or ecologies rarely happens.

The Ako Project did an end run around the starved schools and unresponsive bureaucracies. After Jolly saw the alluring watercolors of fauna and flora by Deborah Ross, she asked the artist if she would illustrate her children's books about lemurs. Ross said yes; Jolly then contacted her old friend, the lemur biologist Hantanirina Rasamimanana. They raised money; the project was off and running.[59] In exciting, beautiful, funny, and scary stories, distributed outside the school bureaucracies, the Ako Project nurtures empathy and knowledge about the extraordinary biodiversity of Madagascar *for the Malagasy*.

The Ako Project is the generative fruit of a colleagueship and friendship over decades.[60] In 1983 Alison Jolly met Hanta Rasamimanana, a scientist seventeen years her junior. They bonded as mothers doing fieldwork in challenging conditions, primatologists riveted by ring-tailed lemurs, lovers of Malagasy people and nature, and participants in global and local politics, with differently situated vulnerability and authority. Born in the capital city and part of the generation sponsored by the Soviet Union under Didier Ratsiraka's socialism, Rasamimanana

3.6. Painting for *Tsambiki Ilamba Fotsy/Bounce the White Sifaka*. UNICEF Madagascar and the Lemur Conservation Foundation. Text by Alison Jolly and Hanta Rasamimanana. Art by Deborah Ross. Courtesy of Margaretta Jolly and Deborah Ross.

trained in animal husbandry at the Veterinary Academy in Moscow. She earned a PhD at the Muséum National d'Histoire Naturelle in Paris, and she has a master's in primate conservation. She is professor of zoology and scientific education at l'École Normale Supérieure of Antananarivo. Studying ringtails, Rasamimanana has published on feeding behavior, energy expenditure, and lemur female precedence and supreme authority in their societies ("dominance"). Her responsibilities in the scientific academy of Madagascar have been multiple, and she initiated a master's degree in primate conservation run in Mahajanga and the Comoros. An adviser on the Madagascar National Curriculum, she heads the Ako Project teacher support program and wrote the Malagasy teacher's guides based on workshops she ran in rural areas.[61]

In the summer of 2013, Rasamimanana was the program chair for the Fifth International Prosimian Congress, held at the Centre ValBio Research Campus in Ranomafana National Park, where Alison Jolly's friend and colleague Patricia Wright and so many others had worked for decades to strengthen biodiversity and primate research *in* Madagascar and *by* Malagasy scientists.[62] Eighty of the two hundred participants in 2013 were from Madagascar. Half of the two hundred present were students, the core of the next generation of scientists dedicated to holding open space and time for lemurs and their associates in vulnerable forest webs. Writing in her conservation diaries shortly before her death, Jolly celebrated what this congress meant: "The big change is that most papers are by Malagasy speaking on their own biodiversity, eager to advance their own careers in conservation. A contrast to the continuing bewilderment of so many other Malagasy as to why anyone would want to visit forests! And a huge swing from all the meetings in the past dominated by foreigners."[63]

In all their attachments, working with book and poster artists, together the scientists and storytellers Jolly and Rasamimanana brought the Ako Project into the world. In this project and in their work and play across many crises in Madagascar and its conservation history, they have nurtured new generations of Malagasy naturalists and scientists, including small children, field station guides, and school and university students. Without innocence and with relentless commitment, Jolly and Rasamimanana have practiced, in solidarity, the arts of living on a damaged planet; it matters.

3.7. Cover image for *Never Alone (Kisima Ingitchuna)*. Courtesy of E-line Media, in collaboration with Upper One Games, and the Cook Inlet Tribal Council.

NEVER ALONE (KISIMA INGITCHUNA)

My third example of science art worlding for living on a damaged planet is making "world games." World games are made with and from indigenous peoples' stories and practices. "But what good are old stories if the wisdom they contain is not shared?"[64] These games both remember and create worlds in dangerous times; they are worlding practices. Indigenous peoples around the earth have a particular angle on the discourses of coming extinctions and exterminations of the Anthropocene and Capitalocene.[65] The idea that disaster will come is not new; disaster, indeed genocide and devastated home places, has already come, decades and centuries ago, and it has not stopped. The resurgence of peoples and of places is nurtured with ragged vitality in the teeth of such loss, mourning, memory, resilience, reinvention of what it means to be native, refusal to deny irreversible destruction, and refusal to disengage from living and dying well in presents and futures. World games require inventive, sympoietic collaborations that bring together such things as computer game platforms and their designers, indigenous storytellers, visual artists, carvers and puppet makers, digital-savvy youngsters, and community activists. The set of computer world games at the moment

I write this sentence is small; there is one. Others, however, are in the collaboration and design phase.[66]

However, even though the models of sympoiesis are expandable, it is critical not to once again raid situated indigenous stories as resources for the woes of colonizing projects and peoples, entities that seem permanently undead. *Never Alone* is not a New Age game for universal oneness, a posthumanist solution to epistemological crises, a general model for collaboration, or a way to finesse the Anthropocene with Native Climate Wisdom. Nor is *Never Alone* a primer for the Chthulucene. If Inupiat "Sila" meets in SF games with the tentacular Chthulucene, it will be a risk-taking proposition, not an innocent translation.[67] *Never Alone* requires a different sort of attention; and perhaps the fact that I continue to die early and often playing the game is less a reflection of my poor gaming skills than a proper reminder that a world game is situated indigenous storytelling in specific histories. The fact that the game is narrated in Inupiaq, with English subtitles, is another reminder where worlding authority lies here. Stories, even stories offered for sale on the Internet, belong to storytellers, who share them, or not, in practices of situated worlding. The conditions for sharing stories must not be set by raiders, academic or otherwise.[68] That does *not* mean the game is restricted to native commentators in native places for native audiences in a perverse caricature of a reservation. It *does* mean the terms of telling, listening, and playing have been relocated decisively.

"*Never Alone (Kisima Ingitchuna)* is the first game developed in collaboration with the Inupiat, an Alaska Native people. Play as a young Inupiat girl and an arctic fox as they set out to find the source of the eternal blizzard which threatens the survival of everything they have ever known."[69] No one acts alone; connections and corridors are practical and material, even if also fabulous, located in what Anglophones tend to dismiss as the spirit world. The girl Nuna's personal courage and skills are also fundamental. These are the arts of living on a damaged planet (Anna Tsing's term). *Never Alone* might be played in a string figure pattern with Ursula Le Guin's *Always Coming Home*.

Game makers define the new genre "world games" as taking place inside ongoing indigenous stories. Makers of *Never Alone (Kisima Ingitchuna)* include Gloria O'Neill, the president and chief executive of the Cook Inlet Tribal Council; dozens of advisers and elders from the Alaska Native community; Alan Gershenfeld, cofounder of E-line Media; E-Line creative director Sean Vesce; the design team's studio in Seattle; young and old people playing the game; and a shared sense of contemporary urgency for the lands and waters with their human and other-than-human beings. "O'Neill said she loved the chance to participate in the video game because the council could be a codeveloper in the process—and because no Native American group had ever played such a role in the history of the video game industry."[70]

The sympoiesis of *Never Alone* has many threads, and one of them is hard for most modernist people, namely the symanimagenic richness of the stories and the game. The girl Nuna and her arctic fox companion go from the home village to face the unprecedented blizzard, find what is causing it, and save the people and the land. Helping each other, girl and fox learn to traverse many obstacles, and even to swim in the belly of a whale, finally escaping into the sky through the blowhole. Those kinds of sym linkages and fabled travels are not an ontological or epistemological problem, or at least not much of one. But the presence and agency of multiple spirit helpers are absolutely central to this worlding, to these stories, and to this sympoiesis in the Arctic of the Anthropocene. Digital information system ontologies, spirit helpers, and biocultural girls and foxes have to play an agile string figure game seriously for "never alone" to have its full meaning.

Working with Brazilian Amerindian hunters, with whom he learned to theorize the radical conceptual realignment he called multinaturalism and perspectivism, Eduardo Viveiros de Castro wrote, "Animism is the only *sensible* version of materialism."[71] I am not talking about people like me—or kids like Nuna—"believing" in the spirit world. Belief is neither an indigenous nor a "chthulucenean" category. Relentlessly mired in both internecine and colonizing disputes of Christianity, including its scholarly and civic secular forms, the category of belief is tied to doctrine, profession, confession, and taxonomies of errors. That is, believing is not sensible.[72] I am talking about material semiotics, about practices of worlding, about sympoiesis that is not only symbiogenetic, but is always a *sensible* materialism. The sensible materialisms of involutionary momentum are much more innovative than secular modernisms will allow. Stories for living in the Chthulucene demand a certain suspension of ontologies and epistemologies, holding them lightly, in favor of more venturesome, experimental natural histories. Without inhabiting symanimagenic sensible materialism, with all its pushes, pulls, affects, and attachments, one cannot play *Never Alone*; and the resurgence of this and other worlds might depend on learning to play.

But, continuing to die early and often in *Never Alone*, I have not forgotten that spirit helpers favor their kin. Animism cannot be donned like a magic cape by visitors. Making kin in the ongoing Chthulucene will be more difficult than that, and even the unwilling heirs of colonizers are poorly qualified to set conditions for recognition of kinship. Plus, many contemporary Inuit, including those committed to cultural renewal, are wary of animism in their own heritage. Staying with the trouble, yearning toward resurgence, requires inheriting hard histories, for everybody, but not equally and not in the same ways.

NAVAJO WEAVING: COSMOLOGICAL PERFORMANCE, MATHEMATICAL RHYTHM, NAVAJO-CHURRO SHEEP, *HÓZHÓ*

Black Mesa, on it life.
There will be life again, this is what they say.
For this reason they are weaving.[73]

For my last model system for sympoiesis, in risky propositions I return to fibers, linking the Crochet Coral Reef to Navajo weaving. Navajo weaving is practiced all over the Navajo Nation, but I will emphasize the weavers of Black Mesa, their sheep, and their alliances.[74] It would be a serious category mistake to call Navajo weaving "art science activism," which was a comfortable enough name for the Crochet Coral Reef. Besides bypassing robust and precise Diné namings, both the categories "art" and "science" continue to do colonizing work in this context. However, it would also be a serious category error to fence Navajo weaving off from ongoing mathematical, cosmological, and creative practice that never fit ongoing colonial definitions of "traditional." Like the Crochet Coral Reef, Navajo weaving, especially with the wool of Churro sheep, ties people to animals through patterns of care and response-ability in blasted places of excess death and threatened ongoingness. As in the Crochet Coral Reef, the play of collective making and personal invention is everywhere in Navajo fiber work. Both the Crochet Coral Reef and Navajo weaving exist in a modernizing ecology of gendered and commodifying structures that elevate "art" over "craft." Both the Crochet Coral Reef and Navajo weaving are done mainly by women, but men also figure in the webs of thinkers/makers.[75] Both the Crochet Coral Reef and Navajo weaving perform worlds with mathematical vitality that remains invisible in the doxa of scholarship on women's fiber practices in both settler and colonized indigenous production. Finally, attuned to a sympoiesis

3.8. Navajo rug, Two Gray Hills. Weaver unknown. Photograph by Donna Haraway. Purchased by Rusten Hogness's father, John Hogness, in the Navajo Nation in the 1960s.

of practical coalitions, both the Crochet Coral Reef and Navajo weaving are at the heart of thinking/making for more livable politics and ecologies in the times of burning and extraction called the Anthropocene and Capitalocene. In face-to-face and hand-to-hand entanglements, the Great Barrier Reef and Black Mesa are crocheted and woven together in cosmological performances to animate the tentacular Chthulucene of a Thousand Names.

A refrain from Navajo prayers often accompanies a weaver's work: "With me there is beauty" (*shil hózhó*); "in me there is beauty" (*shii' hózhó*); "from me beauty radiates" (*shits' áá d óó hózhó*).[76] *Hózhó* is a central concept in Navajo cosmology and daily practice. Usual translations into English are "beauty," "harmony," and "order"; but I think a better translation would emphasize right relations of the world, including human and nonhuman beings, who are *of* the world as its storied and dynamic substance, not *in* the world as a container. Disorder, often figured in the doings of Coyote, disrupts right relations, which must be restored in ceremony and daily life for proper living to be again possible, for the person to be restored in *hózhó* to the People. For the Diné, greed is the greatest source of disorder; it destroys right relations at their root.

Weaving is a useful practice, to be sure, and an economic one; but, fundamentally, weaving is also cosmological performance, knotting proper relationality and connectedness into the warp and weft of the fabric.[77] The geometric patterns of repetition and invention in weaving are performances of Diné stories and knowledge; the patterns propose and embody world-making and world-sustaining relations. The dynamic patterning continues in contemporary weavings, many of which explore new as well as inherited themes, colors, stories, and fibers.[78] Weavings are individual; they are made by a particular woman and embody her style and sensibility, recognizable by knowledgeable members of the community.[79] Names of weavers and weavers' lineages matter, but weavings are not made to be possessed as property. Neither that nor the entanglement of the creative personal and the cosmological is a contradiction. The *sensible* order inherent in the storied cosmos of Changing Woman, the Holy Twins, Spider Woman, and the other world-making Holy People is the pattern for right living. Weaving is neither secular nor religious; it is *sensible*. It performs and manifests the meaningful lived connections for sustaining kinship, behavior, relational action—for *hózhó*—for humans and nonhumans. Situated worlding is ongoing, neither traditional nor modern.

Navajo weaving relied especially on the so-called rough sheep brought to the Americas by the Spanish in the sixteenth century and developed by Navajo herders over a long time as a distinct kind of sheep, named T'aa Dibei or Navajo-Churro sheep, who are particularly well adapted to the lands of Diné bikéyah on the Colorado Plateau.[80] In Western historical temporalities, Navajo matrifocal pastoralism and farming developed in the eighteenth and nineteenth centuries, with sheep as core companions for living and dying in *hózhó*. The art of weaving and care of Churro sheep reciprocally enact Diné relations of natural and cosmic order.

The Diné endured two intense periods of efforts by U.S. officials to exterminate their Churro sheep. The first such genocide, called Hwéeldi and effected in 1863 under Kit Carson for the U.S. War Department, was the Long Walk of all the People who could be forcibly rounded up from Dinetah and marched for hundreds of miles to Bosque Redondo in New Mexico. The Hwéeldi followed a scorched-earth campaign led by Carson against the Navajo. Killing of Navajo animals was a central act of the removal. From the beginning, across the Southwest and West U.S. modernizers saw Spanish-introduced stock as rough and unimproved. Exterminating flocks, cutting down peach orchards, and forcing the removal of people to Fort Sumner/Bosque Redondo were, in effect, normal actions of U.S. colonizing officials pacifying and civilizing an unruly mobile population. The correct name is attempted genocide. Full of suffering and death, this forced march was followed by four years in a prison camp and then the walk back to their lands. The Hwéeldi is remembered in the flesh of land and people; it is an "originary" trauma, of the kind Toni Morrison understood in her novel *Paradise*.[81]

The Diné returned to the Navajo reservation on the Colorado Plateau. Churro sheep had been carefully tended by people who escaped Kit Carson's soldiers in the deep canyons and remote areas of Dinetah, including Big Mountain/Dzil ni Staa/Black Mesa. The boundaries of the reservation extended gradually until the 1930s; and, despite the failure of the U.S. government after the Diné return from Bosque Redondo to provide promised stock, sheep flocks grew much faster than the human population. This growth was partly driven by the trading post system, which turned wool into blankets to realize value and bought these blankets by the pound in a system of perpetual indebtedness. To obtain basic necessities in this system of debt, the Navajo were forced to produce more and more wool from more and more sheep. The traders sold the weavings in the art and tourist market, but purchased the women's weavings as if they were low-value raw wool. Despite the efforts of federal agents, most of the Diné continued to prefer multipurpose, hardy Churro sheep to merinos and other "improved" breeds. Sheep, goats, horses, and cattle were all part of the pattern of Navajo pastoralism, ordered by complex clan and gender relationships. The animals and the people made kin together.[82] Sheep and goats were especially crucial for women's abilities to feed and provision their families, as well as to their authority in the clans.

With intensifying erosion, severe grazing, and sustained drought, by the 1930s the system was increasingly out of harmony, a condition recognized by both whites and Navajos. The second intense efforts of the U.S. government to exterminate Navajo-Churro sheep occurred in this context; like the first originary trauma, this lethal event can be neither forgotten nor effectively mourned. It bears evil fruit to this day. Restoring the land, animals, and people to *hózhó* is an ongoing process that continues to require continuous weaving. The colonial and capitalist structures of both exterminations have not been dismantled. The first Churro sheep extermination was conducted by U.S. military men; the second was also conducted by force, this time by U.S. progressive agricultural authorities within the ideology and apparatus of the New Deal. These officials worked within the ecological concept of carrying capacity, the patriarchal colonial concepts of male-headed households, and the modernizers' concepts of progress. Without asking how colonial economic structures like the unequal wool trade might be a significant cause of both poverty and ecological damage and judging the erosion of Navajo lands to be due to overstocking as a biological sort of fact, U.S. government scientists in the Department of Agriculture and others in 1934 killed most of the women's goats, the primary source of subsistence meat for families. White-settler divisions of the world into nature and culture split Navajo lifeways into colonial apparatuses of ecology and economics, practiced by different sorts of scientific specialists who could not systematically think even with each other, much less with Navajo herders and weavers. In 1935, officials killed vast numbers of sheep. Churro sheep, many known individually by their people, were preferentially killed, often in front of their human families. Evident in photographs, piles of bones from these animal murders were still prominent in the 1970s; and people still dramatically narrated the trauma, even describing particular animals in their flocks.

Following the killing of about a million sheep and goats (without significant compensation to this day), stocking quotas were imposed, and collective ownership of land was not recognized. The census by which stock quotas and permits were allocated recognized only heads of households, who could not be married women, which was a major blow to Diné matrifocal ways of ordering their relations with land, animals, and each other. Transhumance was disrupted as land boundaries were redrawn into Land Management Units, exacerbating erosion as both seasonal and dynamic rain-pattern-sensitive movements for grazing became difficult across such boundaries. Besides an act of scientific colonial arrogance and culpable ignorance, the animal exterminations of the 1930s effected a profound decapitalization of the whole people, whose existing poverty, itself linked to the consequences of the first Hwéeldi, was structurally intensified. With the failure to restore the health of lands, waters, animals, and people in *hózhó*, balanced pastoralism was not reconstructed; resurgence on the Colorado Plateau was wounded. Stock levels and erosion remain a major problem, intensified by deep resentment of forced controls, including colonial conceptual apparatuses within the Navajo Nation.

In a crisis of drought and multispecies lifeways out of balance in the 1930s, the opportunity was missed to bring scientific ecological ideas like carrying capacity into difficult but necessary conversation with Navajo concepts and practices of *hózhó*. Neither carrying capacity nor *hózhó* is a fixed, deterministic concept; both are relational, contextual, tuned to some ways of living and dying and not others. It matters what concepts think concepts, and vice versa; but in this case, colonial structures assured that the important concepts would not be allowed to think each other, would not be allowed perhaps to issue in something that did not yet exist in thought for either people, but might be needed by both. When one system of thinking and practice can only disparage and nullify another in colonial recursions, there can be no sympoiesis and no *hózhó*. The consequences of the failure to invent the needed decolonial conversations ramifies into the present. Since this period, pastoralism has not been able to support the Diné; and poverty is perpetuated by the post–World War II wage-based economy in the context of extreme under- and unemployment, federal subsidies, tourism, and income from uranium and coal mining.[83]

However, there is also an extraordinary story of resurgence and partial healing to be told, one that belongs to the Diné and their allies in the ongoing Chthulucene and the ongoing Diné Bahane' / Story of the People / Navajo Creation Story. By 1970, only about 430 Navajo-Churro sheep survived, scattered across the reservation. The traditional Diné of Black Mesa and others had protected what sheep they could in remote places. Other Churro sheep survived from a research population studied from 1934 to 1967 at the Southwest Range and Sheep Breeding Laboratory at Fort Wingate, New Mexico. When the research project shut down, 165 Churro sheep were auctioned off in 1967 to a rancher in Gonzales, California, who used them in a shoot-in-a-barrel safari enterprise for Hollywood notables. Besides their double coat, long fibers, high-lanolin wool, ability to survive on scrubby pasture, and excellent mothering skills by the ewes, Churro rams frequently have a double set of horns that incite hunting fantasists to pay to turn them into trophies. The story of Navajo-Churro resurgence—with Navajo herders and weavers; an Anglo scientist committed to Churro sheep and their people; Navajo and Anglo students; Hispanic and Anglo ranchers; Tarahumara/Rarámuri Indians of the Sierra Madre Occidental of northern Mexico, who interbred Churro from the Navajo Sheep Project with their own rough sheep to recover genetic diversity; activists on Black Mesa; and more—begins at these crossroads. Over decades Diné herders nurtured remnant flocks in spite of the odds, and Buster Naegle, who had taken over the ranch in Gonzales in 1970 to raise paint horses, donated six ewes and two four-horned rams to Lyle McNeal, an animal scientist then at Cal Poly San Luis Obispo, as seed animals. In ensuing lifelong coalitional work, McNeal founded the Navajo Sheep Project in 1977.[84]

The story of Navajo-Churro restoration is complexly tentacular and fibrous, braided by many actors and full of obstacles as well as successes. Collecting sheep on the reservation from Diné cooperating to help rebuild the flocks, Lyle McNeal donated some of the first rams born from his seed flock in the 1980s to Women in Resistance on Black Mesa. Keeping his nucleus flock and operations alive involved thirteen moves in four states over twenty-five years with many adventures with the law, especially private property law. With Diné Churro sheep herders and weavers including Glenna Begay, Lena Nez, and others, Carol Halberstadt, a poet, activist, and lover of wool from Massachusetts, cofounded Black Mesa Weavers for Life and Land as a fair trade cooperative association to better the economic and social conditions of Black Mesa Diné through supporting sheep herding, wool buys, and weaving.[85] A Navajo-Churro flock has been established at the Diné College in Tsaile, Arizona, for teaching. Diné be'iína/The Navajo Lifeway was founded in 1991 to nurture community-based partnerships to restore economy and culture. The college hosts the Dibé be'iína / Sheep Is Life celebration every summer.[86] Churro are central to cultural renewal through weaving and taking care of sheep. Reconnecting generations broken by boarding schools and forced stock exterminations and encouraging Navajo language use among the young are also tied to these sheep.[87] Kosher Navajo-Churro sheep jerky, guard llamas, the American Livestock Breeds Conservancy, the Navajo-Churro Sheep Association, the Agricultural Research Service National Center for Genetic Resources Preservation, the Slow Food Foundation for Biodiversity, Two Grey Hills Trading Post, the Teec Nos Pos Chapter and its regional wool-processing facility, the Ganados del Valle Hispanic agricultural development corporation, Tierra

Wool and Los Ojos Handweavers, the Crownpoint Auction, and Heifer International are all involved in diverse configurations.[88]

Not least, the sheep themselves are active participants in the interlaced relational worlds. Like all sheep, they recognize hundreds of faces; they know their people and their land.[89] Weaving is cosmological performance, relational worlding, with human and nonhuman fibers from the Holy People, ordinary human beings, plants, soils, waters, and sheep. The critters are critical to taking care of country, to environmental justice, to robust ecosystems for humans and nonhumans, to *hózhó*. It matters which beings recognize beings.

So the sheep lead back to Black Mesa and to a concluding sympoiesis with the activists—the thinkers/makers—of the Black Mesa Water Coalition (BMWC). Supporting the weavers, herders, and sheep of the region, BMWC partners with Diné be'iína and holds wool buys; they even partner with a sheep-farming outfit in Maine called Peace Fleece.[90] BMWC is thoroughly entangled with sheep and their people across damaged lands and blasted histories. But my reason for tying the threads of cosmological performance and continuous weaving together through BMWC is grounded in coal, water, indigenous environmental justice movements, and surging coalitions for Just Transition toward still possible worlds in urgent times. Probably still possible. Barely still possible. Still possible *if* we render each other capable of worlding and reworlding for flourishing. I want to propose the Black Mesa Water Coalition as a sympoietic model for learning to stay with the trouble together, for *hózhó*.

The BMWC was founded in 2001 by a group of young intertribal, interethnic people, mostly students at the time, committed to addressing water depletion, natural resource exploitation, and health in Navajo and Hopi communities.[91] Quickly focusing on Peabody Energy, they were central to the actions that closed down the Black Mesa Mine and Mohave Generating Station in 2006. But that was the beginning, not the end. The coalition sees Black Mesa as a critical place for learning to transition out of coal-based economies and ecologies and into abundant solar and other renewable power, situated on damaged lands, as a needed practice for multispecies environmental justice. Black Mesa itself is not just any place; within Navajo cosmology Black Mesa is the mother encircled by the four sacred mountains. The waters are the mother's blood, and coal is her liver. That condensed Diné geo-anatomy is only an indication of the corporeal relational cosmology of place that is utterly illegible to Peabody Energy—and to settler colonialism more broadly, to this day. My colleague Anna Tsing talks about "worlds worth fighting for"; Black Mesa is such a world.[92]

The BMWC's Just Transition Initiative, beginning in 2005, is a comprehensive vision and practice for building on the strengths of local people, culture, and land, in alliance with many partners, to make resurgence on Black Mesa and beyond a reality. Pilot projects for restoring regional watersheds and for economic development, the vision and work toward a Black Mesa Solar Project, the Food Security Project, the Navajo Wool Market Project, the Green Economy Project, and the Climate Justice Solutions Project are all part of the BMWC's work. These activists aim to develop a strong regional, integrated environmental and social justice movement led by indigenous communities and organizations, as well as to ally with the worldwide Climate Justice Alliance.[93] These are big, important ideas and actions; these kinds of continuous weaving are at the heart of staying with the trouble in a damaged world. Continuing to be led by young adults within a multigenerational web, the BMWC proposes the sort of resurgence that can face the originary, repeating traumas of history without denial and without cynicism or despair. In my idiom, the Black Mesa Water Coalition is a strong tentacle in the surging Chthulucene.

Conclusion: Tying Off the Threads

We relate, know, think, world, and tell stories through and with other stories, worlds, knowledges, thinkings, yearnings. So do all the other critters of Terra, in all our bumptious diversity and category-breaking speciations and knottings. Other words for this might be materialism, evolution, ecology, sympoiesis, history, situated knowledges, cosmological performance, science art worldings, or animism, complete with all the contaminations and infections conjured by each of these terms. Critters are at stake in each other in every mixing and turning of the terran compost pile. We are compost, not posthuman; we inhabit the humusities, not the humanities. Philosophically and materially, I am a compostist, not a posthumanist. Critters—human and not—become-with each other, compose and decompose each other, in every scale and register of time and stuff in sympoietic tangling, in ecological evolutionary developmental earthly worlding and unworlding.

This chapter began with Lynn Margulis's proposition of symbiogenesis and segued into the biologies that make an extended evolutionary synthesis necessary to thinking well about multispecies living and dying on earth at every scale of time and space. The involutionary momentum of a vanishing bee and its faithful orchid enfolded the EcoEvoDevo biologies into four naturalsocial ecologies of a damaged planet. Actual places, these are worlds worth fighting for; and each has nourished brave, smart, generative coalitions of artists/scientists/activists across dangerous historical divisions. The biologies, arts, and politics need each other; with involutionary momentum, they entice each other to thinking/making in sympoiesis for more livable worlds that I call the Chthulucene.[94]

Isabelle Stengers's sense of cosmopolitics gives me courage.[95] Including human people, critters are in each other's presence, or better, inside each other's tubes, folds, and crevices, insides and outsides, and not quite either. The decisions and transformations so urgent in our times for learning again, or for the first time, how to become less deadly, more response-able, more attuned, more capable of surprise, more able to practice the arts of living and dying well in multispecies symbiosis, sympoiesis, and symanimagenesis on a damaged planet, must be made without guarantees or the expectation of harmony with those who are not oneself—and not safely other, either. Neither One nor Other, that is who we all are and always have been. All of us must become more ontologically inventive and sensible within the bumptious holobiome that earth turns out to be, whether called Gaia or a Thousand Other Names.

CHAPTER 4

Making Kin

Anthropocene, Capitalocene, Plantationocene, Chthulucene

There is no question that anthropogenic processes have had planetary effects, in inter/intra-action with other processes and species, for as long as our species can be identified (a few tens of thousand years); and agriculture has been huge (a few thousand years). Of course, from the start the greatest planetary terraformers (and reformers) of all have been and still are bacteria and their kin, also in inter/intra-action of myriad kinds (including with people and their practices, technological and otherwise).[1] The spread of seed-dispersing plants millions of years before human agriculture was a planet-changing development, and so were many other revolutionary evolutionary ecological developmental historical events.

People joined the bumptious fray early and dynamically, even before they/we were critters who were later named *Homo sapiens*. But I think the issues about naming relevant to the Anthropocene, Plantationocene, or Capitalocene have to do with scale, rate/speed, synchronicity, and complexity. The constant questions when considering systemic phenomena have to be, When do changes in degree become changes in kind? and What are the effects of bioculturally, biotechnically, biopolitically, historically situated people (not Man) relative to, and combined with, the effects of other species assemblages and other biotic/abiotic forces?

No species, not even our own arrogant one pretending to be good individuals in so-called modern Western scripts, acts alone; assemblages of organic species and of abiotic actors make history, the evolutionary kind and the other kinds too.

But is there an inflection point of consequence that changes the name of the "game" of life on earth for everybody and everything? It's more than climate change; it's also extraordinary burdens of toxic chemistry, mining, nuclear pollution, depletion of lakes and rivers under and above ground, ecosystem simplification, vast genocides of people and other critters, et cetera, et cetera, in systemically linked patterns that threaten major system collapse after major system collapse after major system collapse. Recursion can be a drag.

Anna Tsing in a recent paper called "Feral Biologies" suggests that the inflection point between the Holocene and the Anthropocene might be the wiping out of most of the refugia from which diverse species assemblages (with or without people) can be reconstituted after major events (like desertification, or clear cutting, or, or, . . .).[2] This is kin to the World-Ecology Research Network coordinator Jason Moore's arguments that cheap nature is at an end; cheapening nature cannot work much longer to sustain extraction and production in and of the contemporary world because most of the reserves of the earth have been drained, burned, depleted, poisoned, exterminated, and otherwise exhausted.[3] Vast investments and hugely creative and destructive technology can drive back the reckoning, but cheap nature really is over. Anna Tsing argues that the Holocene was the long period when refugia, places of refuge, still existed, even abounded, to sustain reworlding in rich cultural and biological diversity. Perhaps the outrage meriting a name like Anthropocene is about the destruction of places and times of refuge for people and other critters. I along with others think the Anthropocene is more a boundary event than an epoch, like the K-Pg boundary between the Cretaceous and the Paleogene.[4] The Anthropocene marks severe discontinuities; what comes after will not be like what came before. I think our job is to make the Anthropocene as short/thin as possible and to cultivate with each other in every way imaginable epochs to come that can replenish refuge.

Right now, the earth is full of refugees, human and not, without refuge.

So I think a big new name, actually more than one name, is warranted—hence Anthropocene, Plantationocene,[5] and Capitalocene (Andreas Malm's and Jason Moore's term before it was mine).[6] I also insist that we need a name for the dynamic ongoing symchthonic forces and powers of which people are a part, within which ongoingness is at stake. Maybe, but only maybe, and only with intense commitment and collaborative work and play with other terrans, flourishing for rich multispecies assemblages that include people will be possible. I am calling all this the Chthulucene—past, present, and to come.[7] These real and possible timespaces are not named after SF writer H. P. Lovecraft's misogynist racial-nightmare monster Cthulhu (note spelling difference), but rather after the diverse earthwide tentacular powers and forces and collected things with names like Naga, Gaia, Tangaroa (burst from water-full Papa), Terra, Haniyasu-hime, Spider Woman, Pachamama, Oya, Gorgo, Raven, A'akuluujjusi, and many many more. "My" Chthulucene, even burdened with its problematic Greek-ish rootlets, entangles myriad temporalities and spatialities and myriad intra-active entities-in-assemblages—including the more-than-human, other-than-human, inhuman, and human-as-humus. Even rendered in an American English-language text like this one, Naga, Gaia, Tangaroa, Medusa, Spider Woman, and all their kin are some of the many thousand names proper to a vein of SF that Lovecraft could not have imagined or embraced—namely, the webs of speculative fabulation, speculative feminism, science fiction, and scientific fact.[8] It matters which stories tell stories, which concepts think concepts. Mathematically, visually, and narratively, it matters which figures figure figures, which systems systematize systems.

All the thousand names are too big and too small; all the stories are too big and too small. As Jim Clifford taught me, we need stories (and theories) that are just big enough to gather up the complexities and keep the edges open and greedy for surprising new and old connections.[9]

One way to live and die well as mortal critters in the Chthulucene is to join forces to reconstitute refuges, to make possible partial and robust biological-cultural-political-technological recuperation and recomposition, which must include mourning irreversible losses. Thom van Dooren and Vinciane Despret taught me that.[10] There are so many losses already, and there will be many more. Renewed generative flourishing cannot grow from myths of immortality or failure to become-with the dead and the extinct. There is a lot of work for Orson Scott Card's Speaker for the Dead.[11] And even more for Ursula Le Guin's worlding in *Always Coming Home*.

I am a compostist, not a posthumanist: we are all compost, not posthuman. The boundary that is the Anthropocene/Capitalocene means many things, including that immense irreversible destruction is really in train, not only for the 11 billion or so people who will be on earth near the end of the twenty-first century, but for myriads of other critters too. (The incomprehensible but sober number of around 11 billion will only hold if current worldwide birth rates of human babies remain low; if they rise again, all bets are off.) The edge of extinction is not just a metaphor; system collapse is not a thriller. Ask any refugee of any species.

The Chthulucene needs at least one slogan (of course, more than one); still shouting "Cyborgs for Earthly Survival," "Run Fast, Bite Hard," and "Shut Up and Train," I propose "Make Kin Not Babies!" Making—and recognizing—kin is perhaps the hardest and most urgent part.[12] Feminists of our time have been leaders in unraveling the supposed natural necessity of ties between sex and gender, race and sex, race and nation, class and race, gender and morphology, sex and reproduction, and reproduction and composing persons (our debts here are due especially to Melanesians, in alliance with Marilyn Strathern and her ethnographer kin).[13] If there is to be multispecies ecojustice, which can also embrace diverse human people, it is high time that feminists exercise leadership in imagination, theory, and action to unravel the ties of both genealogy and kin, and kin and species.

Bacteria and fungi abound to give us metaphors; but, metaphors aside (good luck with that!), we have a mammalian job to do, with our biotic and abiotic sympoietic collaborators, colaborers. We need to make kin symchthonically, sympoetically. Who and whatever we are, we need to make-with—become-with, compose-with—the earth-bound (thanks for that term, Bruno Latour–in-Anglophone-mode).[14]

We, human people everywhere, must address intense, systemic urgencies; yet so far, as Kim Stanley Robinson put it in *2312*, we are living in times of "The Dithering" (in this SF narrative, lasting from 2005 to 2060—too optimistic?), a "state of indecisive agitation."[15] Perhaps the Dithering is a more apt name than either the Anthropocene or Capitalocene! The Dithering will be written into earth's rocky strata, indeed already is written into earth's mineralized layers. Symchthonic ones don't dither; they compose and decompose, which are both dangerous and promising practices. To say the least, human hegemony is not a symchthonic affair. As ecosexual artists Beth Stephens and Annie Sprinkle say on a sticker they had made for me, composting is so hot!

My purpose is to make "kin" mean something other/more than entities tied by ancestry or genealogy. The gently defamiliarizing move might seem for a while to be just a mistake, but then (with luck) appear as correct all along. Kin making is making persons, not necessarily as individuals or as humans. I was moved in college by Shakespeare's punning between *kin* and *kind*—the kindest were not necessarily kin as family; making kin and making kind (as category, care, relatives without ties by birth, lateral relatives, lots of other echoes) stretch the imagination and can change the story. Marilyn Strathern taught me that "relatives" in British English were originally "logical relations" and only became "family members" in the seventeenth century—this is definitely among the factoids I love.[16] Go outside English, and the wild multiplies.

I think that the stretch and recomposition of kin are allowed by the fact that all earthlings are kin in the deepest sense, and it is past time to practice better care of kinds-as-assemblages (not species one at a time). Kin is an assembling sort of word. All critters share a common "flesh," laterally, semiotically, and genealogically. Ancestors turn out to be very interesting strangers; kin are unfamiliar (outside what we thought was family or gens), uncanny, haunting, active.[17]

Too much for a tiny slogan, I know! Still, try. Over a couple hundred years from now, maybe the human people of this planet can again be numbered 2 or 3 billion or so, while all along the way being part of increasing well-being for diverse human beings and other critters as means and not just ends.

So, make kin, not babies! It matters how kin generate kin.[18]

CHAPTER 5

Awash in Urine

DES and Premarin in Multispecies Response-ability

Cyborg Littermates

Cyborgs are kin, whelped in the litter of post–World War II information technologies and globalized digital bodies, politics, and cultures of human and not-human sorts. Cyborgs are not machines in just any sense, nor are they machine-organism hybrids. In fact, they are not hybrids at all. They are, rather, imploded entities, dense material semiotic "things"—articulated string figures of ontologically heterogeneous, historically situated, materially rich, virally proliferating relatings of particular sorts, not all the time everywhere, but here, there, and in between, with consequences. Particular sorts of historically situated machines signaled by the words *information* and *system* play their part in cyborg living and dying. Particular sorts of historically situated organisms, signaled by the idioms of labor systems, energetics, and communication, play their part. Finally, particular sorts of historically situated human beings, becoming-with the practices and artifacts of technoscience, play their part. Characterized by partial connections, the parts do not add up to any whole; but they do add up to worlds of nonoptional, stratified, webbed, and unfinished living and dying, appearing and disappearing. Cyborgs are constitutively full of multiscalar, multitemporal, multimaterial critters of both living and nonliving persuasions.[1] Cyborgs matter in terran worlding.

But cyborgs are critters in a queer litter, not the Chief Figure of Our Times. Queer here means not committed to reproduction of kind and having bumptious relations with futurities. Irreducible to cyborgs, the litter interests me, the particular kin and kind nursed on the fluid and solid effluvia of terra in the late twentieth and early twenty-first centuries. I return to "my" cyborg in this litter in order to relay the string figures—the speculative fabulations, the scientific facts, the science fictions, and the speculative feminisms—to whatever sorts of tentacular grippers will receive the pattern to keep living and dying well possible in "our times." Made up of an aging California dog, pregnant mares on the western Canadian prairies, human women who came to be known as DES daughters, lots of menopausal U.S. women, and assorted other players in the story of "synthetic" and "natural" estrogens, the litter for this chapter is decanted from bodies awash in a particular pungent fluid—urine. Waste and resource, out-of-place urine from particular female bodies is the salty ocean needed for my tale. Leaks and eddies are everywhere. These leaks and eddies might help open passages for a praxis of care and response—response-ability—in ongoing multispecies worlding on a wounded terra.

DES for Hot Peppers

In October 2011 my twelve-year-old canine friend and lifelong sports partner Cayenne, aka Hot Pepper, started taking a notorious, industrially produced, nonsteroidal, synthetic estrogen called DES (diethylstilbesterol) to deal with urinary leakage.[2] Perhaps I should not write she "started taking," but, rather, "I started feeding her as an occasional late-night treat, following her last pee, a luscious, slippery, Earth Balance® margarine-coated capsule of DES." Plato gave us the tones in the inextricable ambiguities of his pharmakon: cure and poison; care, curare; remedy, toxin; treat, threat. Aging spayed bitches like Cayenne and post-menopausal women like me often could use a hormonal tightening of slack smooth muscles in the urethra to keep socially unacceptable leaks plugged up. The term *estrogen deficiency* is a tough one for feminists like me, marinated at a young age in the women's health movements and feminist science studies, to pronounce. But the fact is that a few extra dabs of estrogens do some handy jobs in aging mammalian female bodies—at a price, of course, in many currencies of living and dying. Granted, the adrenal glands still secrete some estrogens for those of us with missing or dried-up ovaries, but output is pretty low and smooth muscle can get pretty flaccid.

But giving this beloved, elder, nonreproducing dog to whom I am responsible even very low-dose and infrequent diethylstilbesterol caused acute DES Anxiety Syndrome in me. My blood pressure rose higher than the high canine blood pressure that motivated changing urine-plugging drugs for Cayenne in the first place. Even if I could keep my critique of biocapital in a sealed flask, my feminist biopolitical juices started oozing from every pore, leaking all over my obligations to our dog. Rusten, my male human spouse, was drawn deeply into this mammalian female well of worry, and not just because neither one of us much wanted to sleep in the urinary wet spot if estrogen-deprived urethral smooth muscle were left unattended in the nocturnal hours in our species-queer connubial bed. Cross-species kinship has consequences. Our now shared DES Anxiety Syndrome had to be treated immediately by our excellent primary care veterinarian, who did the service of presenting us with scientific studies and her own history of practice with low-dose, minimum-frequency DES for elder dogs; this was the "talking cure" we needed—a high dose of reason, evidence, and story, taken weekly with uncanny and unruly molecules. Still, my vet herself is too young to have been infected with my kind of terror of DES. Besides, she can't possibly be the daughter of a woman who took DES sometime between 1940 and 1971, when a report in the *New England Journal of Medicine* tied DES to a nasty vaginal clear cell adenocarcinoma in girls and young women who had been exposed to this drug in utero. Otherwise, my vet would surely do more than remember what I am afraid of. A talking cure might not be enough.

Even though a double-blind study done at the University of Chicago in the early 1950s showed no benefit from DES for sustaining pregnancies in women, and even though by the late 1960s six of seven leading human gynecological textbooks stated that DES did not prevent miscarriages, the drug continued to be prescribed frequently over three decades for averting miscarriage and also for an almost comical (except it was not funny) host of other "indications," both on and off label. Ultimately, probably 2 million women in the United States alone took DES during pregnancy. Probably every reader of this chapter knows some of the offspring of these pregnancies, but may or may not know their often hidden suffering. I do—both people and their suffering, or a little bit of it—and the extraordinary psychologist-scholar-friend who told me about her DES history when I told her about Cayenne performed just the generative acts of "becoming-with" that have occupied my soul since writing *When Species Meet*, or really since the "Cyborg Manifesto." My human friend, this human DES daughter, was already an avid, if dog-allergic, admirer of Cayenne; she is one of the humans who, when she visits, induces enthusiastic canine play-solicitation performances from my very nonpromiscuous dog. But suddenly and oddly, their unexpected DES kinship threw them transversally, not genealogically, into a litter together differently. Separated by allergies, they were joined in the flesh by a disreputable nonsteroidal estrogen. It is clear to me that human Sheila has assigned herself to keep a baleful queer sisterly eye on those gelcaps I give dog Cayenne. That kindly critical lateral eye will complement the regular blood tests Cayenne will now have to endure to keep track of the health of her blood-forming cells and immune functions.[3] A good sphincter can be hard to find.

For very good reasons tied to the history of the women's health movements and to action, finally, by agencies like the U.S. Food and Drug Administration, these days DES is a controlled substance you can get only (or mainly) for nonhumans. In the 1990s, the only approved indication for DES in human beings was treatment of advanced prostate cancer in men and of advanced breast cancer in postmenopausal women, and that use has been superseded. The last U.S. manufacturer of DES, Eli Lilly, stopped making and marketing the no-longer-profitable drug in 1997. That's why Cayenne and I found ourselves at a homeopathic and compounding pharmacy in 2011.[4]

Diethylstilbesterol was first synthesized in 1938 in a laboratory of the Oxford University in the waning heroic days of the history of endocrinology. Those were the days when one might still find eminent biochemists prowling around nonhuman animal slaughter floors collecting many pounds of ovaries, pancreas, testes, adrenal glands, kidneys, pituitaries (try collecting pounds or kilograms of pituitary gland!), and other organs and tissues from many species to ferret back to the lab to extract and then chemically and physiologically characterize the first few precious micrograms of natural steroids or other potent hormones. Stalking the newly dead in graves of the European Renaissance for human bodies to dissect has a long uncanny laboratory history into the present. Current-day laboratory mice and their archived and curated parts would probably be our best informants for today's stories of organs without bodies and life after death. The 1930s were still the days when biochemical laboratories were accustomed to distilling tiny amounts of chemical gold from the dross of vats of urine and other bodily fluids, human and not.

DES did not come from these material sources, but it inhabits the same cross-hatched histories, where what counts as natural or artificial was (and is) constantly morphing in the study and production of things called "sex hormones." No wonder biologist feminists like me find our politics and psyches relentlessly and variously material in ways that Foucault hardly dreamed of. It's those laboratory wet spots in nighttime knowledge making that get feminists roused.

So in my town, you purchase the expensive, carcinogenic, immune-suppressing, anemia-inducing, smooth muscle–plumping molecules known as DES in snowy powder form in gelcaps from a homeopathic and compounding pharmacy, Lauden Integrative Pharmacy. "Compounding pharmacy" sounds so early twentieth-century to my ear, but I can see that when Big Pharma no longer makes or sells a still useful molecule, one that is no better than she should be, the up-to-the-minute, seemingly old-fashioned drugstore gets the leavings. Lauden Integrative Pharmacy sells lots of homeopathic substances for both human and more-than-human animals. I paid for Cayenne's DES capsules at a counter draped in the colors, posters, and icons of "Western" and "Eastern" alternative medicine, both ancient and modern.

To say this scene is emblematic of the mixed structures and affects of biomedical technoscience is an understatement. Lauden formulates many of the chemotherapeutic and other drugs prescribed by the veterinary specialist clinic where Cayenne and I were under the care of a fine consulting internist and cardiac specialist for her early mitral valve disease (MVD). MVD is the reason my fast and sporty Hot Pepper's moderately high blood pressure was not acceptable, and so a new diagnosis of MVD is the reason we changed prescriptions from a drug she had gobbled happily for a few years, called Propolin® (PPA or phenylpropanolamine, "for oral use in dogs only"), to DES. PPA does a fine job pumping up urethral smooth muscle and keeping urine in its hygienic reservoirs for properly timed release in assigned places. But unfortunately PPA is indiscriminate and tightens arterial smooth muscle too, thereby raising blood pressure—not a good idea in dogs with early heart disease. For better and for worse, estrogens are more discriminating in the tissues they home in on. Anyone with breasts—or breast cancer—knows this.

But it's the multispecies business in the compounding pharmacy that really drew my attention. When I get an anxiety syndrome, I am thrown into compulsive scholarly antics, and DES Anxiety Syndrome was no exception. Urine, urethras, damaged heart valves, "abnormal pregnancy outcomes," and cancer-ravaged breasts and uteruses have provided the cross-species organic stuff of the story. So far, my tale has emphasized a litter of critters made up of dogs, humans, and slaughtered animals, mainly pigs, sheep, and cows. It's that last category that will take me into the last stanza of the DES recitative and plump out the litter a bit before we get to my next starring estrogen molecules for remaking kin and kind.

DES was the molecule used in the first experimental, scientific demonstration of successful hormonal growth promotion in cattle in the history of those animal-human relations called agriculture.[5] Although in 1947 researchers at Purdue University demonstrated DES-induced growth promotion in heifers, Purdue did not pursue patent protection for the cattle and sheep work that its investigators carried out. These agricultural scientists used DES because implants had already been formulated for use in poultry, those feathered workhorses in so much of the history of factory farming. The FDA banned DES for growth promotion in chickens and lambs in 1959 and in all animal feed in 1979. But from 1954 to the early 1970s DES was used widely as a growth promoter in the beef industry. Agricultural industry and university agricultural science (especially Iowa State College) were close partners in research for this use. The agricultural-industrial complex was in its postwar adolescent growth spurt. Iowa State and W. Burroughs filed for a patent on oral DES for cattle in 1953, granted in 1956. In 1972 the FDA removed oral DES from the market for use in cattle (1973 for implants). DES residues found in bovine livers and human DES daughters converged to take the drug off the legal agricultural market, although stories of illegal use still surface.

But the core story here is not DES as such; the big story is the relentless rise of hormonal growth promoters of the next molecular generations that are integral to the ecosystem-destroying, human and animal labor-transforming, multispecies soul-mutilating, epidemic-friendly, corn monocrop-promoting, cross-species heartbreaking, feedlot cattle industries. All of a sudden, I cannot forget that in 1947 heifers too became DES daughters, and the bovine sons followed in droves. Cyborg's enhanced litter is outsized. Daughter of a dog family known for prowess in cattle herding before the times of DES, my dog dribbling urine spots leads inexorably to feedlots, slaughterhouses, and unmet agricultural animal, human, and ecological well-being and advocacy obligations around the world. Response-ability yet to come, again. In companion species worlding, becoming-with makes strong demands on the littermates.

Conjugating Kin with Premarin

Conjugating is about yoking together; conjugal love is yoked love; conjugated chemical compounds join together two or more constituents. People conjugate in public spaces; they yoke themselves together transversally and across time and space to make significant things happen. Students conjugate verbs to explore the yoked inflections of person, number, gender, kind, voice, mood, position, tense, and aspect in a field of material-semiotic meaning making. To learn about recursive yoking, conjugate "to conjugate." Now, do that with estrogens. Conjunctivitis is an irritation of the mucous membrane lining the inner surface of the eyelid. What might conjunctivitis mean in the odoriferous fluid mixtures of conjugated estrogens, such as the motley of naturally occurring but nonhuman estrogens purified from pregnant mare's urine to make very profitable pills for Big Pharma? And also to give lots of human women the means to decide whether or not they will bear children, endure hot flashes, lose bone mass, or add to or subtract from their risk of cancer or heart disease? Or, to find that "our bodies ourselves" includes mares and their foals (and a few stallions), with all the political and ethical consequences of that conjugation? Conjugated estrogens are about yoking molecules and species to each other in consequential ways. In the Moby Thesaurus, one mouth-watering synonym of conjugate is conglobulate; that is what I will try to do with horses, humans, urine, and hearts conjugated with Premarin.

Once upon a time, when I thought I needed estrogen during menopause—to stave off familial heart disease, of all things—I relied on the animal-industrial complex, repeated pregnancies and long-term confinement of mares, and natural conjugated estrogens called Premarin (compounded with a progestin into HRT, Hormone Replacement Therapy) extracted from equine urine.[6] Now I give my dog a synthetic estrogen with a terrible human and bovine history to control her urinary incontinence, for the sake of her heart—and her indoor way of life. (It works.) Oh, Cayenne, dog of my heart, a human taste for irony will not get us through these companion species relationships, these meals of situated molecules and required response-ability yet to come. Somehow, a feminist science studies scholar and lifelong animal lover, my menopausal self failed to know much about the pregnant mares and their disposable foals.

Did I forget, never know, not look—or just not care? What kind of conjunctivitis was that? Social movements for animal flourishing had noticed those horses and made a very effective fuss about it, and these movements were full of feminist women and men. Why not me too? Was it only after it turned out that HRT probably harmed my heart rather than guarded it that the horses came into my ken? I don't remember. Marx understood all about how privileged positions block knowledge of the conditions of one's privilege. So did the innovators of feminist standpoint theory, the founders of the women's health movements, and the thinkers and activists shaping movements for animal flourishing—that is, my friends, comrades, and colleagues—well before I was in menopause. Still, I managed not to know about the conditions of work for those adult horses for a very long time, much less know about the fate of the excess foals. I ate equine conjugated estrogens; I drank pooled mares' urine, literally; but I did not conjugate well with the horses themselves. Shame is a prod to lifelong rethinking and recrafting one's accountabilities!

A collaboration between a Canadian pharmaceutical company and an endocrinologist at McGill University led to the development in 1930 of the first orally active, water-soluble, conjugated estrogen, called Emmenin®.[7] Emmenin® was extracted from the urine of Canadian women in late stages of their pregnancies, but supply considerations set the researchers and company to looking for a more copious and available mammalian source. Even if they were paid and desperate, pregnant women would not stay attached to collection bags for long, nor did they pee nearly enough to supply their sisters with hormones. German researchers at the time were studying water-soluble estrogens in the urine of pregnant zebras and horses in the Berlin zoo, and by 1939 the pharmaceutical company Ayerst had established a method to get a stable concentrate from pregnant mares' urine. The result of an extraction and concentration process with more than one hundred steps, Premarin was ready to be marketed in Canada in 1941. The horses, confined in stalls for months at a time attached to collection bags, were originally contract workers on Quebec farms, and the product was manufactured in Montreal. Eventually, high demand issuing from the growing practice of prescribing hormones for menopause, coupled with the history of successive buyouts among the pharmaceutical companies, resulted in production moving to the expansive Canadian western prairies, with a new processing plant in Manitoba.

About a decade after I started menopause in the late 1980s, by 1997 Premarin became the number-one prescribed drug in the United States, reaching the sales figure of $2 billion by 2002.[8] Used in over three thousand scientific investigations by 2011, this drug complex remains the most studied estrogen therapy in the world. Definitively by 2002, strong data gathered in the context of the Women's Health Initiative showed that not only did estrogens not prevent heart disease; they were also positively correlated with increased incidences of blood clots, strokes, heart attack, and breast cancer. Sales of Premarin dropped fast—by a lot. Redundant equine workers went to slaughter—lots of them. Dependent contract farmers were put out of business. Drug companies scrambled. Women worried; I know.

However reduced in volume, harvesting of pregnant mare urine remains a worldwide business, and Premarin remains a much-prescribed and profitable product. Today Pfizer, which bought out Wyeth-Ayerest in 2009, contracts with about two dozen horse ranches, mainly in western Canada. In 2003, there were over four hundred Wyeth-Ayerest contracted PMU farms in Manitoba. With industry reorganization in the wake of the crisis in Premarin prescriptions after 2002, profit per PMU mare went up between 2003 and 2007—a lot. The North American Equine Ranching Information Council (NAERIC) is a committed, sophisticated industry group that presents in its best light the history and contemporary practices of PMU farming.[9] The NAERIC website includes a "four seasons" description, with beautiful pictures, of the annual life cycle of the horses on idyllic-looking farms said to be thoroughly regulated and inspected for animal welfare. This site narrates that, from autumn through early spring, mares are confined in their own "comfortable" stalls attached to a "lightweight, flexible pouch that is suspended from the ceiling by rubber suspension lines" that allow a full range of motion, including lying down. Horses have access to sufficient water—a major change from the period before reform, when the demand for concentrated urine trumped equine thirst, with predictable medical consequences for the horses. Put together by international veterinary and welfare groups which inspect equine ranches, and available online from the NAERIC site, the "Equine Veterinarians' Consensus Report on the Care of Horses on PMU Ranches" concluded that numerous reforms after an investigation in 1995 led to major improvements in the lives of the horses. "The public should be assured that the care and welfare of the horses involved in the production of an estrogen replacement medication is good, and is closely monitored."[10]

Onsite analysis on several farms by HorseAid in 1999 found conditions much less satisfactory than NAERIC claims, even if one grants that the guidelines, which leave the question of exercise to the discretion of pressed farmers with no indoor exercise facilities in a northern plains winter, are good enough for horses.[11] Confined mares stand around too much, eat too much, get fat, and develop bad feet—sounds like a lot of working females across species to me. With very little room to make costly changes in care practices, contract farmers are at the low end of the financial food chain generated by pregnant mares' urine, just as they are for broiler chickens or other animal industrial products.

By 2011, about 2000 NAERIC foals per year were born on twenty-six PMU farms, including draft, light horse, and sport breeds. The foals were sold mostly to families and show barns. About forty-nine thousand horses have been registered with NAERIC since 1998. Better-bred foals are more profitable, so fewer are slaughtered or enter the rescue and adoption apparatus. These days, ranchers collecting pregnant mares' urine "rely on selling foals as much as they rely upon the urine collected from the pregnant mares. Many of these farms utilize websites and forms of promotion identical to non-Premarin-related horse breeders, and, in nearly all ways, are indistinguishable from the average breeder of equines."[12]

The reforms promoted by NAERIC came into being because of activist animal rights, women's health, and horse advocacy groups. Beginning with a hands-on, on-site study in 1986, HorseAid was the first animal rights organization to investigate conditions on PMU farms and the risks to women from HRT medication, publishing its damning results first in print in 1988 and then in 1994 on the Internet, with graphic images and details about farm and industry practices and human medical data.[13] In 1995, seven years after HorseAid's 1988 report, NAERIC formed to advocate for reform and humane treatment of PMU horses. But reform was not and is not the ultimate goal for HorseAid or organizations like the International Fund for Horses. Both groups continue to argue for shutting down all PMU farming, where months-long confinement of pregnant horses, however "comfortable" (i.e., in a box 8 feet wide by 3.5

feet wide by 5 feet high for six months), and slaughtering mares who fail to become pregnant, continue.[14] The availability of a wider range of laboratory-synthesized and plant-derived hormones makes arguments to end PMU production harder to evade. Taking account of all PMU farms in 2002, HorseAid estimated that about fifteen thousand "excess" foals went to slaughter. Reflecting Premarin sales declines since 2002, cuts in contracts to farms, and a more market-oriented foal production, the number now is much lower, but could be zero.[15]

HorseAid was always clear about its advocacy both for women's health and the well-being of horses, and its reports also paid attention to the difficulties of farms and farmers in an agribusiness system in which making a living by farming has become brutal.That fact, of course, does not address what would make raising horses on the northern prairies viable for economic and ecological human-animal well-being, and that should not be an idle goal.

Viral Response-ability

There is no innocence in these kin stories, and the accountabilities are extensive and permanently unfinished. Indeed, responsibility in and for the worldings in play in these stories requires the cultivation of viral response-abilities, carrying meanings and materials across kinds in order to infect processes and practices that might yet ignite epidemics of multispecies recuperation and maybe even flourishing on terra in ordinary times and places. Call that utopia; call that inhabiting the despised places; call that touch; call that the rapidly mutating virus of hope, or the less rapidly changing commitment to staying with the trouble. My slogan from the 1980s, "Cyborgs for Earthly Survival," still resonates, in a cacophony of sound and fury emanating from a very big litter whelped in shared but nonmimetic suffering and issuing in movements for flourishing yet to come.

In my DES story, tracking Cayenne's urine spots to out-of-the-way places brought us into a still-expanding conglobulation of interlinked research, marketing, medical and veterinary, activist, agricultural, and scholarly body- and subject-making apparatuses. Digital and molecular species vied for attention with urethras and vaginas. Females in trouble seemed to luxuriate everywhere; even the industrially synthesized molecules seemed to respond to the lure of (always nonreproductive, in this story) sexual tropisms, despite decades of astute feminist wariness of so-called sex hormones. Cyborgs laughed. Do cyborgs get mitral valve disease or go through menopause? Of course they do, just like their kin. The relations of intimate care yoking together one woman and one dog rampaged virally into all sorts of publics. Sheer contagion. Companion species infect each other all the time. Bodily ethical and political obligations are infectious, or they should be. Before my dog and I could get out of the story, we were in the nonoptional company of—and accountable to—heifers in labs, beef cattle in feedlots, pregnant women in all sorts of places, daughters and sons and granddaughters and grandsons of once pregnant women, angry and well-informed women's health movement activists, dogs with heart disease, and bevies of other spayed leaky bitches and their people in vet clinics and on beds.

In my Premarin story, all the players seemed to be marinating in vats of Canadian equine urine, the only thing that seemed to hold together the virally exploding, vulnerable species of the tale. One registered trademark's travels through bodies brought together, in the need to craft response-ability, quite a motley of mortal beings: fetal calves stripped of amniotic fluid, urinating pregnant Canadian women, pregnant mares and their foals and consorts in Manitoba and beyond, activists in horse rescue and women's health, economically strapped contract farmers, a California menopausal woman worried about familial heart disease in the company of a lucrative market-ready crowd of other menopausal Americans, and German zebras in zoos in the 1930s. Big Pharma, Big Agribusiness, and Big Science provided drama and villains aplenty, but also plenty of reason to damp down the certainty of villainy and explore the complexities of cyborg worlding.

Each diner exposed to high risks of familial heart failure, eating dangerous and notorious estrogens in later life seems, finally, to be what here, in this tale, conjugates—yokes together—the cyborg author and the dog of her heart. *Cum panis*, companion species, females of two species (along with their microbiomes with species in the zillions) at table together, in different decades, slurping drafts of dubious estrogens in self-care and care of the other. Why tell stories like this, when there are only more and more openings and no bottom lines? Because there are quite definite response-abilities that are strengthened in such stories.

It is no longer news that corporations, farms, clinics, labs, homes, sciences, technologies, and multispecies lives are entangled in multiscalar, multitemporal, multimaterial worlding; but the details matter. The details link actual beings to actual response-abilities. Each time a story helps me remember what I thought I knew, or introduces me to new knowledge, a muscle critical for caring about flourishing gets some aerobic exercise. Such exercise enhances collective thinking and movement too. Each time I trace a tangle and add a few threads that first seemed whimsical but turned out to be essential to the fabric, I get a bit straighter that staying with the trouble of complex worlding is the name of the game of living and dying well together on terra. Having eaten Premarin makes me more accountable to the well-being of ranchers, northern prairie ecologies, horses, activists, scientists, and women with breast cancer than I would otherwise be. Giving my dog DES makes me accountable to histories and ongoing possibilities differently than if we never shaped kinships with the attachment sites of this molecule. Perhaps reading this chapter has consequences for response-ability too. We are all responsible to and for shaping conditions for multispecies flourishing in the face of terrible histories, but not in the same ways. The differences matter—in ecologies, economies, species, lives.

CHAPTER 6

Sowing Worlds

A Seed Bag for Terraforming with Earth Others

"Do you realize," the phytolinguist will say to the aesthetic critic, "that they couldn't even read Eggplant?" And they will smile at our ignorance, as they pick up their rucksacks and hike on up to read the newly deciphered lyrics of the lichen on the north face of Pike's Peak.
—Ursula K. Le Guin, "The Author of the Acacia Seeds"

The political slogan I wore in the Reagan Star Wars era of the 1980s read, "Cyborgs for Earthly Survival!" The terrifying times of George H. W. Bush and the secondary Bushes made me switch to slogans purloined from tough schutzhund dog trainers, "Run Fast, Bite Hard!" and "Shut Up and Train!" Today my slogan reads, "Stay with the Trouble!" But in all these knots and especially now, wherewhenever that potent and capacious placetime is—we need a hardy, soiled kind of wisdom. Instructed by companion species of the myriad terran kingdoms in all their placetimes, we need to reseed our souls and our home worlds in order to flourish—again, or maybe just for the first time—on a vulnerable planet that is not yet murdered.[1] We need not just reseeding, but also reinoculating with all the fermenting, fomenting, and nutrient-fixing associates that seeds need to thrive. Recuperation is still possible, but only in multispecies alliance, across the killing divisions of nature, culture, and technology and of organism, language, and machine.[2] The feminist cyborg taught me that; the humanimal worlds of dogs, chickens, turtles, and wolves taught me that; and in fugal, fungal, microbial, symbiogenetic counterpoint, the acacia trees of Africa, the Americas, Australia, and the Pacific Islands, with their congeries of associates reaching across taxa, teach me that. Sowing worlds is about opening up the story of companion species to more of its relentless diversity and urgent trouble.

To study the kind of situated, mortal, germinal wisdom we need, I turn to Ursula K. Le Guin and Octavia Butler.[3] It matters what stories we tell to tell other stories with; it matters what concepts we think to think other concepts with. It matters wherehow Ouroboros swallows its tale, again. That's how worlding gets on with itself in dragon time. These are such simple and difficult koans; let us see what kind of get they spawn. A careful student of dragons, Le Guin taught me the carrier bag theory of fiction and of naturalcultural history.[4] Her theories, her stories, are capacious bags for collecting, carrying, and telling the stuff of living. "A leaf a gourd a shell a net a bag a sling a sack a bottle a pot a box a container. A holder. A recipient."[5]

So much of earth history has been told in the thrall of the fantasy of the first beautiful words and weapons, of the first beautiful weapons *as* words and vice versa. Tool, weapon, word: that is the word made flesh in the image of the sky god. In a tragic story with only one real actor, one real world-maker, the hero, this is the Man-making tale of the hunter on a quest to kill and bring back the terrible bounty. This is the cutting, sharp, combative tale of action that defers the suffering of glutinous, earth-rotted passivity beyond bearing. All others in the prick tale are props, ground, plot space, or prey. They don't matter; their job is to be in the way, to be overcome, to be the road, the conduit, but not the traveler, not the begetter. The last thing the hero wants to know is that his beautiful words and weapons will be worthless without a bag, a container, a net.

Nonetheless, no adventurer should leave home without a sack. How did a sling, a pot, a bottle suddenly get in the story? How do such lowly things keep the story going? Or maybe even worse for the hero, how do those concave, hollowed-out things, those holes in Being, from the get-go generate richer, quirkier, fuller, unfitting, ongoing stories, stories with room for the hunter but which weren't and aren't about him, the self-making Human, the human-making machine of history? The slight curve of the shell that holds just a little water, just a few seeds to give away and to receive, suggests stories of becoming-with, of reciprocal induction, of companion species whose job in living and dying is not to end the storying, the worlding. With a shell and a net, becoming human, becoming humus, becoming terran, has another shape—the side-winding, snaky shape of becoming-with.

Le Guin quickly assures all of us who are wary of evasive, sentimental holisms and organicisms: "Not, let it be said at once, [am I] an unaggressive or uncombative human being. I am an aging, angry woman laying about me with my handbag, fighting hoodlums off . . . It's just one of those damned things you have to do in order to go on gathering wild oats and telling stories."[6] There is room for conflict in Le Guin's story, but her carrier bag narratives are full of much else in wonderful, messy tales to use for retelling, or reseeding, possibilities for getting on now, as well as in deep earth history. "It sometimes seems that that [heroic] story is approaching its end. Lest there be no more telling of stories at all, some of us out here in the wild oats, amid the alien corn, think we'd better start telling another one, which maybe people can go on with when the old one's finished . . . Hence it is with a certain feeling of urgency that I seek the nature, subject, words of the other story, the untold one, the life story."[7]

Octavia Butler knows all about the untold stories, the ones that need a restitched seedbag and a traveling sower to hollow out a place to flourish after the catastrophes of that Sharp Story. In *Parable of the Sower*, the U.S. teenage hyperempath Lauren Oya Olamina grows up in a gated community in Los Angeles. Important in New World Santeria and in Catholic cults of the Virgin Mary, in Yoruba Oya, mother of nine, is the Orisha of the Niger River, with its nine tributaries, its nine tentacles gripping the living and the dead. She is among the chthonic entities of a thousand names, generators of persisting times called the Chthulucene. Wind, creation, and death are Oya's attributes and powers for worlding. Olamina's gift and curse were her inescapable abilities to feel the pain of all living beings, a result of a drug taken by her addicted mother during pregnancy. After the murder of her family, the young woman traveled from a devastated and dying society with a motley band of survivors to sow a new community rooted in a religion called Earthseed. In the story arc of what was to be a trilogy (*Parable of the Trickster* was not completed before her death), Butler's SF worlding imagined Earthseed ultimately flourishing on a new home world among the stars. But Olamina started the first Earthseed community in Northern California, and it is there and at other sites on Terra where my own explorations for reseeding our home world must stay. This home is where Butler's lessons apply with special ferocity.

In the Parable novels, "God is change," and Earthseed teaches that the seeds of life on earth can be transplanted and can adapt and flourish in all sorts of unexpected and always dangerous places and times. Note "can," not necessarily "may" or "should." Butler's entire work as an SF writer is riveted on the problem of destruction and wounded flourishing—not simply survival—in exile, diaspora, abduction, and transportation—the earthly gift-burden of the descendants of slaves, refugees, immigrants, travelers, and of the indigenous too. It is not a burden that stops with settlement. In the SF mode,[8] my own writing works and plays only on earth, in the mud of cyborgs, dogs, acacia trees, ants, microbes, fungi, and all their kin and get. With the twist in the belly that etymology brings, I remember too that *kin*, with the *g-k* exchange of Indo-European cousins, becomes *gen* on the way to *get*. Terran spawn all, we are sidewinding as well as arboreal kindred—blown get—in infected and seedy generation after generation, blowsy kind after blowsy kind.

Planting seeds requires medium, soil, matter, mutter, mother. These words interest me greatly for and in the SF terraforming mode of attention. In the feminist SF mode, matter is never "mere" medium to the "informing" seed; rather, mixed in terra's carrier bag, kin and get have a much richer congress for worlding. *Matter* is a powerful, mindfully bodied word, the matrix and generatrix of things, kin to the riverine generatrix Oya. It doesn't take much digging or swimming to get to matter as source, ground, flux, reason, and consequential stuff—the matter of the thing, the generatrix that is simultaneously fluid and solid, mathematical and fleshly—and by that etymological route to one tone of matter as timber, as hard inner wood (in Portuguese, *madeira*). Matter as timber brings me to Le Guin's *The Word for World Is Forest*, published in 1976 as part of her Hainish fabulations for dispersed native and colonial beings locked in struggle over imperialist exploitation and the chances for multispecies flourishing. That story took place on another planet and is very like the tale of colonial oppression in the name of pacification and resource extraction that takes place on Pandora in James Cameron's 2009 blockbuster film *Avatar*. Except one particular detail is very different; Le Guin's *Forest* does not feature a repentant and redeemed "white" colonial hero. Her story has the shape of a carrier bag that is disdained by heroes. Also, even as they condemn their chief oppressor to live, rather than killing him after their victory, for Le Guin's "natives"

6.1. An ant of the species *Rhytidoponera metallica* in western Australia holding a seed of *Acacia neurophylla* by the elaiosome during seed transport. © Benoit Guenard, 2007.

the consequences of the freedom struggle bring the lasting knowledge of how to murder *each other*, not just the invader, as well as how to recollect and perhaps relearn to flourish in the face of this history. There is no *status quo ante*, no salvation tale, like that on Pandora. Instructed by the struggle on *Forest*'s planet of Athshea, I will stay on Terra and imagine that Le Guin's Hainish species have not all been of the hominid lineage or web, no matter how dispersed. Matter, mater, mutter make me—make us, that collective gathered in the narrative bag of the Chthulucene—stay with the naturalcultural multispecies trouble on earth, strengthened by the freedom struggle for a postcolonial world on Le Guin's planet of Athshea. It is time to return to the question of finding seeds for terraforming for a recuperating earthly world of difference, wherewhen the knowledge of how to murder is not scarce.

My carrier bag for terraforming is full of acacia seeds, but as we shall see, that collection brings its full share of trouble too. I begin with the decapitated corpse of an ant found by scientist-explorers next to Seed 31 in a row of degerminated acacia seeds at the end of an ant-colony tunnel in Le Guin's story "'The Author of the Acacia Seeds' and Other Extracts from the *Journal of the Association of Therolinguistics*." The therolinguists were perplexed in their reading of the touch-gland exudate script that the ant seemed to have written in her biochemical ink on the aligned seeds. The scientists were uncertain both about how to interpret the script and about who the ant was—an intruder killed by the colony's soldiers? A resident rebel writing seditious messages about the queen and her eggs? A myrmexian tragic poet?[9] The therolinguists could not apply rules from human languages to their task, and their grasp of animal communication was (is) still raggedly fragmentary, full of guesses across profound naturalcultural difference. From the scientific and hermeneutic study of other animal languages recorded in difficult expeditions of discovery, therolinguists held that "language is communication" and that many animals use an active collective kinesthetic semiotics, as well as chemosensory, visual, and tactile language. They might have been troubled about their reading of this unexpected ant's exudate text, but they felt confident that at least they were engaging therolingistic acts and would someday learn to read them.

Plants, however, they speculated, "do not communicate" and so have no language. Something else is going on in the vegetative world, perhaps something that should be called art.[10] Phytolinguistics pursued along these lines by the scientists and explorers was just beginning and would surely require entirely new modes of attention, field methodology, and conceptual invention. The president of the Therolinguists Association waxed lyrical: "If a noncommunicative, vegetative art exists, we must re-think the very elements of our science, and learn a whole new set of techniques. For it is simply not possible to bring the critical and technical skills appropriate to the study of weasel murder-mysteries, or Battrachian erotica, or the tunnel-sagas of the earthworm, to bear on the art of the redwood or the zucchini."[11]

In my view, the president got it right about the need to question the tissues of one's knowings and ways of knowing in order to respond to nonanthropocentric difference. But a closer look at that decapitated ant and the degerminated acacia seeds should have told those still zoocentric scientists that their sublime aestheticization of plants led them astray about earth-making companion species. Plants are consummate communicators in a vast terran array of modalities, making and exchanging meanings among and between an astonishing galaxy of associates across the taxa of living beings. Plants, along with bacteria and fungi, are also animals' lifelines to communication with the abiotic world, from sun to gas to rock. To pursue this matter, I need to leave Le Guin's story for now and instead draw on the stories told by students of symbiosis, symbiogenesis, and ecological evolutionary developmental biology.[12]

Acacias and ants can do almost all of the work for me. With fifteen hundred species (about a thousand of which are indigenous to Australia), the genus *Acacia* is one of the largest genera of trees and shrubs on earth. Different acacias flourish in temperate, tropical, and desert climates across oceans and continents. They are crucial species maintaining the healthy biodiversity of complex ecologies, housing many lodgers, and nourishing a motley guest list of diners. Relocated from wherever they originated, acacias were the darlings of human colonial foresters and still are the stock-in-trade of landscapers and plant breeders. In those histories, some acacias become the overgrowing destroyers of endemic ecologies that are the special responsibility of restoration biologists and just plain citizens of recuperating places.[13] In part and in whole, acacias show up in the most unexpected places. They give the bounty of the gorgeous hardwoods like Hawaiian koa, which are cut down in greedy, exterminating, global-capitalist excess. Acacias also make the humble polysaccharide gums, including gum arabic from *Acacia senegal*, which show up in human industrial products like ice cream, hand lotion, beer, ink, jelly beans, and old-fashioned postage stamps. Those same gum exudates are the immune system of the acacias themselves, helping to seal wounds and discourage opportunistic fungi and bacteria. Bees make a prized honey from acacia flowers, among the few honeys that will not crystallize. Many animals, including moths, human beings, and the only known vegetarian spider, use acacias for food. People rely on acacias for seed pastes, fritters made from pods, curries, shoots, toasted seeds, and root beer.

Acacias are members of the vast family of legumes. That means that, among their many talents, in association with fungal mycorrhizal symbionts (which host their own bacterial endosymbionts), many acacias fix the nitrogen crucial to soil fertility, plant growth, and animal existence.[14] In defending themselves from grazers and pests, acacias are veritable alkaloid chemical factories, making many compounds that are psychoactive in animals like me. I can only imagine with my hominid brain what these compounds feel like to critters like insects. From giraffes' points of view, acacias sport lovely leafy salads on their crowns, and the acacias respond to assiduous giraffe pruning by producing the picturesque African savannah flattop tree landscape prized by human photographers and tourist enterprises, not to mention life-saving shade and rest for many critters.

Supported within this big narrative netbag, I am ready to add a few details of my own to Le Guin's ongoing carrier bag story of the decapitated ant and her acacia-seed writing tablet. The therolinguists were worried about the message they tried to decipher in the writing, but I am riveted by what drew ant and acacia seed together in the first place. How did they know each other? How did they communicate? Why did the ant paint her message on that shiny surface? The degerminated seed is the clue. *Acacia verticulata*, an Australian shrub related to the coastal wattle so worrisome to Southern Californian ecologists, makes seeds that are dispersed by ants. The wily acacias draw the ants' attention with a showy attachment stalk coiled around every seed. The ants carry the decorated seeds to their nests, where they consume the fat-rich attachment stalks, called elaisosomes, at their leisure. In time, the seeds germinate out of the nice womb provided by the ant tunnels, and the ants have the nutritious, calorie-dense food they need to fuel all those stories of their hard-working habits. In evolutionary-ecological terms, these ants and acacias are necessary to each other's reproductive business.

Some ant-acacia associations are much more elaborate than that, reaching into the internal tissues of each participant, shaping genomes and developmental patterning of the structures and functions of both companion species. Several Central American acacias make large, hollow thornlike structures called stipules that provide shelter for several species of *Pseudomyrmex* ants. "The ants feed on a secretion of sap on the leaf-stalk and small, lipid-rich [and protein-rich] food-bodies at the tips of the leaflets called Beltian bodies. In return, the ants add protection to the plant against herbivores."[15] There is nothing like a dedicated bevy of angry, biting ants to make a day's foraging uncomfortable and the leaf-grazer of whatever species move on to less infested pantries. In the 2005 BBC Science and Nature five-part special with David Attenborough, in the episode called "Intimate Relations," we see these matters in exquisite, sensuous detail. We also witness that "some ants 'farm' the trees that give them shelter, creating areas known as 'Devil's gardens.' To make sure these grow without competition, they kill off other seedlings in the surrounding vegetation."[16] The ants accomplish this task by gnawing methodically through branches and shoots and then injecting formic acid into the conductive tissue of the offending plants. Similar ant-acacia mutualisms occur in Africa. For example, the Whistling Thorn acacias in Kenya provide shelter for ants in the thorns and nectar in extrafloral nectaries for their symbiotic ants, such as *Crematogaster mimosae*. In turn, the ants protect the plant by attacking large mammalian herbivores and stem-boring beetles that damage the plant. The more one looks, the more the name of the game of living and dying on earth is a convoluted multispecies affair that goes by the name of symbiosis, the yoking together of companion species, at table together.

Ants and acacias are both highly diverse, well-populated groups. They are sometimes world travelers and sometimes homebodies that cannot flourish away from natal countries and natal neighbors. Homebody or traveler, their ways of living and dying have consequences for terrraforming, past and present. Ants and acacias are avid for association with critters of all sorts of sizes and scales, and they are opportunistic in their approaches to living and dying in both evolutionary and organismic or whole group timeplaces. These species in all their complexities and ongoingness both do great harm and sustain whole worlds, sometimes in association with human people, sometimes not. The devil is truly in the details of response-able naturecultures inhabited by accountable companion species. They—we—are here to live and die with, not just think and write with. But also that, also here to sow worlds with, to write in ant exudates on acacia seeds to keep the stories going. No more than Le Guin's carrier bag story—with the crusty elderly lady ready to use her purse to whack evildoers and the author avid for the mess as well as the order of her bumptious critters, human and not—is my story of these worldly wise symbionts a tale of rectitude and final peace. With Le Guin, I am committed to the finicky, disruptive details of good stories that don't know how to finish. Good stories reach into rich pasts to sustain thick presents to keep the story going for those who come after.[17] Emma Goldman's understanding of anarchist love and rage make sense in the worlds of ants and acacias. These companion species are a prompt to shaggy dog stories—growls, bites, whelps, games, snufflings, and all. Symbiogenesis is not a synonym for the good, but for becoming-with each other in response-ability.

Finally, and not a moment too soon, sympoesis enlarges and displaces autopoesis and all other self-forming and self-sustaining system fantasies. Sympoesis is a carrier bag for ongoingness, a yoke for becoming-with, for staying with the trouble of inheriting the damages and achievements of colonial and postcolonial naturalcultural histories in telling the tale of still possible recuperation. Le Guin's therolinguists, even bound in their animal hides, had the vision of these scary and inspiring possibilities: "And with them, or after them, may there not come that even bolder adventurer—the first geolinguist, who, ignoring the delicate, transient lyrics of the lichen, will read beneath it the still less communicative, still more passive, wholly atemporal, cold, volcanic, poetry of the rocks; each one a word spoken, how long ago, by the earth itself, in the immense solitude, the immenser community, of space."[18] Communicative and mute, the old lady and her purse will be found in Earthseed communities on terra and throughout timespace. Mutter, matter, mother.

CHAPTER 7

A Curious Practice

Interesting research is research conducted
under conditions that make beings interesting.
—Vinciane Despret

To think with an enlarged mentality means that one trains
one's imagination to go visiting.
—Hannah Arendt, *Lectures on Kant's Political Philosophy*

Vinciane Despret thinks-with other beings, human and not. That is a rare and precious vocation. Vocation: calling, calling with, called by, calling as if the world mattered, calling out, going too far, going visiting. Despret listened to a singing blackbird one morning—a living blackbird outside her particular window—and that way learned what importance sounds like. She thinks in attunement with those she thinks with—recursively, inventively, relentlessly—with joy and verve. She studies how beings render each other capable in actual encounters, and she theorizes—makes cogently available—that kind of theory and method. Despret is not interested in thinking by discovering the stupidities of others, or by reducing the field of attention to prove a point. Her kind of thinking enlarges, even invents, the competencies of all the players, including herself, such that the domain of ways of being and knowing dilates, expands, adds both ontological and epistemological possibilities, proposes and enacts what was not there before. That is her worlding practice. She is a philosopher and a scientist who is allergic to denunciation and hungry for discovery, needy for what must be known and built together, with and for earthly beings, living, dead, and yet to come.

Referring both to her own practice for observing scientists and also to the practices of ethologist Thelma Rowell observing her Soay sheep, Despret affirmed "a particular epistemological position to which I am committed, one that I call a virtue: the virtue of politeness."[1] In every sense, Despret's cultivation of politeness is a curious practice. She trains her whole being, not just her imagination, in Arendt's words, "to go visiting." Visiting is not an easy practice; it demands the ability to find others actively interesting, even or especially others most people already claim to know all too completely, to ask questions that one's interlocutors truly find interesting, to cultivate the wild virtue of curiosity, to retune one's ability to sense and respond—and to do all this politely! What is this sort of politeness? It sounds more than a little risky. Curiosity always leads its practitioners a bit too far off the path, and that way lie stories.

The first and most important thing at risk in Despret's practice is an approach that assumes that beings have pre-established natures and abilities that are simply put into play in an encounter. Rather, Despret's sort of politeness does the energetic work of holding open the possibility that surprises are in store, that something *interesting* is about to happen, but only if one cultivates the virtue of letting those one visits intra-actively shape what occurs. They are not who/what we expected to visit, and we are not who/what were anticipated either. Visiting is a subject- and object-making dance, and the choreographer is a trickster. Asking questions comes to mean both asking what another finds intriguing and also how learning to engage *that* changes everybody in unforeseeable ways. Good questions come only to a polite inquirer, especially a polite inquirer provoked by a singing blackbird. With good questions, even or especially mistakes and misunderstandings can become interesting. This is not so much a question of manners, but of epistemology and ontology, and of method alert to off-the-beaten-path practices. At the least, this sort of politeness is not what Miss Manners purveys in her advice column.

There are so many examples of Despret learning and teaching polite inquiry. Perhaps the most famous is her visit to the Negev desert field site of the Israeli ornithologist Amotz Zahavi, where she encountered Arabian babblers who defied orthodox accounts of what birds should be doing, even as the scientists also acted off-script *scientifically*. Specifically, Zahavi asked in excruciating detail, what matters to babblers? He could not do good science otherwise. The babblers' practices of altruism were off the charts, and they seemed to do it, according to Zahavi, for reasons of competitive prestige not well accounted for by theories like kin selection. Zahavi let the babblers be interesting; he asked them interesting questions; he saw them dance. "Not only were these birds described as *dancing* together in the morning sunrise, not only were they eager to offer presents to one another, not only would they *take pride* in caring for each other's nestlings or in defending an endangered comrade, but also, according to Zahavi's depiction, their relations relied on trust."[2]

What Despret tells us she came to know is that the specific practices of observation, narration, and the liveliness of the birds were far from independent of each other. This was not just a question of worldviews and related theories shaping research design and interpretations, or of any other purely discursive effect. What scientists actually do in the field affects the ways "animals see their scientists seeing them" and therefore how the animals respond.[3] In a strong sense, observers and birds rendered each other capable in ways not written into preexisting scripts, but invented or provoked, more than simply shown, in practical research. Birds and scientists were in dynamic, moving relations of attunement. The behavior of birds and their observers were made, but not made up. Stories are essential, but are never "mere" stories. Zahavi seemed intent on making experiments *with* rather than *on* babblers. He was trying to look at the world *with* the babblers rather than *at* them, a very demanding practice. And the same demands were made of Despret, who came to watch scientists but ended up in a much more complex tangle of practices. Birds and scientists do something, and they do it together. They become-with each other.

The world in the southern Israeli desert was composed by adding competencies to engage competencies, adding perspectives to engage perspectives, adding subjectivities to engage subjectivities, adding versions to understand versions. In short, this science worked by addition, not subtraction. Worlds enlarged; the babblers and the scientists—Despret included—inhabited a world of propositions not available before. "Both humans and babblers create narratives, rather than just telling them. They create/disclose new scripts."[4] Good questions were posed; surprising answers made the world richer. Visiting might be risky, but it is definitely not boring.

Despret's work is full of literal collaborations, with people and with animals, not simply metaphors of thinking with each other. I admit I am drawn most by the collaborations that entangle people, critters, and apparatuses. No wonder that Despret's work with sociologist Jocelyne Porcher and the farmers, pigs, and cows in their care sustains me. Despret and Porcher visited cow and pig breeders on nonindustrial French farms, where the humans and animals lived in daily interaction that led sober, nonromantic, working breeders to say such things as, "We don't stop talking with our animals."[5] The question that led Despret and Porcher to the farmers circled around their efforts to think through what it means to claim that these domestic food-producing animals are *working*, and *working with* their people. The first difficulty, not surprisingly, was to figure out how to ask questions that interested the breeders, that engaged them in their conversations and labors with their animals. It was decidedly not interesting to the breeders to ask how animals and people are the same or different in general. These are people who make particular animals live and die and who live, and die, by them. The task was to engage these breeders in constructing the questions that mattered to them. The breeders incessantly "uprooted" the researchers' questions to address the queries that concerned them in their work.

The story has many turns, but what interested me most was the insistence of the breeders that their animals "know what we want, but we, we don't know what they want."[6] Figuring out what their animals want, so that people and cows could together accomplish successful breeding, was the fundamental conjoined work of the farm. Farmers bad at listening to their animals, bad at talking to them, and bad at responding were not good farmers in their peers' estimation. The animals paid attention to their farmers; paying equally effective attention to the cows and pigs was the job of good breeders. This is an extension of subjectivities for both people and critters, "becoming what the other suggests to you, accepting a proposal of subjectivity, acting in the manner in which the other addresses you, actualizing and verifying this proposal, in the sense of rendering it true."[7] The result is bringing into being animals that nourish humans, and humans that nourish animals. Living and dying are both in play. "Working together" in this kind of daily interaction of labor, conversation, and attention seems to me to be the right idiom.

Continually hungry for more of Despret's visiting with critters, their people, and their apparatuses—hungry for more of her elucidations of "anthropo-zoo-genesis"[8]—I have a hard time feeling satisfied with only human people on the menu. That prejudice took a tumble when I read *Women Who Make a Fuss: The Unfaithful Daughters of Virginia Woolf,* which Isabelle Stengers and Vinciane Despret wrote together with an extraordinary collective of bumptious women.[9] "Think we must!" cries this book, in concert with the famous line from Virginia Woolf's *Three Guineas*. In Western worlds, and elsewhere too, women have hardly been included in the patrilines of thinking, most certainly including the patrilines making decisions for (yet another) war. Why should Virginia Woolf, or any other woman, or men for that matter, be faithful to such patrilines and their demands for sacrifice? Infidelity seems the least we should demand of ourselves!

This all matters, but the question in this book is not precisely that, but rather what thinking can possibly mean in the civilization in which we find ourselves. "But how do we take back up a collective adventure that is multiple and ceaselessly reinvented, not on an individual basis, but in a way that passes the baton, that is to say, affirms new givens and new unknowns?"[10] We must somehow make the relay, inherit the trouble, and reinvent the conditions for multispecies flourishing, not just in a time of ceaseless human wars and genocides, but in a time of human-propelled mass extinctions and multispecies genocides that sweep people and critters into the vortex. We must "dare 'to make' the relay; that is to create, to fabulate, in order not to despair. In order to induce a transformation, perhaps, but without the artificial loyalty that would resemble 'in the name of a cause,' no matter how noble it might be."[11]

Hannah Arendt and Virginia Woolf both understood the high stakes of training the mind and imagination to go visiting, to venture off the beaten path to meet unexpected, non-natal kin, and to strike up conversations, to pose and respond to interesting questions, to propose together something unanticipated, to take up the unasked-for obligations of having met. This is what I have called cultivating response-ability. Visiting is not a heroic practice; making a fuss is not the Revolution; thinking with each other is not Thought. Opening up versions so stories can be ongoing is so mundane, so earth-bound. That is precisely the point. The blackbird sings its importance; the babblers dance their shining prestige; the storytellers crack the established disorder. That is what "going too far" means, and this curious practice is not safe. Like Arendt and Woolf, Despret and her collaborators understand that we are dealing with "the idea of a world that could be habitable."[12] "The very strength of women who make a fuss is not to represent the True, rather to be witnesses for the possibility of other ways of doing what would perhaps be 'better.' The fuss is not the heroic statement of a grand cause . . . It instead affirms the need to resist the stifling impotence created by the 'no possibility to do otherwise, whether we want it or not,' which now reigns everywhere."[13] It is past time to make such a fuss.

Despret's curious practice has no truck with loyalty to a cause or doctrine; but it draws deeply from another virtue that is sometimes confused with loyalty, namely, "thinking from" a heritage. She is tuned to the obligations that inhere in starting from situated histories, situated stories. She retells the parable of the twelve camels in order to tease out what it means to "start from," that is, to "remain obligated with respect to that *from* which we speak, think, or act. It means to let ourselves learn from the event and to create from it." In a sort of cat's cradle with powerful fables, Despret received the parable from Isabelle Stengers, and then she relayed it to me in early 2013. I relay it back to her here. To inherit is an act "which demands thought and commitment. An act that calls for our transformation by the very deed of inheriting."[14]

In his will, the father in this story left his three quarrelsome sons a seemingly impossible inheritance: eleven camels to be divided in a precise way, half to the eldest son, a quarter to the second son, and a sixth to the third. The perverse requirements of the legacy provoked the confused sons, who were on the verge of failing to fulfill the terms of the will, to visit an old man living in the village. His savvy kindness in giving the sons a twelfth camel allowed the heirs to create a solution to their difficult heritage; they could make their inheritance active, alive, generative. With twelve camels, the fractions worked, and there was one camel left over to give back to the old man.

Despret notes that the tale she read left actual camels out of the enlargement and creativity of finding what it means to "start from." Those storied camels were conventional, discursive, figural beasts, whose only function was to give occasion for the problematic sons to grow in patriarchal understanding, recapitulating more than a little the history of philosophy that Despret—and I—inherited. But by listening, telling, and activating that particular story her way, she makes something that was absent present. She made an interesting, curious fuss without denouncing anybody. Therefore, another heritage emerges and makes claims on anyone listening, anyone attuned. It isn't just philosophy that has to change; the mortal world shifts. Long-legged, big-lipped, humped camels shake the dust from their hot, hard-worked hides and nuzzle the storyteller for a scratch behind the ears. Despret, and because of her, we, inherit camels now, camels with their people, in their markets and places of travel and labor, in their living and dying in worlds-at-stake, like the contemporary Gobi Desert.[15] We start from what is henceforth a dilated story that makes unexpected demands to cultivate response-ability. If we are to remain faithful to starting from the transformed story, we can no longer not know or not care that camels and people are at stake to each other—across regions, genders, races, species, practices. From now on, call that philosophy, a game of cat's cradle, not a lineage. We are obligated to speak from situated worlds, but we no longer need start from a humanist patriline and its breath-taking erasures and high-wire acts. The risk of listening to a story is that it can obligate us in ramifying webs that cannot be known in advance of venturing among their myriad threads. In a world of anthropozoogenesis, the figural is more likely than not to grow teeth and bite us in the bum.

Despret's philosophical ethology starts from the dead and missing as well as from the living and visible. She has studied situated human beings' mourning practices for their dead in ways strongly akin to her practice of philosophical ethology; in both domains, she attends to how—in practice—people can and do solicit the absent into vivid copresence, in many kinds of temporality and materiality. She attends to how practices—activated storytelling—can be on the side of what I call "ongoingness": that is, nurturing, or inventing, or discovering, or somehow cobbling together ways for living and dying well with each other in the tissues of an earth whose very habitability is threatened.[16] Many kinds of failure of ongoingness crumble lifeways in our times of onrushing extinctions, exterminations, wars, extractions, and genocides. Many kinds of absence, or threatened absence, must be brought into ongoing response-ability, not in the abstract but in homely storied cultivated practice.

To my initial surprise, this matter brought Despret and me together with racing pigeons, also called carrier pigeons (in French *voyageurs*) and with their avid fanciers (in French *colombophiles*, lovers of pigeons). I wrote an essay for Despret after an extraordinary week with her and her colleagues in the chateau at Cerisy in July 2010, in which I proposed playing string figure games with companion species for cultivating multispecies response-ability.[17] I sent Despret a draft containing my discussion of the wonderful art-technology-environmental-activist project by Beatriz da Costa called PigeonBlog, as well as a discussion of the communities of racing pigeons and their fanciers in Southern California.

Pigeon racing is a working-class men's sport around the world, one made immensely difficult in conditions of urban war (Baghdad, Damascus), racial and economic injustice (New York, Berlin), and displaced labor and play of many kinds across regions (France, Iran, California).

I care about art-design-activist practices that join diverse people and varied critters in shared, often vexed public spaces. "Starting from" *this* caring, not from some delusional caring in general, landed me in innovative pigeon lofts, where, it turned out, Despret, attuned to practices of commemoration, had already begun to roost. In particular, by leading me to Matali Crasset's *Capsule*, built in 2003 in the leisure park of Caudry, she shared her understanding of the power of holding open actual space for ongoing living and working in the face of threatened absence as a potent practice of commemoration.[18] The Beauvois association of carrier pigeon fanciers asked Crasset, an artist and industrial designer, to build a prototype pigeon loft that would combine beauty, functionality for people and birds, and a pedagogic lure to draw future practitioners into learning demanding skills. Actual pigeons had to thrive inhabiting this loft; actual colombophiles had to experience the loft working; and actual visitors to the ecological park, which was rehabilitating exhausted farmland into a variegated nature reserve for recuperating critters and people, had to be infected with the desire for a life transformed with avian voyageurs. Despret understood that the prototype, the memorial, had to be *for* both the carrier pigeons and their people—past, present and yet to come.[19]

Neither the critters nor the people could have existed or could endure without each other in ongoing, curious practices. Attached to ongoing pasts, they bring each other forward in thick presents and still possible futures; they stay with the trouble in speculative fabulation.

CHAPTER 8

The Camille Stories

Children of Compost

And then Camille came into our lives, rendering present the cross-stitched generations of the not-yet-born and not-yet-hatched of vulnerable, coevolving species. Proposing a relay into uncertain futures, I end *Staying with the Trouble* with a story, a speculative fabulation, which starts from a writing workshop at Cerisy in summer 2013, part of Isabelle Stengers's colloquium on *gestes spéculatifs*. Gestated in SF writing practices, Camille is a keeper of memories in the flesh of worlds that may become habitable again. Camille is one of the children of compost who ripen in the earth to say no to the posthuman of every time.

I signed up for the afternoon workshop at Cerisy called Narration Spéculative. The first day the organizers broke us down into writing groups of two or three participants and gave us a task. We were asked to fabulate a baby, and somehow to bring the infant through five human generations. In our times of surplus death of both individuals and of kinds, a mere five human generations can seem impossibly long to imagine flourishing with and for a renewed multispecies world. Over the week, the groups wrote many kinds of possible futures in a rambunctious play of literary forms. Versions abounded. Besides myself, the members of my group were the filmmaker Fabrizio Terranova and psychologist, philosopher, and ethologist Vinciane Despret. The version

8.1. *Mariposa* mask, Guerrero, Mexico, 62 cm× 72.5 cm × 12.5 cm, before 1990, Samuel Frid Collection, UBC Museum of Anthropology, Vancouver. Installation view, *The Marvellous Real: Art from Mexico, 1926–2011* exhibition (October 2013—March 2014), UBC Museum of Anthropology. Curator Nicola Levell. Photograph by Jim Clifford.

I tell here is itself a speculative gesture, both a memory and a lure for a "we" that came into being by fabulating a story together one summer in Normandy. I cannot tell exactly the same story that my cowriters would propose or remember. My story here is an ongoing speculative fabulation, not a conference report for the archives. We started writing together, and we have since written Camille stories individually, sometimes passing them back to the original writers for elaboration, sometimes not; and we have encountered Camille and the Children of Compost in other writing collaborations too.[1] All the versions are necessary to Camille. My memoir for that workshop is an active casting of threads from and for ongoing, shared stories. Camille, Donna, Vinciane, and Fabrizio brought each other into copresence; we render each other capable.

The Children of Compost insist that we need to write stories and live lives for flourishing and for abundance, especially in the teeth of rampaging destruction and impoverization. Anna Tsing urges us to cobble together the "arts of living on a damaged planet"; and among those arts are cultivating the capacity to reimagine wealth, learn practical healing rather than wholeness, and stitch together improbable collaborations without worrying overmuch about conventional ontological kinds.[2] The Camille Stories are invitations to participate in a kind of genre fiction committed to strengthening ways to propose near futures, possible futures, and implausible but real nows. Every Camille Story that I write will make terrible political and ecological mistakes; and every story asks readers to practice generous suspicion by joining in the fray of inventing a bumptious crop of Children of Compost.[3] Readers of science fiction are accustomed to the lively and irreverent arts of fan fiction. Story arcs and worlds are fodder for mutant transformations or for loving but perverse extensions. The Children of Compost invite not so much fan fiction as sym fiction, the genre of sympoiesis and symchthonia—the coming together of earthly ones. The Children of Compost want the Camille Stories to be a pilot project, a model, a work and play object, for composing collective projects, not just in the imagination but also in actual story writing. And on and under the ground.

Vinciane, Fabrizio, and I felt a vital pressure to provide our baby with a name and a pathway into what was not yet but might be. We also felt a vital pressure to ask our baby to be part of learning, over five generations, to radically reduce the pressure of human numbers on earth, currently set on a course to climb to more than 11 billion by the end of the twenty-first century CE. We could hardly approach the five generations through a story of heteronormative reproduction (to use the ugly but apt American feminist idiom)! More than a year later, I realized that Camille taught me how to say, "Make Kin Not Babies."[4]

Immediately, however, as soon as we proposed the name of Camille to each other, we realized that we were now holding a squirming child who had no truck with conventional genders or with human exceptionalism. This was a child born for sympoiesis—for becoming-with and making-with a motley clutch of earth others.[5]

Imagining the World of the Camilles

Luckily, Camille came into being at a moment of an unexpected but powerful, interlaced, planetwide eruption of numerous communities of a few hundred people each, who felt moved to migrate to ruined places and work with human and nonhuman partners to heal these places, building networks, pathways, nodes, and webs of and for a newly habitable world.[6]

Only a portion of the earthwide, astonishing, and infectious action for well-being came from intentional, migratory communities like Camille's. Drawing from long histories of creative resistance and generative living in even the worst circumstances, people everywhere found themselves profoundly tired of waiting for external, never materializing solutions to local and systemic problems. Both large and small individuals, organizations, and communities joined with each other, and with migrant communities like Camille's, to reshape terran life for an epoch that could follow the deadly discontinuities of the Anthropocene, Capitalocene, and Plantationocene. In system-changing simultaneous waves and pulses, diverse indigenous peoples and all sorts of other laboring women, men, and children—who had been long subjected to devastating conditions of extraction and production in their lands, waters, homes, and travels—innovated and strengthened coalitions to recraft conditions of living and dying to enable flourishing in the present and in times to come. These eruptions of healing energy and activism were ignited by love of earth and its human and nonhuman beings and by rage at the rate and scope of extinctions, exterminations, genocides, and immiserations in enforced patterns of multispecies living and dying that threatened ongoingness for everybody. Love and rage contained the germs of partial healing even in the face of onrushing destruction.

None of the Communities of Compost could imagine that they inhabited or moved to "empty land." Such still powerful, destructive fictions of settler colonialism and religious revivalism, secular or not, were fiercely resisted. The Communities of Compost worked and played hard to understand how to inherit the layers upon layers of living and dying that infuse every place and every corridor. Unlike inhabitants in many other utopian movements, stories, or literatures in the history of the earth, the Children of Compost knew they could not deceive themselves that they could start from scratch. Precisely the opposite insight moved them; they asked and responded to the question of how to live in the ruins that were still inhabited, with ghosts and with the living too. Coming from every economic class, color, caste, religion, secularism, and region, members of the emerging diverse settlements around the earth lived by a few simple but transformative practices, which in turn lured—became vitally infectious for—many other peoples and communities, both migratory and stable. The communities diverged in their development with sympoietic creativity, but they remained tied together by sticky threads.

The linking practices grew from the sense that healing and ongoingness in ruined places requires making kin in innovative ways. In the infectious new settlements, every new child must have at least three parents, who may or may not practice new or old genders. Corporeal differences, along with their fraught histories, are cherished. New children must be rare and precious, and they must have the robust company of other young and old ones of many kinds. Kin relations can be formed at any time in life, and so parents and other sorts of relatives can be added or invented at significant points of transition. Such relationships enact strong lifelong commitments and obligations of diverse kinds. Kin making as a means of reducing human numbers and demands on the earth, while simultaneously increasing human and other critters' flourishing, engaged intense energies and passions in the dispersed emerging worlds. But kin making and rebalancing human numbers had to happen in risky embodied connections to places, corridors, histories, and ongoing decolonial and postcolonial struggles, and not in the abstract and not by external fiat. Many failed models of population control provided strong cautionary tales.

Thus the work of these communities was and is intentional kin making across deep damage and significant difference. By the early twenty-first century, historical social action and cultural and scientific knowledges—much of it activated by anticolonial, antiracist, proqueer

8.2. Make Kin Not Babies. Sticker, 2 × 3 in., made by Kern Toy, Beth Stephens, Annie Sprinkle, and Donna Haraway.

feminist movement—had seriously unraveled the once-imagined natural bonds of sex and gender and race and nation, but undoing the widespread destructive commitment to the still-conceived natural necessity of the tie between kin making and a treelike biogenetic reproductive genealogy became a key task for the Children of Compost.

The decision to bring a new human infant into being is strongly structured to be a collective one for the emerging communities. Further, no one can be coerced to bear a child or punished for birthing one outside community auspices.[7] The Children of Compost nurture the born ones every way they can, even as they work and play to mutate the apparatuses of kin making and to reduce radically the burdens of human numbers across the earth. Although discouraged in the form of individual decisions to make a new baby, reproductive freedom of the person is actively cherished.

This freedom's most treasured power is the right and obligation of the human person, of whatever gender, who is carrying a pregnancy to choose an animal symbiont for the new child.[8] All new human members of the group who are born in the context of community decision making come into being as symbionts with critters of actively threatened species, and therefore with the whole patterned fabric of living and dying of those particular beings and all their associates, for whom the possibility of a future is very fragile. Human babies born through individual reproductive choice do not become biological symbionts, but they do live in many other kinds of sympoiesis with human and nonhuman critters. Over the generations, the Communities of Compost experienced complex difficulties with hierarchical caste formations and sometimes violent clashes between children born as symbionts and those born as more conventional human individuals. Syms and non-syms, sometimes literally, did not see eye to eye easily.

The animal symbionts are generally members of migratory species, which critically shapes the lines of visiting, working, and playing for all the partners of the symbiosis. The members of the symbioses of the Children of Compost, human and nonhuman, travel or depend on associates that travel; corridors are essential to their being. The restoration and care of corridors, of connection, is a central task of the communities; it is how they imagine and practice repair of ruined lands and waters and their critters, human and not.[9] The Children of Compost came to see their shared kind as humus, rather than as human or nonhuman. The core of each new child's education is learning how to live in symbiosis so as to nurture the animal symbiont, and all the other beings the symbiont requires, into ongoingness for at least five human generations. Nurturing the animal symbiont also means being nurtured in turn, as well as inventing practices of care of the ramifying symbiotic selves. The human and animal symbionts keep the relays of mortal life going, both inheriting and inventing practices of recuperation, survival, and flourishing.

Because the animal partners in the symbiosis are migratory, each human child learns and lives in nodes and pathways, with other people and their symbionts, in the alliances and collaborations needed to make ongoingness possible. Literally and figurally, training the mind to go visiting is a lifelong pedagogical practice in these communities. Together and separately, the sciences and arts are passionately practiced and enlarged as means to attune rapidly evolving ecological naturalcultural communities, including people, to live and die well throughout the dangerous centuries of irreversible climate change and continuing high rates of extinction and other troubles.

A treasured power of individual freedom for the new child is to choose a gender—or not—when and if the patterns of living and dying evoke that desire. Bodily modifications are normal among Camille's people;

and at birth a few genes and a few microorganisms from the animal symbiont are added to the symchild's bodily heritage, so that sensitivity and response to the world as experienced by the animal critter can be more vivid and precise for the human member of the team. The animal partners are not modified in these ways, although the ongoing relationships with lands, waters, people and peoples, critters, and apparatuses render them newly capable in surprising ways too, including ongoing EcoEvoDevo biological changes.[10] Throughout life, the human person may adopt further bodily modifications for pleasure and aesthetics or for work, as long as the modifications tend to both symbionts' well-being in the humus of sympoiesis.

Camille's people moved to southern West Virginia in the Appalachian Mountains on a site along the Kanawha River near Gauley Mountain, which had been devastated by mountaintop removal coal mining. The river and tributary creeks were toxic, the valleys filled with mine debris, the people used and abandoned by the coal companies. Camille's people allied themselves with struggling multispecies communities in the rugged mountains and valleys, both the local people and the other critters.[11] Most of the Communities of Compost that became most closely linked to Camille's gathering lived in places ravaged by fossil fuel extraction or by mining of gold, uranium, or other metals. Places eviscerated by deforestation or agriculture practiced as water and nutrient mining and monocropping also figured large in Camille's extended world.

Monarch butterflies frequent Camille's West Virginia community in the summers, and they undertake a many-thousand-mile migration south to overwinter in a few specific forests of pine and oyamel fir in central Mexico, along the border of the states of Michoacán and México.[12] In the twentieth century, the monarch was declared the state insect of West Virginia; and the Sanctuarío de la Biosfera Mariposa Monarca (Monarch Butterfly Biosphere Reserve), a UNESCO World Heritage Site after 2008, was established in the ecoregion along the Trans-Mexican Volcanic Belt of surviving woodlands.

Throughout their complex migrations, the monarchs must eat, breed, and rest in cities, *ejidos*, indigenous lands, farms, forests, and grasslands of a vast and damaged landscape, populated by people and peoples living and dying in many sorts of contested ecologies and economies. The larvae of the leap-frogging spring eastern monarch migrations from south to north face the consequences of genetic and chemical technologies of mass industrial agriculture that make their indispensable food—the leaves of native, local milkweeds—unavailable along most of the routes. Not just the presence of any milkweed, but the seasonal appearance of local milkweed varieties from Mexico to Canada, is syncopated in the flesh of monarch caterpillars. Some milkweed species flourish in disturbed land; they are good pioneer plants. The common milkweed of central and eastern North America, *Asclepias syriaca*, is such an early successional plant. Milkweeds thrive on roadsides and between crop furrows, and these are the milkweeds that are especially susceptible to herbicides like Monsanto's glyphosphate-containing herbicide, Roundup. Another milkweed is also important to the eastern migration of monarchs, namely the climax prairie species native to grasslands in later successional stages. With the nearly complete destruction of climax prairies across North America, this milkweed, *Asclepias meadii*, is fiercely endangered.[13]

Throughout the spring, summer, and fall, a large variety of early, midseason, and late flowering plants, including milkweed blossoms, produce the nectar sucked greedily by monarch adults. On the southern journey to Mexico, the future of the North American eastern migration is threatened by loss of the habitats of nectar-producing plants to feed the nonbreeding adults flying to overwinter in their favorite roosting trees in mountain woodlands. These woodlands in turn face natural-cultural degradation in complex histories of ongoing state, class, and ethnic oppression of campesinos and indigenous peoples in the region, for example, the Mazahuas and Otomi.[14]

Unhinged in space and time and stripped of food in both directions, larvae starve and hungry adults grow sluggish and fail to reach their winter homes. Migrations fail across the Americas. The trees in central Mexico mourn the loss of their winter shimmying clusters, and the meadows, farms, and town gardens of the United States and southern Canada are desolate in summer without the flitting shimmer of orange and black.

For the child's symbionts, Camille 1's birthing parent chose monarch butterflies of North America, in two magnificent but severely damaged streams, from Canada to Mexico, and from the state of Washington, along California, and across the Rocky Mountains. Camille's gestational parent exercised reproductive freedom with wild hope, choosing to bond the soon-to-be-born fetus with both the western and eastern currents of this braid of butterfly motion. That meant that Camille of the first generation, and further Camilles for four more human generations at least, would grow in knowledge and know-how committed to the ongoingness of these gorgeous and threatened insects and their human and nonhuman communities all along the pathways and nodes of their migrations and residencies in *these* places and corridors, not all the time everywhere. Camille's community understood that monarchs as a widespread global species are not threatened; but two grand currents of a continental migration, a vast connected sweep of myriad critters living and dying together, were on the brink of perishing.

The child-bearing parent who chose the monarch butterfly as Camille's symbiont was a single person with the response-ability to exercise potent, noninnocent, generative freedom that was pregnant with consequences for ramifying worlds across five generations. That irreducible singularity, that particular exercise of reproductive choice, set in train a several-hundred-year effort, involving many actors, to keep alive practices of migration across and along continents for all the migrations' critters. The Communities of Compost did not align their children to "endangered species" as that term had been developed in conservation organizations in the twentieth century. Rather, the Communities of Compost understood their task to be to cultivate and invent the arts of living with and for damaged worlds in place, not as an abstraction or a type, but as and for those living and dying in ruined places. All the Camilles grew rich in worldly communities throughout life, as work and play with and for the butterflies made for intense residencies and active migrations with a host of people and other critters. As one Camille approached death, a new Camille would be born to the community in time so that the elder, as mentor in symbiosis, could teach the younger to be ready.[15]

The Camilles knew the work could fail at any time. The dangers remained intense. As a legacy of centuries of economic, cultural, and ecological exploitation both of people and other beings, excess extinctions and exterminations continued to stalk the earth. Still, successfully holding open space for other critters and their committed people also flourished, and multispecies partnerships of many kinds contributed to building a habitable earth in sustained troubled times.

The Camille Stories

The story I tell below tracks the five Camilles along only a few threads and knots of their lifeways, between the birth of Camille 1 in 2025 and the death of Camille 5 in 2425. The story I tell here cries out for collaborative and divergent story-making practices, in narrative, audio, and visual performances and texts in materialities from digital to sculptural to everything practicable. My stories are suggestive string figures at best; they long for a fuller weave that still keeps the patterns open, with ramifying attachment sites for storytellers yet to come. I hope readers change parts of the story and take them elsewhere, enlarge, object, flesh out, and reimagine the lifeways of the Camilles.

The Camille Stories reach only to five generations, not yet able to fulfill the obligations that the Haudenosaunee Confederacy imposed on themselves and so on anyone who has been touched by the account, even in acts of unacknowledged appropriation, namely, to act so as to be response-able to and for those in the seventh generation to come.[16] *The Children of Compost beyond the reach of the Camille Stories might become capable of that kind of worlding, which somehow once seemed possible, before the Great Acceleration of the Capitalocene and the Great Dithering.*

Over the five generations of the Camilles, the total number of human beings on earth, including persons in symbiosis with vulnerable animals chosen by their birth parent (syms) and those not in such symbioses (non-syms), declined from the high point of 10 billion in 2100 to a stable level of 3 billion by 2400. If the Communities of Compost had not proved from their earliest years so successful and so infectious among other human people and peoples, the earth's population would have reached more than 11 billion by 2100. The breathing room provided by that difference of a billion human people opened up possibilities for ongoingness for many threatened ways of living and dying for both human and nonhuman beings.[17]

CAMILLE 1

Born 2025. Human numbers are 8 billion.
Died 2100. Human numbers are 10 billion.

In 2020, about three hundred people with diverse class, racial, religious, and regional heritages, including two hundred adults of the four major genders practiced at the time[18] and one hundred children under the age of eighteen, built a town where the New River and Gauley River flowed together to form the Kanawha River in West Virginia. They named the settlement New Gauley to honor the lands and waters devastated by mountaintop removal coal mining. Historians of this time have suggested that the period between about 2000 and 2050 on earth should be called the Great Dithering.[19] The Great Dithering was a time of ineffective and widespread anxiety about environmental destruction, un-

8.3. Monarch butterfly caterpillar *Danaus plexippus* on a milkweed pod. Photograph by Singer S. Ron, U.S. Fish and Wildlife Service.

mistakable evidence of accelerating mass extinctions, violent climate change, social disintegration, widening wars, ongoing human population increase due to the large numbers of already-born youngsters (even though birth rates most places had fallen below replacement rate), and vast migrations of human and nonhuman refugees without refuges.

During this terrible period, when it was nonetheless still possible for concerted action to make a difference, numerous communities emerged across the earth. The English-language name for these gatherings was the Communities of Compost; the people called themselves compostists. Many other names in many languages also proposed the string figure game of collective resurgence. These communities understood that the Great Dithering could end in terminal crises; or radical collective action could ferment a turbulent but generative time of reversals, revolt, revolution, and resurgence.

For the first few years, the adults of New Gauley did not birth any new children, but concentrated on building culture, economy, rituals, and politics in which oddkin would be abundant, and children would be rare but precious.[20] The kin-making work and play of the community built capacities critical for resurgence and multispecies flourishing. In particular, friendship as a kin-making practice throughout life was elaborated and celebrated. In 2025, the community felt ready to birth their first new babies to be bonded with animal symbionts. The adults judged that most of their already-born children, who had helped found the community, were ready and eager to be older siblings to the coming symbiont youngsters. Everybody believed that this kind of sympoiesis had not been practiced anywhere on earth before. People knew it would not be simple to learn to live collectively in intimate and worldly care-taking symbiosis with another animal as a practice of repairing damaged places and making flourishing multispecies futures.

Camille 1 was born among a small group of five children, and per[21] was the only youngster linked to an insect. Other children in this first cohort became symbionts with fish (American eel, *Anguilla rostrata*), birds (American kestrel, *Falco sparverius*), crustaceans (the Big Sandy crayfish, *Cambarus veteranus*), and amphibians (streamside salamander, *Ambystoma barbouri*).[22] Beginning with vulnerable bats, mammal symbioses were undertaken in the second wave of births about five years later. It was often easier to identify migratory threatened insects, fish, mammals, and birds as potential symbionts for new children than reptiles, amphibians, and crustaceans. The preference for migratory symbionts was often relaxed, especially since corridor conservation of all kinds was ever more urgent as rising temperatures due to climate change forced many usually nonmigratory species outside their previous ranges. Although their first loves remained traveling critters and far-flung pathways—mostly because their own small human communities were made geographically and culturally more worldly through cultivating the linkages required to take care of their partners in symbiosis—some members of the Communities of Compost committed themselves to critters in tiny remnant habitats, as well as to those whose finicky ecological requirements and love of home tied them tightly to particular places only.[23]

Over the first hundred years, New Gauley welcomed 100 new births with babies joined to animal symbionts, 10 births to single parents or couples who declined the three-parent model and whose offspring did not receive these sorts of symbionts, 200 deaths, 175 in-migrants, and 50 out-migrants. The scientists of the Communities of Compost found it impossible to establish successful animal-human symbioses with adults; the critical receptive times for humans were fetal development, nursing, and adolescence. During the times that they contributed cellular or molecular materials for modifying the human partner, the animal partners also had to be in a period of transformation, such as hatching, larval ecdysis, or metamorphosis. The animals themselves were not modified with human material; their roles in the symbioses were to teach and to flourish in every way possible in dangerous and damaged times.

Almost everywhere, the Communities of Compost committed themselves to maintaining their size or to growing through immigration, while keeping their own new births at a level compatible with the earth's overall human numbers eventually declining by two-thirds. If new in-migrants accepted the basic practices of the Communities of Compost, upon request they received permanent residency and citizenship rights as compostists in inventive and usually raucous kin-making ceremonies. Nonresident visitors were always welcome; hospitality was regarded as both a basic obligation and source of mutual renewal. Visitors' lengths of stay could become a contentious matter and was even known to break up kin affiliations and sometimes entire compostist communities.

If many more in-migrants wanted to join Communities of Compost than were possible to accommodate, new settlements formed with mentors from the seed towns. In-migrants in the early centuries often came from ruined areas elsewhere, and their seeking both refuge and belonging in the Communities of Compost—themselves committed to the arts of living in damaged places—was an act of both desperation and faith. The original founders of the Communities of Compost quickly realized that in-migrants from desperate situations brought with them not only trauma, but also extraordinary insight and skill for the work to be done. Resettlement in still other ruined sites and establishment of alliances and collaborations with people and other critters in those areas required the best abilities of the mentors and the in-migrants. Plant symbionts were not joined to babies in the Communities of Compost for several generations, although recognizing profuse sympoiesis—world making—with plants was fundamental for all compostists.

New Gauley decided to emphasize in-migration of people over new births for the first three generations, and after that time there was both more flexibility and the need to recalibrate births and deaths. In- and out-migration tended to equalize as more places on earth restored conditions for modest resurgence and reasons for seeking new homes rested much less on war, exploitation, genocide, and ecological devastation, and much more on adventure, curiosity, desire for new kinds of abundance and skill, and the old habits of human beings to move, including hunter-gatherers, pastoralists, and farm-and-town-living people. Opportunistic social species tend to move around a lot; human beings outside captivity have always been extraordinary ecosocial opportunists, travelers, and path makers. Added to that, by 2300, more than a billion human beings on earth had themselves been born into new kinds of symbiotic relationships with other critters, in addition to the much older multispecies associations that characterized human people as well as every other sort of living being throughout ecological, evolutionary, developmental, historical, and technological histories.

Before birth, Camille 1 was given a suite of pattern-forming genes expressed on monarch surfaces over their transformations from caterpillar to winged adult. Camille 1 also received genes allowing per to taste in the wind the dilute chemical signals crucial to adult monarchs selecting diverse nectar-rich flowers and the best milkweed leaves for depositing their eggs. Camille 1's gut and mouth microbiomes were enhanced to allow per to safely savor milkweed plants containing the toxic alkaloids that the monarchs accumulate in their flesh to deter predators. As an infant, Camille 1's oral satisfactions with fragrant mammalian milk were laced with the bitter tastes of cardiac glycosides, tastes that the human parent nursing per dared not share. In per's maturing mindful body, Camille 1 had to learn to become in symbiosis with an insect composed as five caterpillar instars before metamorphosis into a flying adult, which in turn experienced seasonally alternating sexually excited phases and sexually quiet diapause. The symbiogenetic join of Camille and monarchs also had to accommodate the diverse parasitic and beneficial associates of the butterfly holobiont, as well as pay attention to the genetics of the migrating populations.[24]

The compostists did not attempt to introduce into Camille 1's already complicated symbiotic reformatting any of the genes and timing patterns that the butterflies use to utterly disassemble and recompose their entire being in the chrysalis before emerging as winged imagos. Nor did the parents attempt to alter Camille's visual capacities and neural arrangements to perceive physically in the butterfly color spectrum, or to see as if Camille had the compound eyes of an insect. Mimesis was not the point of the alterations, but fleshly suggestions braided through innovative pedagogical practices of naturalsocial becoming-with that could help the symbiosis thrive through five human generations committed to healing damaged human and nonhuman lives and places. In its most reductive expression, the point was to give the butterflies and their people—to give the Migrations—a chance to have a future in a time of mass extinctions.

By five years of age, Camille 1's skin was brilliantly banded and colored in yellow and black like a late-stage monarch caterpillar, increasing in intensity until age ten. But by initiation into adult responsibilities at age fifteen, Camille 1's skin had the muted tones and patterns of the monarch chrysalis. As an adult, Camille 1 gradually acquired the pattern and coloration of a vibrant orange and black adult butterfly. Camille 1's adult body was more androgynous in appearance than that of sexually dimorphic monarch adults.

All of the symbiont children developed both visible traits and subtle sensory similarities to their animal partners in early childhood. Although they should not have been surprised, the consequences of this developmental fact blindsided the adult compostists, as the first serious conflicts in New Gauley erupted in the learning groups of the young. Five youngsters who were bonded to animal symbionts, two children born to dissenting parents, and so not bonded with such symbionts, and five in-migrant children without symbionts made up the first cohort of little ones. The symbiotic young were struggling to integrate mindful bodies unimaginable to their parents. In addition, each symbiosis was the only one of its kind in these early generations.

Camille 1 formed fierce friendships, especially with Kess, the youngster bonded with the American kestrels; but each symbiotic child was acutely aware of their irreducible difference. Kess and Camille gravitated to each other partly because they knew kestrels ate butterflies, and both of their threatened animal symbionts flourished best in fields, meadows, roadsides, pastures, and mixed woodlands full of a myriad of flowering plants. From the beginning, the symbiont children developed a complex subjectivity composed of loneliness, intense sociality, intimacy with nonhuman others, specialness, lack of choice, fullness of meaning, and sureness of future purpose. This landscape of converging and diverging feelings tended to grade into arrogance and exceptionalism toward the nonsymbiotic children, and even toward their parents and other nonsymbiotic adults of New Gauley. Because symbionts were still rare in the overall population of an area in the initial generations after the first Communities of Compost were established, in vulnerable moments nonsymbiotic children and adults could and did feel the symbionts were freaks, both more-than- and other-than-human, and seriously threatening. Remembering that humanity meant humus, and not Anthropos or Homo, did not come easily in the webs of Western cultures that predominated in New Gauley. Determined to help youngsters through the mazes of self-preoccupation, social enthusiasm, playfulness, pride in each other, fear, competition, and bullying that they had known in school, the New Gauley adults and their young were faced with quite another challenge in the emerging community of both symbiotic and nonsymbiotic children.

New Gauley compostists soon found that storytelling was the most powerful practice for comforting, inspiring, remembering, warning, nurturing compassion, mourning, and becoming-with each other in their differences, hopes, and terrors. Of course, the Communities of Compost emphasized a deep and wide range of approaches to educating both young and old, and the sciences and arts were especially elaborated and cherished. For youngsters and adults of most species in the communities, play was the most powerful and diverse activity for rearranging old things and proposing new things, new patterns of feeling and action, and for crafting safe enough ways to tangle with each other in conflict and collaboration.[25] The practice of friendship and the practice of play, both ritualized and celebrated in small and large ways, were the core kin-forming apparatuses. Libraries in many formats and materialities abounded to evoke curiosities and sustain knowledge projects for learning to live and die well in the work of healing damaged places, selves, and other beings. Decolonial multispecies studies (including diverse and multimodal human and nonhuman languages) and an indefinitely expandable transknowledging approach called EcoEvoDevoHistoEthnoTechnoPsycho (Ecological Evolutionary Developmental Historical Ethnographic Technological Psychological studies) were essential layered and knotted inquiries for compostists.[26]

Compostists eagerly found out everything they could about experimental, intentional, utopian, dystopian, and revolutionary communities and movements across times and places. One of their great disappointments in these accounts was that so many started from the premises of starting over and beginning anew, instead of learning to inherit without denial and stay with the trouble of damaged worlds. Although hardly free of the sterilizing narrative of wiping the world clean by apocalypse or salvation, the richest humus for their inquiries turned out to be SF—science fiction and fantasy, speculative fabulation, speculative feminism, and string figures. Blocking the foreclosures of utopias, SF kept politics alive.

So storytelling was the seed bag for flourishing for compostists, and Camille 1 was fed on stories. Because the brave young princess loved the toxic forest beings, especially the despised and feared insects called the Ohmu, Camille's favorite story was *Nausicaä of the Valley of the Wind*. Like a turbo butterfly, Nausicaä could fly over the forest, fields, and towns on her agile personal jet-powered glider. The young Camille 1 could never resist that vivid sensation. Hayao Miyazaki's manga and anime story is set on a postapocalyptic earth menaced by the toxic forest's critters, who were defending themselves and taking revenge for the natural world's relentless destruction at the hands of militarized, power-mad, technological humans. Evil rulers continued to promise ultimate destruction in their drive to exterminate the toxic forest and extract the last drams of resources for the walled cities of privilege and exception. Through her study of the forest's ecology, understanding of the physiology of the mushroomlike infected poisonous trees, and love for the dangerous mutant giant insects and their larvae, Nausicaä triumphed in her efforts to save both the people and the forest. She discovered that the trees purified the toxins and drop by drop were forming a vast underground aquifer of pure water that could regenerate the biodiverse earth. Attuned to the languages of the plants, fungi, and animals, Nausicaä could calm the incomprehension and fear of the people who were poisoned by the toxic emanations of the disturbed forest. She could propose peace between humans and other-than-humans because she befriended the toxic forest, a practice that reached deep into the young Camille 1's psyche. In the dramatic concluding scenes of the story, at great risk to herself, Nausicaä rescued a threatened larval Ohmu and so stopped the stampede of its giant adult conspecifics in their rage at the humans' capturing and wounding of the youngster.

Camille 1 learned that there were many inspirations for Miyazaki's story,[27] including a Phaecian princess from Homer's *Odyssey* named Nausicaä, who loved nature and music, cultivated a fervid imagination, and disdained possessions. Along with European medieval accounts of witches' mastery of the winds, Master Windkey from Ursula Le Guin's *Earthsea* also infused the Nausicaä tale. The adult Camille 1 thought that the most generative inspiration, however, was a Japanese story from the Heian period, called "The Princess Who Loved Insects."[28] The princess did not beautify herself by blackening her teeth or plucking her eyebrows, and she scorned the idea of a husband. All her passions were for the caterpillars and creeping crawling critters disdained by others.[29]

Nausicaä had a companion animal, really a symbiont, a fierce and gentle little fox squirrel. In per's memoir, the elder Camille 1 described

Nausicaä of the Valley of the Wind as a fable of great danger and great companionship. Unlike conventional heroes, Nausicaä accompanied by animals, is a girl child and healer, whose courage matures in thick connection with many others and many kinds of others. Nausicaä cannot act alone, and also her personal response-ability and actions have great consequences for herself and for myriad human and nonhuman beings. Nausicaä's connections and corridors are practical and material, as well as fabulous and enspirited in bumptious animist fashion. Hers are the arts of living on a damaged planet. This twentieth-century Japanese anime child sustained Camille 1 in symbiosis with the monarchs for a lifetime.

CAMILLE 2

Born in 2085. Human numbers are 9.5 billion.
Died in 2185. Human numbers are 8 billion.

At initiation at age fifteen, as a coming-of-age gift the second Camille decided to ask for chin implants of butterfly antennae, a kind of tentacular beard, so that more vivid tasting of the flying insects' worlds could become the heritage of the human partner too, helping in the work and adding to the corporeal pleasures of becoming-with.[30] Proud of this vibrant sign of the lived symbiosis now in its second generation, once the procedures were complete the adolescent Camille 2 undertook a trip to the overwintering habitat of the eastern migration to meet with indigenous people and campesinos who were rehabilitating damaged lands and waters along the transvolcanic belt between the states of México and Michoacán.

Camille 1 had been Camille 2's mentor, and over the first fifteen years of the new child's life Camille 1 tried to prepare the second-generation New Gauley human-butterfly symbiont for sustained visiting as the guest of the diverse communities of Michoacán. But Camille 1's lifework had been almost entirely along the corridors and in the towns, fields, mines, woods, coasts, mountains, deserts, and cities of the great eastern and western monarch migrations that are north of Mexico, into southern Canada on the east and into Washington and the Northern Rockies on the west. Camille 1 worked, played, and struggled primarily with midwestern and southern farmers, agribusiness scientists, energy companies and their ruthless lawyers, mine workers, unemployed peo-

8.4. Monarch butterfly resting on fennel in the Pismo Butterfly Grove near the entrance to Pismo Beach State Park, November 15, 2008. Photograph by docentjoyce / Wikimedia Commons.

ple, nature lovers, gardeners, corridor ecologists, insect specialists, and climate scientists, as well as artists performing with and for nonhuman critters. Although based in New Gauley and attuned especially to the ravaged landscapes and peoples of coal country there and across the continent,[31] Camille 1 also sojourned with the insects and their people in the winter homes of the western migration of the monarchs, especially along the Monterey Bay of Central California. So per understood the biological, cultural, historical worlds of these clusters of monarchs clinging to their local Monterey pines and crucial (if never really accepted by ecological nativists) Australian gum trees.

Both sym and non-sym people across this vast expanse of land experienced the sympoieses of the Children of Compost primarily through their *biological* semiotic materialities. Of course, as an important component of per's education and working alliances as both child and adult, Camille 1 had studied with Native American, First Nation, and Métis teachers, who explained and performed diverse practices and knowledges for conjoined human and other-than-human becoming and exchange. But

Camille 1 had never deeply questioned the settler practices or categories of nature, culture, and biology that made per's own transformative sympoiesis with monarchs comprehensible to perself. For practical and political, as well as ontological and epistemological, reasons, Camille 1 recognized the urgent need both to deepen and to change the terms of the exchanges and collaborations with people, peoples, and other-than-human critters in the southern migrations and residencies of the monarchs in Mexico.

And so, although well read in decolonial and postcolonial literatures and engaged in lifelong correspondence with Mexican comrades, several of whom had personally made the journey to New Gauley and to California's Santa Cruz to meet with the Communities of Compost there, Camille 1 died before encountering monarch sympoiesis in material semiotic forms other than symbiogenesis. Then, on the first Día de los Muertos of per's first sojourn in Mexico after coming of age, Camille 2 was introduced to the Monarcas returning to their winter mountain home in November as the souls of the Mazahua dead. The Monarcas did not *represent* the souls of the dead; they *were* syms of the living butterflies and the human dead, in multinaturalist worldings that Camille 2 had studied but could barely recognize and did not know how to greet.

There were Communities of Compost in Michoacán and all over Mexico that had birthed symbiogenetic children for the work of rehabilitating ruined lands and waters through the coming generations. Mexicans were as much at home in the extended apparatuses of naturalcultural, biological knowledges and practices as their northern comrades. But none of the Mexican Communities of Compost, no matter their settler, mixed, or native heritages, had joined a human baby to migrating endangered critters that were themselves the visiting ancestors. The consequences of Camille 2's powerful introduction to the Monarca sympoiesis that joined the overwintering winged insects and the visiting ancestors reshaped the fabric of work for ecojustice that joined New Gauley and the communities of the transvolcanic belt for the coming three hundred years. Searching for a term they could share, Camille 2 and per's Mazahua hosts decided to name this kind of becoming-with "symanimagenesis." The corridors, migrations, and contact zones of the monarchs collect up many ways of living and dying across the Americas!

Thus the Mazahuas of central Mexico (México, Michoacán, and Queretaro) became vital to Camille's story from the second generation on. Decolonial work on all sides of the border had to be intrinsic to every form of sympoiesis with the monarchs.[32] Struggling with the consequences for the Communities of Compost of practices of conservation and restoration inherited from Anglo and Spanish settler colonialism and of both the Mexican and U.S. states' ongoing suppression of and extraction from indigenous peoples, Camille 2 no longer could not know about the Mazahuas' land and water struggles, migrations to near and distant cities for ill-paid work, customary and illegal forest cutting, charcoal making, history of acting to preserve the trees and woodlands before the butterfly migration became an international issue, history also of old and new indigenous and outsider exploitation of the forest and the watershed, and resistance to U.S. and other foreign scientists and Mexican state regulations and bureaucrats that criminalize local subsistence practices in the butterfly preserve that became a World Heritage Site.

In Camille 2's first weeks in Mazahua monarch country, the women of the communities in and around the butterfly reserve took per in hand.[33] When the young sym arrived in autumn of 2100, the radical Mazahua women were celebrating the ninety-sixth anniversary of the founding of their movement. In 2004, "symbolically armed with farming tools and wooden rifles, they formed el Ejército de Mujeres Zapatistas en Defensa del Agua and pledged a strategy of non-violence."[34] The Zapatista movement had begun with a transformative armed uprising in Chiapas on January 1, 1994, but the Zapatistas' most important contribution to the ongoing strategies of the Mazahuan communities was their development of vigorous nonviolence in a broad, multigenerational opposition front.[35] Communities of Compost all over the world studied Zapatista-held municipalities called *caracoles* (snail shells).

Initiating the visiting New Gauley sym into intense relationships with the living dead was the Mazahua women's first priority. Politely resisting the temptation to thread their fingers continuously though the sensitive tentacular organs of their oddly bearded visitor, they were charmed by the butterfly antennae on Camille 2's chin and by per's vividly colored adult skin patterns that had gradually replaced the dramatic caterpillar skin banding of the younger child. The Mazahuas were confident that these markings signified that the adolescent would become an apt student of their own human-butterfly worldings, and so a useful ally. To join in the work of human and other-than-human rehabilitation and multispecies environmental justice in indigenous territory, which had been dominated and drained for centuries by the state and other outside forces, Camille 2 had to study the resurgence of peoples that ignited in the early years of the twenty-first century.

On the night the monarchas returned in November 2100, the women taught Camille 2 a poem, "Soy Mazahua," composed by Julio Garduño Cervantes, which remained vital to their work for the dead and the living. The beautifully dressed people at the fiesta were singing the poem in the midst of extravagant fireworks, feasting, and greeting returning kin. The poem commemorated a Mazahua leader who had been murdered in 1980 on the way home from a cemetery on el Día de los Muertos. All over Mexico, this killing outraged indigenous pueblos and ignited a movement that was stronger than ever by the time Camille 2 visited. The women of El Ejército de Mujeres Zapatistas en Defensa del Agua taught that butterflies drank the tears of those who mourned the murdered, raped, and disappeared ones of every land.[36]

I AM MAZAHUA

You have wanted to deny my existence
But I do not deny yours.
But I exist. I am Mazahua.
. . .
I am made of this land, the air, water and sun.
And together we repeat, We are Mazahuas
. . .
You have enslaved my ancestors and stolen their lands. You have murdered them.
. . .
I build the house but you live in it.
You are the criminal but I am in prison.
We made the revolution but you took advantage of it.
My voice rises and joins with a thousand others.
And together we repeat, We are Mazahuas
Our hands sowed for everyone.
Our hands will struggle for everyone.
I am Mazahua.

—Julio Garduño Cervantes

Camille 2 found it difficult at first to grasp how active the dead were across this region and how critical to the work of compostists to restore damaged land and its human and nonhuman beings.[37] Camille 2 had to learn to let go of colonialist notions of religion and secularism to begin to appreciate the sheer semiotic materiality of those who came before. Until sympoiesis with the dead could be acknowledged, sympoiesis with the living was radically incomplete. Visiting urbanites from Mexico City were no better at seriously engaging the epistemological, ontological, and practical demands of this aspect of indigenous cosmopolitics than Camille 2 was. Modernity and its category work proved terribly durable for hundreds of years after the withering critique conducted in the late twentieth and early twenty-first centuries had made explicit adherence to the tenets of philosophical and political modernity unthinkable for serious people, including scientists and artists. Modernity was driven underground, but remained undead. Making peace with this vampire ancestor was an urgent task for the Communities of Compost.[38]

In the early weeks of Camille 2's sojourn, the teenage Mazahua women of El Ejército de Mujeres Zapatistas en Defensa del Agua undertook the task of teaching the sym about the struggle for water and forest eco-justice in the region. Throughout the late twentieth and twenty-first centuries, in immense water transfer projects, Mexico City drew this precious fluid from the lakes, rivers, and aquifers that reached across the mountains and into the basins occupied by indigenous peoples and other critters to the north and west.[39] This practice was immensely destructive, a fact highlighted in a 2015 report of the Union of Scientists Committed to Society, which argued, "Transferring water, moving large volumes from one basin to another, is not only unsustainable and affects the environment in the medium and long term, but also causes forced displacements that destroy towns and communities and marginalize people, who are obliged to move to the poverty belts of the large cities."[40] By the time of the birth of Camille 1, the Cutzamala System already pumped 127 billion gallons per year to Mexico City and its twenty-seven municipalities, while the Mazahua communities did not have potable water. El Ejército de Mujeres Zapatistas en Defensa del Agua carried out relentless demonstrations and actions in Mexico City and elsewhere and won many partial victories. The struggle continued throughout Camille 2's lifetime and after. In sympoiesis, the monarch critters, human and other-than-human, drank from the healing tears of the living and the dead.[41]

8.5. Mural painted by youth in La Hormiga, Putumayo, in southwestern Colombia depicting landscapes before and after aerial fumigation during the U.S.-Colombia "War on Drugs." Photograph by Kristina Lyons.

CAMILLE 3

Born in 2170; human numbers are 8.5 billion.
Died in 2270; human numbers are 6 billion.

By this generation, two-thirds of the residents of the Communities of Compost around the world were symbionts engaged in intense work and play for sustaining vulnerable beings across the hardest centuries of planetary crisis and widespread human and other-than-human suffering. A significant number of syms had decided to leave their compostist communities, surrendering residency rights for citizenship in other political formations. Some humans, both in-migrants and non-sym offspring born in the new towns, became solid compostists without ever wanting to engage personally in symbiogenetic kin making. Allied with diverse non-sym peoples, compostist practices of living and dying flourished everywhere, and the people in the emerging epoch of partial healing felt deeply entangled with the ongoing tentacular Chthulucene. There had been great losses of kinds of living beings, as rapid climate change and interlocked ecosystem collapses swept the earth; and the mass extinction event of the Capitalocene and Anthropocene was not over.

Still, by the time Camille 3 was fifty years old, it was clear that human numbers, while still too heavy in most places for the damaged natural, social, and technical systems of earth to sustain, were declining within a deliberate pattern of heightened environmental justice. That pattern emphasized a preference for the poor among humans, a preference for biodiverse naturalsocial ecosystems, and a preference for the most vulnerable among other critters and their habitats. Much of the most inventive work over the 150 years since the first compostist communities appeared was developing the linkages of this pattern. That work required both powerful recognition and strengthening of inherited Chthulucene practices that had not been fully obliterated in the Capitalocene and Anthropocene, and also newly invented ways of linking the three critical preferences to each other. The wealthiest and highest-consuming human populations reduced new births the most, with the support of the Communities of Compost; but human births everywhere were deliberately below replacement rates, so as to slowly and effectively reach levels that made sense for distributed and diverse humanity as humus, rather than as end points of nature and culture. Practices of making kin, not babies, had taken hold inside and outside the Communities of Compost.

Against all expectations in the early twenty-first century, after only 150 years sympoieses, both symbiogenetic and symanimagenic, seemed to be making a difference in holding open time and space for many of earth's most vulnerable, including the monarch migrations and their diverse human people and peoples. Forests in the transvolcanic belt of Mexico were resurgent, and water had been restored to the pillaged aquifers. People had worked out robust peace with the critters and scientists of the Biosphere Reserve, as greater control was relocated to campesino and indigenous organizations for environmental justice. Migrations north of Mexico now could count on larval and adult food as nonmonocropping organic agriculture, ubiquitous gardens, and species-rich roadside verges filled the landscape. Devastation of habitats for people and other critters caused by Big Energy and Big Capital was not finished, but the tide had definitively turned.

Humus-friendly technological innovation, creative rituals and celebrations, profound economic restructuring, reconfiguration of political control, demilitarization, and sustained work for connecting corridors and for ecological, cultural, and political restoration had all made an impact and were growing in force. While Camille 3 could not forget the monarchs, per's attention was turned to the fact that syms needed to take stock of themselves collectively in unprecedented ways.

The major events of Camille 3's life were travels to gatherings around the world of sym and non-sym humans in the face of general recognition that both humanity and animality had been fundamentally transformed by compostist practices. Of course, many peoples of earth had never divided living beings into human and animal; nonetheless, they had all ordered things differently from what was plainly happening everywhere by 2200. It was also inescapably clear by 2200 that the changes were not the same everywhere. Sym worlding was not one thing, and it was diverging and adapting exuberantly in EcoEvoDevoHistoTechnoEthnoPsycho fashion. The recognition was turbulent, exhilarating, and dangerous. The crises of bullying among the children in Camille 1's generation were nothing to the terrors of transition in the third generation of the Communities of Compost, who would in a few more generations become the majority of people on earth. Inventing earthwide cosmopolitics between

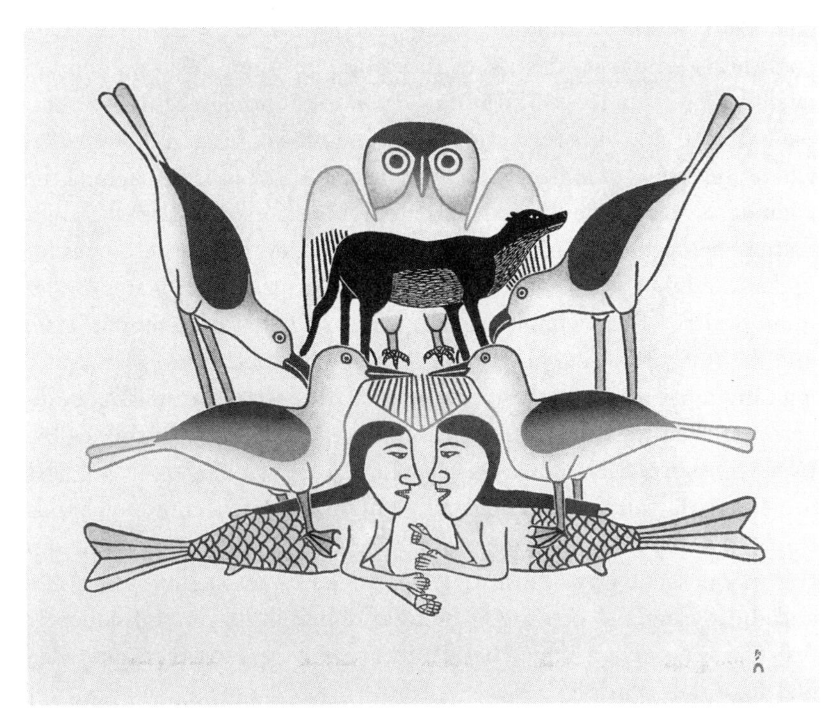

8.6. Kenojuak Ashevak, *Animals of Land and Sea,* 1991, stonecut with stencil, 25 × 30 in. (paper). Courtesy of Dorset Fine Arts.

and among syms and non-syms was the daunting task of Camille 3's generation.

Peoples from every fold of earth had long been both generated and nourished by stories, myths, performances, powers, and embodiments of entities not divided into categories recognizable to most conventional Western philosophy and politics. Such stories and embodiments were also deeply embedded in the practices and accounts of both recent and long-established peoples living throughout what was once called the West. Camille 3's generation found biologies and storytelling to be the richest veins for weaving the needed fabrics to bind syms and non-syms together.

Coming-of-age stories in SF, from Joanna Russ's Alyx in *The Adventures of Alyx* to Julie Czerneda's decidedly nonhumanoid young female entity Esen-alit-Quar in her Web Shifters series, entranced many in Camille 3's generation, and these tales had been carefully maintained in the compostist archives. Camille 3 was especially drawn to Philip Pullman's twentieth-century stories about the young girl Lyra Belacqua and her daemon, Pantalaimon.[42] The animal daemons of human beings were capable of multiple transformations until the time of human adolescence, when the animal stabilized into a single morph. Pullman imagined the daemons as expressions of a tripartite human person, rather than as a distinct beings; but compostist syms had other and better resources for understanding the join, which did not rely on the trinary soul/body/mind heritage that Pullman used in his war against both monotheism and the Authority. Compostists understood the daemons in less ontologically purified ways, more entangled with situated animisms in diverse modern and traditional, past and present worlds. The bonds between human and daemon were very close to the symbiogenetic linkages forged for the young in compostist communities. Severing such bonds threatened persons at the deepest core of their being. Living-with was the only possible way to live-well. Heartened by these stories between and during endless meetings among quibbling and not infrequently fearful syms and non-syms, the critters of earth were forging planetwide ontological revolutions for making kin.

CAMILLE 4

Born in 2255. Human numbers are 6.5 billion.
Died in 2355. Human numbers are 3.5 billion.

After decades of heartening progress, new viral diseases afflicting soil-based fungal symbionts of food plants needed by many species in the subfamily Danainae to fight off their own protozoan parasites emerged earthwide too quickly for response.[43] The monarchs joined the myriad kinds of beings that disappeared in the ongoing great extinctions that had been unloosed by the Plantationocene, Anthropocene and Capitalocene. Near the end of life, witnessing the loss of the great monarch migrations across the Americas, and with them the loss of the patterns of living and dying they sustained, Camille 4 knew that mentoring Camille 5 would take other paths than those that had guided Camille 3's teaching of Camille 4.

In a diary archived by the compostists, Camille 4 wrote about per's wrenching feelings in 2340 at the age of eighty-five, while watching the

8.7. The relations of infectious disease and parasitism are not the enemies of earth's critters; the killing of ongoingness—double death—is the crime. "When monarch butterflies are heavily infected with the protozoan parasite *Ophryocystis elektroscirrha*, they can sometimes get stuck to their chrysalis. In this case, a paper wasp takes advantage of the situation." Photograph and description by Jaap de Roode, Emory University.

celebrations for fifteen-year-old Camille 5's initiation into full sym responsibilities. Involved for decades in research on the ecologies of the insects in their human and nonhuman holobiomes, Camille 4 had been studying the reports from all over the earth about the rapid population crashes of moths and butterflies, especially among the Danainae. Although widespread and diverse, monarchs would be among the first to disappear; and no one yet knew why. It was not clear that total extinction would result, but it was clear the migrations were doomed. Holding open space would involve much different work for Camille 5 than for the previous generations, and it was Camille 4's hard task to take the young sym through another sort of initiation before dying in 2355. Of course, Camille 4 had a great deal of experience to draw from with other syms who had lost their critters.

8.8. Make Kin Not Babies. Design by Elaine Gan for the exhibit *dump!* at Kunsthal Aarhus, 2015.

There were thousands of Speakers for the Dead around the earth by 2300, each tasked with bringing critters who had been irretrievably lost into potent presence for giving knowledge and heart to all of those continuing to work for the still diverse earth's robust and partial recuperation.[44] Over three hundred years, the Communities of Compost had built a potent earthwide network of refugia and foci of resurgent naturalcultural diversity. The Speakers for the Dead teach practices of remembering and mourning that enlist extinct human and nonhuman critters in the ongoing work of cutting the shackles of Double Death, which strangled a vast proportion of ways of living and dying in the Plantationocene, Anthropocene and Capitalocene.[45]

For help in preparing Camille 5 to take up the tasks of a Speaker for the Dead, Camille 4 turned to the Canadian-Nunavut, nontraditional, young Inuk woman throat singer Tanya Tagaq and her 2014 album *Animism*, which had been so powerful in strengthening Inuit and also other situated resurgence in the twenty-first century. Tagaq practiced what a twenty-first-century anthropologist, Susan Harding, tentatively called "experimental animism."[46] In *Animism*, Tagaq and her partners, violinist Jesse Zubot and drummers Jean Martin and DZ Michael Red, performed a musical argument for and about continuities, transformations, contradictions, and SF visual and acoustic kinetic interconversions of human and animal beings in situated worlds. Hunting, eating, living-with, dying-with, and moving-with in the turbulent folds and eddies of a situated earth: these were the affirmations and controversies of Tagaq's singing and website texts and interviews. Tagaq embraced oppositions and conflicts, not to purify them, but to live inside complexities of shared flesh, casting herself for some worlds and not others. At her Polaris Music Prize performance in September 2014, the names of murdered and missing Aboriginal women scrolled behind Tagaq. The last track of *Animism* was titled "Fracking"; the first was "Caribou." She wore seal fur cuffs during her Polaris performance; she affirmed the natural world and hunting by her people. Her risk-taking animism performed materialist worlds—gone, here, and to come. Proclaiming, "I want to live in worlds that are not supposed to be," she affirmed that such worlds already are, have been, and will be.[47] The music was utterly contemporary, and many mobile identities were in play and at risk. The work reached out to unexpected techniques and audiences; and it was unapologetically rooted in specific places, peoples, and critters.

Tagaq's practices of transforming sound, flesh, and kind were animist in old and new Inuit terms and in the related sense proposed by the anthropologist Eduardo Viveiros de Castro. Viveiros de Castro studied with Brazilian Amerindians, with whom he learned to theorize the radical conceptual realignment he called multinaturalism and perspectivism. "Animism is the only *sensible* version of materialism."[48] It matters which concepts conceptualize concepts. Materialist, experimental animism is not a New Age wish nor a neocolonial fantasy, but a powerful proposition for rethinking relationality, perspective, process, and reality without the dubious comforts of the oppositional categories of modern/traditional or religious/secular. Human-animal knots do something different in this world.

It matters which worlds world worlds. It matters who eats whom and how. It remains a material question for cosmopolitical critters in the Communities of Compost. For these reasons, Camille 4 invoked Tanya Tagaq to share her power more than two hundred years after her death.

CAMILLE 5

Born 2340. Human numbers are 4 billion.
Died in 2425. Human numbers are 3 billion.
One billion human-critter symbionts inhabit the earth in 2425.
Two billion humans are not syms.
Over 50 percent of all critter species living in 2015
have vanished by 2425.
Millions of kinds of critters are syms with humans.
The animal sym partners remain unaltered by human genes.
The human syms take on ever more properties
of their animal partners.
Many humans are syms with extinct partners.

STARHAWK'S SONG, TAUGHT BY THE SPEAKERS FOR THE DEAD

Breathe deep.
Feel the pain
where it lives deep in us
for we live, still,
in the raw wounds
and pain is salt in us, burning.
Flush it out.
Let the pain become a sound,
a living river on the breath.
Raise your voice.
Cry out. Scream. Wail.
Keen and mourn
for the dismembering of the world.[49]

And so the fifth Camille inherited a powerful task from per's mentor—to become a Speaker for the Dead, to bring into ongoing presence, through active memory, the lost lifeways, so that other symbiotic and sympoietic commitments would not lose heart. Crucial to the work was not to forget the stink in the air from the burning of the witches, not to forget the murders of human and nonhuman beings in the Great Catastrophes named the Plantationocene, Anthropocene, Capitalocene, to "keen and mourn for the dismembering of the world." Moving through mourning to representing, to the practice of vital memory, was the work of the Speakers for the Dead. Their task was to strengthen the healing that was gaining momentum across the earth. The fourth and fifth Camilles both

8.9. Monarch butterfly caterpillars regularly have to share their milkweed food plants with oleander aphids (*Aphis nerii*). © Jaap de Roode, Emory University.

traveled widely, drawing from their heritage of monarch symbioses, to teach and learn how to practice healing and ongoingness in the cyclones of continuing damage and partial resurgence.

Before taking up the tasks of the Speaker for the Dead, remembering the bonds formed more than two hundred years earlier, Camille 5 again sought help from the evolving, historically situated, symanimagenic kin-making practices of twenty-fourth-century Mazahuas. Camille 5 began per's work with a year-long residence in Michoacán studying with the overlapping indigenous-scientific-activist communities that continued to heal damaged lives and lands there. The Mazahuas also mourned the loss of the living monarchs, and they were deeply worried how this extinction would reshape their own symanimagenic relations with their dead. By Camille 5's generation, there were millions of vanished species, vanished kinds of critters both human and other-than-human, and so much for the Speakers for the Dead to do to replenish heart and mind for and with those who continue to stay with the trouble. And to stay with the ragged joy of ordinary living and dying, up to and beyond the year 2400. The monarchs' people decided that this Speaker for the Dead should itself be a new kind of sym, joining the symbiogenetic Camille 5 with symanimagenic persons from the transvolcanic belt. Such persons had been friends and coworkers before; now they were to undertake another experimental and risky sympoiesis for unfurling times.

The Speakers for the Dead are also tasked with bringing into mind and heart the new things of earth, not only the symbionts and symanimants and their communities and corridors, but also the emerging kinds of beings and ways of life of an always evolving home world. The Speakers for the Dead seek and release the energies of the past, present, and future Chthulucene, with its myriad tentacles of opportunistic, dangerous, and generative sympoiesis. The Children of Compost would not cease the layered, curious practice of becoming-with others for a habitable, flourishing world.

Notes

Introduction

1 *Critters* is an American everyday idiom for varmints of all sorts. Scientists talk of their "critters" all the time; and so do ordinary people all over the U.S., but perhaps especially in the South. The taint of "creatures" and "creation" does not stick to "critters"; if you see such a semiotic barnacle, scrape it off. In this book, "critters" refers promiscuously to microbes, plants, animals, humans and nonhumans, and sometimes even to machines.

2 Less simple was deciding how to spell Chthulucene so that it led to diverse and bumptious chthonic dividuals and powers and not to Chthulhu, Cthulhu, or any other singleton monster or deity. A fastidious Greek speller might insist on the "h" between the last "l" and "u"; but both for English pronunciation and for avoiding the grasp of Lovecraft's Cthulhu, I dropped that "h." This is a metaplasm.

Chapter 1: Playing String Figures with Companion Species

1 In languages attuned to partial translation, in U.S. English string figures are called cat's cradle; in French, *jeux de ficelle*; in Navajo, *na'atł'o'*. See Haraway, "SF: Science Fiction, Speculative Fabulation, String Figures, So Far."

2 For a mathematical joke-exposition of Terrapolis, see Haraway, *SF: Speculative Fabulation and String Figures*.

3 From Proto-Germanic and Old English, *guman* later became *human*, but both come soiled with the earth and its critters, rich in humus, *humaine*, earthly beings as

opposed to the gods. In Hebrew, Adam is from *adamah* or "ground." The historical linguistic gender tone of *guman*, like *human* and *man*, is masculine/universal; but in SF worlding *adam*, *guman*, *adamah* become more a microbiome of fermenting critters of many genders and kinds, i.e., companion species, at table together, eating and being eaten, messmates, compost. Puig de la Bellacasa, "Ethical Doings in Naturecultures," discusses a transformational biopolitics, care of the earth and its many species, including people, through care of the soil in the permaculture movement.

4 Strathern, *Reproducing the Future*, 10; Strathern, *The Gender of the Gift*.

5 Whitehead, *Adventures*.

6 Stengers, *Cosmopolitics I* and *II*.

7 Despret, "The Body We Care For"; Despret, "The Becoming of Subjectivity in Animal Worlds." Despret gave me "rendering capable" and much more. "Becoming-with" is developed in Haraway, *When Species Meet*, 16–17, 287.

8 On agential realism and intra-action, see Barad, *Meeting the Universe Halfway*.

9 For the old-style ethnology, see Jayne, *String Figures*.

10 Hogness, "California Bird Talk."

11 Naabeehó Bináhásdzo (the Navajo Nation, the legal geographically defined territory for the semi-autonomous nation), or Diné Bikéyah (the People's name for Navajoland), is located in the Four Corners area of the southwestern United States, surrounded by Colorado, Arizona, Utah, and New Mexico. For Navajo scholarship on their history, written in the web of Diné creation stories and the discipline of academic history, see Denetdale, *Reclaiming Diné History*. There are several Internet sources for Navajo string games and string figures, with varied stories and names, e.g., "Diné String Games" and the large "Library of Navajo String Games." For an extraordinary video of an elder Navajo woman, Margaret Ray Bochinclonny, playing string games, see "Navajo String Games by Grandma Margaret." Margaret Ray's grandson, Terry Teller, explains Navajo string figure star constellations at "So Naal Kaah, Navajo Astronomy." Navajo string games are played mainly in winter, Spider Woman's storytelling season.

12 Anderson, *Creatures of Empire*.

13 Rock doves have probably been in relations of codomestication with people for about ten thousand years, and they are recorded five thousand years ago in Mesopotamia on cuneiform tablets. Throughout this chapter, I will use the common term *pigeons* and *rock doves* interchangeably unless otherwise noted. There are several dozens of living and fossil species in the Columbidae family that includes *C. livia domestica*, including over thirty living Old World pigeon or dove species. Some Columbidae species have expansive natural ranges, and others have specialized needs met only in small ranges. The greatest variety of the family is in the Indomalaya and Australasian ecozones. Domestic pigeons have diversified into many dozens of formal and informal kinds and breeds, as well as the ubiquitous feral pigeons making a living from Istanbul to Tokyo to London to Los Angeles to Berlin to Cairo to Cape Town to Buenos Aires. For an updated "List of Pigeon Breeds," see the Wikipedia entry in the bibliography. A Google images search for

pigeon breeds yields a visual feast. Domestic breeds are thought to have originated in the Middle East and Central Asia. For some of the breeds from this region, including tumblers, spinners, and rollers, see "Turkish Tumblers.com." The BBC did a show about how the pigeon fanciers of Baghdad kept their birds and their sport alive after 2003 during the Iraq war; the detailed corporeal practices of love and care of the men for their pigeons are palpable. See Muir, "The Pigeon Fanciers of Baghdad." For sociological ethnography of racing-pigeon worlds, see Jerolmack, "Primary Groups and Cosmopolitan Ties"; Jerolmack, "Animal Practices, Ethnicity and Community"; Jerolmack, *The Global Pigeon*. For centuries Iran has been a passionate center of pigeon racing, a practice that continues today despite being illegal (but tolerated) under the current regime because racing events are occasions for gambling. For the Persian/French bilingual ethnographic treatment of this fascinating story, see Goushegir, *Le combat du colombophile*. See also "World Market in Pigeons." For an extraordinary index to articles and other information relating to racing pigeons written mainly by fanciers, see "Racing Pigeon-Post."

14 For research on how pigeons do these things, see Walcott, "Pigeon Homing."

15 For a crime thriller set in the racing-pigeon world, check out Scottoline, *The Vendetta Defense*. Working-class New York dockworkers, who are racing-pigeon men, feature prominently in the famous 1954 film starring Marlon Brando, *On the Waterfront*. See Elizabeth Jones, *Night Flyers*, for a moving historical mystery thriller for girls that tells the story of a twelve-year-old farm girl in North Carolina in World War I who loves, protects, and raises homing pigeons. She agrees to train night flyers for the U.S. Army message service. The birds themselves are vivid, fully fleshed out, relational players in the narrative.

16 U.S. Coast Guard, "Pigeon Search and Rescue Project."

17 For example, see Prior, Schwarz, and Güntürkün, "Mirror-Induced Behavior in the Magpie." The mirror test was developed in 1970 by Gordon Gallop Jr.

18 Epstein, Lanza, and Skinner, "'Self-awareness' in the Pigeon"; Allen, DeLabar, and Drossel, "Mirror Use in Pigeons."

19 See Keio University, "Pigeons Show Superior Self-recognition Abilities"; Toda and Watanabe, "Discrimination of Moving Video Images of Self by Pigeons."

20 Watanabe, Sakamoto, and Wakita, "Pigeons' Discrimination of Paintings by Monet and Picasso."

21 Berokoff, "Attachment" and "Love." In a post called "Let's Hear," providing a fascinating window into structures of gender in marriage in these worlds, Berokoff surveyed other wives of pigeon racers from several continents about how they felt about the sport, the pigeons, their husbands, time, and the labor and pleasure of care for the birds.

22 Berokoff, "Love."

23 Beatriz da Costa died on December 27, 2012. For access to her work, including *PigeonBlog*, see "Beatriz da Costa's Blog and Project Hub." See also da Costa, "Interspecies Coproduction." For a discussion of da Costa's work, especially her last project, *Dying for the Other*, see Haraway, Lord, and Juhasz, "Feminism, Technology, Transformation." See da Costa, *Dying for the Other*.

24 Da Costa, "PigeonBlog," 31. All quotes are from this essay. Giving a sense of the skills needed for this collective project, human team members included: Beatriz da Costa (artist, researcher), Richard Desroisiers (pigeon owner), Rufus Edwards (scientific consultant), Cina Hazegh (artist, researcher), Kevin Ponto (artist, researcher), Bob Matsuyama (pigeon owner), Robert Nideffer (copyeditor), Peter Osterholm (pigeon owner), Jamie Schulte (electronics consultant and dear friend), Ward Smith (videographer). See also da Costa and Philips, *Tactical Biopolitics*. The wonderful SF writer Gwyneth Jones also has an essay in this book that inspires my own storytelling, "True Life Science Fiction."

25 Da Costa, "PigeonBlog," 32.

26 Da Costa, "PigeonBlog," 35.

27 I am so greedy for stories of critters and their people collaborating in work and play that I do not always see the rough edges and ongoing troubles. One PigeonBlog team member informally told me that he sometimes found it hard to watch the pigeons learning to fly with the packs and ruffling their feathers in annoyance while they were fitted. He hoped they took pride in their part; but he reminded me that work and play, whether for art, science, or politics—or for all three—are not innocent activities and the burdens rarely fall symmetrically.

28 For a recent pigeon spy report from Iran, see Hambling, "Spy Pigeons Circle the World." Hambling's speculation about ties of Beatriz da Costa's PigeonBlog project to avian spies at Iran's nuclear facilities is ironic, to say the least. But it does seem that the United States might be losing both high-tech robotic remote-controlled spy drones and well-equipped spy pigeons over Iran. It's enough to make the mullahs suspicious. Me too. See also Denega, *The Cold War Pigeon Patrols*.

29 Da Costa, "PigeonBlog," 36.

30 It is too easy to write as if struggles and positions separating animal rights people from other advocates for animals and animal-human worlding were straightforward and closed; they are not. For one discussion of this issue between differing animal-loving feminists, see Potts and Haraway, "Kiwi Chicken Advocate Talks with Californian Dog Companion."

31 Cornell's Margaret Barker, who ran workshops for Washington, DC, school groups in the late 1990s, provided these optimistic reports. See Youth, "Pigeons."

32 "Mais sans colombophile, sans savoir et savoir-faire des hommes et des oiseaux, sans sélection, sans apprentissage, sans transmission des usages, quand bien même resterait-il des pigeons, plus aucun ne sera voyageur. Ce qu'il s'agit de commémorer n'est donc pas un animal seul, ni une pratique seule, mais bien un agencement de deux 'devenirs avec' qui s'inscrit, explicitement, à l'origine du projet. Autant dire, ce qu'il s'agit de faire exister, ce sont des relations par lesquelles des pigeons transforment des hommes en colombophiles talentueux et par lesquelles ces derniers transforment des pigeons en voyageurs fiables. C'est cela que l'oeuvre commémore. Elle se charge de faire mémoire au sens de prolonger activement. Il y a 'reprise.'" Despret, "Ceux qui insistent." Translation by Donna Haraway. See Crasset, "Capsule," for the story and a picture of the pigeon loft that Matali Crasset designed.

33 Australia was the first continent where Europeans recorded string figures. There

are many names in many Aboriginal languages, for example, *matjka-wuma* in Yirrkala. See Davidson, "Aboriginal Australian String Figures." See also "Survival and Revival of the String Figures of Yirrkala."

34 "Batman's Treaty," "Batman Park," and "Wurundjeri," *Wikipedia*. I leave the *Wikipedia* references unadorned partly to mark my own ignorance, partly in appreciation of a flawed but remarkable tool.

35 Downing, "Wild Harvest—Bird Poo."

Chapter 2: Tentacular Thinking

Epigraph 1: Scott Gilbert, "We Are All Lichens Now." See Gilbert, Sapp, and Tauber, "A Symbiotic View of Life." Gilbert has erased the "now" from his rallying cry; we have always been symbionts—genetically, developmentally, anatomically, physiologically, neurologically, ecologically.

Epigraph 2: These sentences are on the rear cover of Stengers and Despret, *Women Who Make a Fuss*. From Virginia Woolf's *Three Guineas*, "think we must" is the urgency relayed to feminist collective thinking-with in *Women Who Make a Fuss* through Puig de la Bellacasa, *Penser nous devons*.

1 Hormiga, "A Revision and Cladistic Analysis of the Spider Family Pimoidae." See "*Pimoa cthulhu*," *Wikipedia*; "Hormiga Laboratory."

2 "The brand of holist ecological philosophy that emphasizes that 'everything is connected to everything,' will not help us here. Rather, everything is connected to *something*, which is connected to something else. While we may all *ultimately* be connected to one another, the specificity and proximity of connections matters—*who we are bound up with and in what ways*. Life and death happen inside these relationships. And so, we need to understand how particular human communities, as well as those of other living beings, are entangled, and how these entanglements are implicated in the production of both extinctions and their accompanying patterns of amplified death" (Van Dooren, *Flight Ways*, 60).

3 Two indispensable books by my colleague-sibling from thirty-plus years in the History of Consciousness Department at the University of California, Santa Cruz, guide my writing. Clifford, *Routes* and *Returns*.

4 *Chthonic* derives from ancient Greek *khthonios*, of the earth, and from *khthōn*, earth. Greek mythology depicts the chthonic as the underworld, beneath the earth; but the chthonic ones are much older (and younger) than those Greeks. Sumeria is a riverine civilizational scene of emergence of great chthonic tales, including possibly the great circular snake eating its own tail, the polysemous Ouroboros (figure of the continuity of life, an Egyptian figure as early as 1600 BCE; Sumerian SF worlding dates to 3500 BCE or before). The chthonic will accrue many resonances throughout my chapter. See Jacobsen, *The Treasures of Darkness*. In lectures, conversations, and e-mails, the scholar of ancient Middle Eastern worlds at UC Santa Cruz, Gildas Hamel, gave me "the abyssal and elemental forces before they were astralized by chief gods and their tame committees" (personal communication,

June 12, 2014). Cthulhu (note spelling), luxuriating in the science fiction of H. P. Lovecraft, plays no role for me, although it/he did play a role for Gustavo Hormiga, the scientist who named my spider demon familiar. For the monstrous male elder god (Cthulhu), see Lovecraft, *The Call of Cthulhu*.

I take the liberty of rescuing my spider from Lovecraft for other stories, and mark the liberation with the more common spelling of chthonic ones. Lovecraft's dreadful underworld chthonic serpents were terrible only in the patriarchal mode. The Chthulucene has other terrors—more dangerous and generative in worlds where such gender does not reign. Undulating with slippery eros and gravid chaos, tangled snakes and ongoing tentacular forces coil through the twenty-first century CE. Consider: Old English *oearth*, German *Erde*, Greek Gaïa, Roman *terra*, Dutch *aarde*; Old English *w(e)oruld* ("affairs of life," "a long period of time," "the known life," or "life on earth" as opposed to the "afterlife"), from a Germanic compound meaning "age of the human race" (*wer*); Old Norse *heimr*, literally "abode." Then consider Turkish *dünya* and go to *dunyā* (the temporal world), an Arabic word that was passed to many other languages, such as Persian, Dari, Pashto, Bengali, Punjabi, Urdu, Hindi, Kurdish, Nepali, Turkish, Arumanian, and North Caucasian languages. *Dunyā* is also a loanword in Malay and Indonesian, as well as in Greek *δουνιας*—so many words, so many roots, so many pathways, so many mycorrhizal symbioses, even if we restrict ourselves only to Indo-European tangles. There are so many kin who might better have named this time of the Anthropocene that is at stake now. The anthropos is too much of a parochial fellow; he is both too big and too small for most of the needed stories.

5 Eva Hayward proposes the term *tentacularity*; her trans-thinking and -doing in spidery and coralline worlds entwine with my writing in SF patterns. See Hayward, "FingeryEyes"; "SpiderCitySex"; and "Sensational Jellyfish." See Morgan, "Sticky Tales." UK experimental artist Eleanor Morgan's spider silk art spins many threads resonating with this chapter, tuned to the interactions of animals (especially arachnids and sponges) and humans. Morgan, "Website."

6 Katie King aligns Hayward's "fingery eyes" and "tentacularity" with "networked reenactments" or "transknowledges." "Working out in a multiverse of articulating disciplines, interdisciplines, and multidisciplinarities, such transdisciplinary inspection actually *enjoys* the many flavors of details, offerings, passions, languages, things . . . One index for the evaluation of transdisciplinary work is how well it learns and models *how* to be affected or moved, how well it *opens up* unexpected elements of one's own embodiments in lively and re-sensitizing worlds." King, *Networked Reenactments*, 19. See also King, "A Naturalcultural Collection of Affections." Think we must.

7 *Muddle*, Old Dutch for muddying the waters. I use *muddle* as a theoretical trope and soothing wallow to trouble the trope of visual clarity as the only sense and affect for mortal thinking. Muddles team with company. Empty spaces and clear vision are bad fictions for thinking, not worthy of SF or of contemporary biology. My speculative feminist courage has been fed by Puig de la Bellacasa, "Touching Technologies, Touching Visions."

For a gorgeous animated model of a densely packed living neuron, where proteins muddle on their herky-jerky way to making cells work, see "Protein Packing: Inner Life of a Cell" and Zimmer, "Watch Proteins Do the Jitterbug."

8 Ingold, *Lines*, 116–19.

9 The pile was made irresistible by Puig de la Bellacasa, "Encountering Bioinfrastructure."

10 Art science activism infuses this book. In the struggle for multispecies environmental justice in the face of coal company mountaintop removal in her homeworld in West Virginia, with her wife Annie Sprinkle (environmental activist, radical adult film director and performer, former sex worker), UC Santa Cruz artist Beth Stephens made the "sexiest nature documentary ever," *Goodbye Gauley Mountain: An Ecosexual Love Story*. The quote is from a review by Russ McSpadden, "Ecosexuals of the World Unite!" In love and rage (Emma Goldman), think we must (Virginia Woolf) for a habitable planet.

11 Throughout this essay I use the Latinate words *terran* and *terra*, even while I swim in Greek names and stories, including the material-semiotic story of Gaia and Bruno Latour's "Gaia stories/geostories." Terra is especially legible in SF, but Gaia is important in SF too. My favorite is John Varley's Gaea Trilogy, *Titan* (1979), *Wizard* (1980), and *Demon* (1984). Varley's Gaea is an old woman, who/which is a living being in the shape of a 1,300-kilometer-diameter Stanford torus, inhabited by many different species, in orbit around the planet Saturn. For a fan site, see "Gaea, the Mad Titan." Latour's Earthbound ("Terriens" in his French) and Stengers's intrusive Gaia would recognize Varley's irascible, unpredictable Gaea. Gaia is more legible in systems theories than Terra, as well as in "New Age" cultures. Gaia comes into her/its own in the Anthropocene, but Terra sounds a more earthy tone for me. However, Terra and Gaia are not in opposition, nor are the Earthbound, who are given to us in loving, risk-taking, powerful writing by Bruno Latour, in opposition to Terrans. Rather, Gaians and Terrans are in a queer planetwide litter of chthonic ones who must be re-membered urgently. It is in that sense that I hear together Isabelle Stengers's "cosmopolitics" and my verbally miscegenated "Terrapolis." We are making string figures together.

12 Allied to this kind of argument is Barad, *Meeting the Universe Halfway*. Outside (and inside) the odd thing named the West, there are myriad histories, philosophies, and practices—some civilizational, some urban, some neither—that propose living and dying in other knots and patterns that do not presume isolated, much less binary, unities and polarities that then need to be brought into connection. Variously and dangerously configured relationality is just what is. Flawed but powerful systems theories are the best technoscientific models we have so far for many Gaian relationalities.

An American evolutionary biologist, David Barash, writes compellingly about convergences (not identities and not resources that can be hijacked to cure Western ills) between ecological sciences and various Buddhist streams, schools, and traditions that emphasize connectedness. Barash emphasizes that ways of living, dying, acting, and nurturing response-ability are embedded in these matters (*Bud-*

dhist Biology). What if Western evolutionary and ecological sciences had been developed from the start within Buddhist instead of Protestant ways of worlding? Why do I find it so jarring that David Barash is a committed neo-Darwinian in evolutionary theory? See Barash, *Natural Selections*. The need for complexity theories tuned to paradox is obvious!

Based on his extensive study of Chinese knowledges and sciences, Joseph Needham asked a similar question to Barash's many years ago about embryology and biochemistry in *The Grand Titration: Science and Society in East and West*. Needham's organicism and Marxism are both crucial for this story, something to remember in thinking about how to configure what I will explore in this chapter under the sign *Capitalocene*. On Needham, see Haraway, *Crystals, Fabrics, and Fields*. What happens if we cultivate response-ability for the Capitalocene inside the netbags of sympoiesis, Buddhism, ecological evolutionary developmental biology (EcoEvoDevo), Marxism, Stengersian cosmopolitics, and other strong pulls against the modernizing foolishness of some analyses of capitalism? What happens if the relentless zero-sum games of neo-Darwinism give way to an extended evolutionary synthesis?

13 Dempster, "A Self-Organizing Systems Perspective on Planning for Sustainability." See 27–32 for a concise comparison of autopoietic and sympoietic systems. Table 1, p. 30, juxtaposes defining characteristics for autopoietic and sympoietic systems, such as: self-produced boundaries/lacking boundaries; organizationally closed/organizationally ajar; external structural coupling/internal and external structural coupling; autonomous units/complex amorphous entities; central control/distributed control; evolution between systems/evolution within systems; growth and development orientation/evolutionary orientation; steady state/potentially dramatic, surprising change; predictable/unpredictable.

Katie King told me about the Dempster thesis as we tried to sort out our overlapping but not identical pleasures and resistances to autopoiesis and sympoiesis. See King, "Toward a Feminist Boundary Object-Oriented Ontology . . . or Should It Be a Boundary Object-Oriented Feminism?"

14 Stengers, "Relaying a War Machine?," 134.

15 Strathern, *The Relation*; *Partial Connections*; and *Kinship, Law and the Unexpected*.

16 Strathern, *Reproducing the Future*, 10.

17 Baila Goldenthal (1925–2011) painted an extraordinary series of four Cat's Cradle-titled oil-on-wood panels in 1995–96 and an oil-on-canvas in 2008. For her and for me, cat's cradle is an open-ended practice of continuous weaving (see her Weavers Series, 1989–94). "The techniques of under-painting and glazing invoke historical time; the enigma of the game itself reflects the complexity of human relationships." Goldenthal, "Painting/Cats Cradle." Goldenthal relates to cat's cradle games as a metaphor for the game of life, and the intensely present, moving hands invite kinship with other tentacular beings. Her 2008 *Cat's Cradle/String Theory* is the cover image for *Nuclear Abolition Forum*, no. 2 (2013), an issue titled "Moving beyond Nuclear Deterrence to a Nuclear Weapons Free World." Metamorphosis, fragility, temporality, disintegration, revelation—these are everywhere

in her work. A student of the Kabbalah and of South Asian Indian culture and philosophy, Goldenthal worked in oils, bronze, leaded glass, paper, photography, printmaking, film, and ceramics. She accomplished powerful work in sculpture and in two-dimensional formats. Goldenthal, "Resumé." Among my favorites is her *Desert Walls* of the mid-1980s, where she worked in photography and collage with tile, brick, straw, plaster, metal, and glass to evoke the visual enigmas of cliffs and rock walls of the U.S. desert Southwest.

18 Arendt, *Eichmann in Jerusalem*; Hartouni, *Visualizing Atrocity*, especially chapter 3, "Thoughtlessness and Evil." I set aside the strict humanism and the specific kind of thinking subject of Arendt's project, as well as her insistence on the essential solitude of thinking. Thinking-with in the SF compost pile of this essay is not an enemy to the profound secular self-examination of Arendt's historically situated human figure, but that is an argument for another day.

19 Arendt characterized thinking as "training one's mind to going visiting." "This distancing of some things and bridging of others is part of the dialogue of understanding, for whose purposes direct experience establishes too close contact and mere knowledge erects artificial barriers." Arendt, "Truth and Politics," 241, quoted in Hartouni, *Visualizing Atrocity*, 75.

20 Puig de la Bellacasa, "Matters of Care in Technoscience"; Puig de la Bellacasa, *Matters of Care*.

21 Title of a conference that Anna Tsing and coworkers organized at the University of California, Santa Cruz, May 8–10, 2014: "Anthropocene: Arts of Living on a Damaged Planet."

22 All quotations are from Tsing, *The Mushroom at the End of the World*, 34, 2, 4.

23 Van Dooren, *Flight Ways*.

24 Van Dooren's colleague Deborah Bird Rose is everywhere in this thinking, especially in her treatment of the undoing of the tissues of ongoingness, the killing of generations, which she called "double death" in *Reports from a Wild Country: Ethics for Decolonisation*. See also van Dooren and Rose, "Unloved Others"; van Dooren and Rose, "Storied-Places in a Multispecies City." The Extinction Studies Working Group, anchored in Australia, is a rich sympoietic gathering. See also Environmental Humanities South, anchored in Cape Town, South Africa.

25 Van Dooren, "Keeping Faith with Death"; *Flight Ways*, chapter 5, "Mourning Crows: Grief in a Shared World." This writing is in SF exchange with Vinciane Despret's thinking about learning to be affected. See Despret, "The Body We Care For."

26 Van Dooren, *Flight Ways*, 63–86. Also crucial to grasping thinking and semiotics outside the premises of modernist humanist doctrines, see Kohn, *How Forests Think*.

27 Le Guin, "The Carrier Bag Theory," 166. Le Guin's essay (1986) shaped my thinking about narrative in evolutionary theory and of the figure of woman the gatherer in *Primate Visions*. Le Guin learned about the Carrier Bag Theory of Evolution from Elizabeth Fisher, *Women's Creation*, in that period of large, brave, speculative, worldly stories that burned in feminist theory in the 1970s and 1980s. Like speculative fabulation, speculative feminism was, and is, an SF practice. For a fuller

SF game with both Le Guin and Octavia Butler, see chapter 6, "Sowing Worlds: A Seed Bag for Terraforming with Earth Others." First published in Grebowicz and Merrick, *Beyond the Cyborg*.

28 Le Guin, "The Carrier Bag Theory," 169.

29 For introduction and elucidation of the "god trick" in science and politics, see Haraway, "Situated Knowledges."

30 Latour, Gifford Lectures, Lecture 3, "The Puzzling Face of a Secular Gaïa." Quotation from lecture manuscript.

31 Latour, "War and Peace in an Age of Ecological Conflicts." Quotation from lecture manuscript. Latour's proportionality in this lecture is bracing:

Humans : business as usual :: the Earthbound : total subversion.

In "Feral Biologies," Anna Tsing uses the word *Holocene* to mean something radically different from Latour; but their basic arguments rub against each other in often edgy agreement, generating some interesting friction. Tsing refers to the Holocene as the timeplaces of possible resurgence after disturbance; the Anthropocene is the timeplace of radical reduction, radical simplification, radical obliteration of the refugia of the Holocene, from which resurgence of species assemblages could occur. Latour's and Tsing's different uses of the same important words illustrate how polysemous possibilities lurk even in closely scrutinized linguistic precincts. Unnecessary oppositions can be easily spun from such different elaborations of words, and the expertise of geologists only adds to language's generativity. I think some of Latour's and Tsing's hot friction comes from his reliance on Carl Schmitt and her love of Ursula K. Le Guin.

32 Latour's "Why Has Critique Run Out of Steam? From Matters of Fact to Matters of Concern" is a major landmark in our collective understanding of the corrosive, self-certain, and self-contained traps of nothing-but-critique. Cultivating response-ability requires much more from us. It requires the risk of being for some worlds rather than others and helping to compose those worlds with others. In multistranded SF worlding, Maria Puig de la Bellacasa recomposts Latour's "matters of concern" to ferment an even richer soil in her "Matters of Care in Technoscience."

33 Latour, "War and Peace in an Age of Ecological Conflicts," lecture manuscript.

34 To understand how the modernizing category of "belief" works in the United States in law, politics, and pedagogy, including religion and social science, see Harding, "Secular Trouble." The figure of the never properly belonging, always leaving and returning "Prodigal Daughter" further unpacks the enabling and disabling operations of "belief" in deVries, "Prodigal Knowledge." Tying knowledge practices to professions of belief in both religion and science is perhaps the single most difficult habit of thought to dislodge for Moderns, at least in the United States. Where belief is exacted, the Inquisition is never far behind. SF in the muddle of Terra/Gaia cannot exact belief, but can shape committed thinking companions. The figure for thinking-with in this ecology of practices is not so much "decision" as sympoietic "care" and "discernment." The Prodigal Daughter remains

a wayfarer, much more promising for pathways in troubled times than the paved road toward the feast prepared for the returning, forever-after obedient Prodigal Son and legitimate heir.

35 Latour, "War and Peace in an Age of Ecological Conflicts," lecture manuscript; Schmitt, *The Nomos of the Earth*. For a full exposition of his reliance on Schmitt's *hostis* and political theology, see Latour, Gifford Lectures, Lecture 5, "War of Humans and Earthbound": "If Humans are at war with It [Gaia], what about those whom I have proposed to call the Earthbound? Can they be '*artisans of peace*'?" (unpublished lecture manuscript). Such artisans are what Latour works to nourish here and elsewhere.

His question deserves more space, but a few words about *hostis* are necessary. Latour and I both ate the "host" in the sacrificial Eucharistic feast, and so we know what it means to be in the material-semiotic world where sign and signifier have imploded in meaningful flesh. Neither of us fits very well in secular Protestant semiotics, dominant in the university and in science, and that shapes our approaches to science studies and much else. But note that the "host" that we ate—our communion—is firmly ensconced in the story of the acceptable sacrifice to the Father. Latour and I ate too much and too little when we consumed this host and refused (and still refuse) to disavow it. I have a case of permanent raging indigestion, even as I hold fast to the joy and the implosion of metaphor and world. I need to know more about Latour's digestive comforts and discomforts because I suspect they are at the root of our different lures for changing the story for the Earthbound. In the sacrificial Eucharistic worlding, there are strong kin ties, etymologically and historically, to the host of Schmitt, where we find the guest, hostage, one held in surety for another, generator and collector of debt, host as the one who feeds the traveler as guest, stranger to be respected even if killed, hostiles, host as an array armed for combat in the field of battle (a trial of strength). Not vermin, not trash, not *inimicus*, but those coproducing the engagement of war and so perhaps a new peace rather than extermination. But host has other tones too, ones that lead a little way to the chthonic and tentacular ones in the carrier bag story, where Latour and I may yet luckily be gathered and transformed by some old hag collecting dinner. We might be allowed to stay as guests, as companion species, especially if we are on the menu. The host is the habitat for the parasite, the condition of life and ongoingness for the parasite; this host is in the dangerous world-making contact zones of symbiogenesis and sympoiesis, where newly cobbled together, good-enough orders may or may not emerge from the ever so promiscuous and opportunistic associations of host and parasite. Perhaps Gaia's unchristian abyssal gut, habitat for chthonic powers, is the muddle for SF, where ongoingness remains at stake. This is the world that evokes this chapter's epigraph, "We are all lichens." (On the difficulty of becoming unchristian, see Anidjar, *Blood*. Anidjar also does very interesting things with Schmitt.)

But not so fast, my lichen selves and affiliates! First we have to wrestle with the ill-named Anthropocene. I am not against all trials of strength; after all, I love women's basketball. I just think trials of strength are the old story. Overvalued, they are a bit like the never-ending task of cleaning the toilet—necessary but radically insufficient. On the other hand, there are excellent composting toilets . . . We can outsource some trials of strength to the ever-eager microbes to make more time and space for SF in other muddles.

36. Stengers, *Au temps des catastrophes*. Gaia intrudes in this text from p. 48 on. Stengers discusses the "intrusion of Gaïa" in numerous interviews, essays, and lectures. Discomfort with the ever more inescapable label of the Anthropocene, in and out of sciences, politics, and culture, pervades Stengers's thinking, as well as that of many other engaged writers, including Latour, even as we struggle for another word. See Stengers in conversation with Heather Davis and Etienne Turpin, "Matters of Cosmopolitics."

    Stengers's thinking about Gaia and the Lovelock-Margulis development of the Gaia hypothesis was from the start entwined with her work with Ilya Prigogine, which understood that strong linear coupling in complex systems theory entailed the possibility of radical global system change, including collapse. Prigogine and Stengers, *Order Out of Chaos*. The relation of Gaia to Chaos is an old one in science and philosophy. What I want to do is knot that emergence sympoietically into a worlding of ongoing chthonic powers, which is the material-semiotic time-space of the Chthulucene rather than Anthropocene or Capitalocene. This is part of what Stengers means when she says that her intrusive Gaia was "ticklish" from the start. "Her 'autopoietic' functioning is not her truth but what 'we' [human beings] have to face, and are able to read from our computer models, the face she turns on 'us'" (e-mail from Stengers to Haraway, May 9, 2014).

37 Scientists estimate that this extinction "event," the first to occur during the time of our species, could, as previous great extinction events have, but much more rapidly, eliminate 50 to 95 percent of existing biodiversity. Sober estimates anticipate half of existing species of birds could disappear by 2100. By any measure, that is a lot of double death. For a popular exposition, see Voices for Biodiversity, "The Sixth Great Extinction." For a report by an award-winning science writer, see Kolbert, *The Sixth Extinction*. Reports from The Convention on Biological Diversity are more cautious about predictions and discuss the practical and theoretical difficulties of obtaining reliable knowledge, but they are not less sobering. For a disturbing report from summer 2015, see Ceballos et al., "Accelerated Modern Human-Induced Species Losses."

38 Lovelock, "Gaia as Seen through the Atmosphere"; Lovelock and Margulis, "Atmospheric Homeostasis by and for the Biosphere." For a video of a lecture to employees at the National Aeronautic and Space Agency in 1984, go to Margulis, "Gaia Hypothesis." Autopoiesis was crucial to Margulis's transformative theory of symbiogenesis, but I think if she were alive to take up the question, Margulis would often prefer the terminology and figural-conceptual powers of sympoiesis. I suggest that Gaia is a system mistaken for autopoietic that is really sympoietic. See chapter 3, "Sympoiesis." Gaia's story needs an intrusive makeover to knot with a host of other promising sympoietic tentacular ones for making rich compost, for going on. Gaia or Ge is much older and wilder than Hesiod (Greek poet around the

time of Homer, circa 750 to 650 BCE), but Hesiod cleaned her/it up in the *Theogony* in his story-setting way: after Chaos, "wide-bosomed" Gaia (Earth) arose to be the everlasting seat of the immortals who possess Olympus above (*Theogony*, 116–18, translated by Glenn W. Most, Loeb Classical Library), and the depths of Tartarus below (*Theogony*, 119). The chthonic ones reply, Nonsense! Gaia is one of theirs, an ongoing tentacular threat to the astralized ones of the Olympiad, not their ground and foundation, with their ensuing generations of gods all arrayed in proper genealogies. Hesiod's is the old prick tale, already setting up canons in the eighth century BCE.

39 Although I cannot help but think more rational environmental and socialnatural policies of all sorts would help!

40 Isabelle Stengers, from English compilation on Gaia sent by e-mail January 14, 2014.

41 I use "thing" in two senses that rub against each other: (1) the collection of entities brought together in the Parliament of Things that Bruno Latour called our attention to, and (2) something hard to classify, unsortable, and probably with a bad smell. Latour, *We Have Never Been Modern*.

42 Crutzen and Stoermer, "The 'Anthropocene'"; Crutzen, "Geology of Mankind"; Zalasiewicz et al., "Are We Now Living in the Anthropocene?" Much earlier dates for the emergence of the Anthropocene are sometimes proposed, but most scientists and environmentalists tend to emphasize global anthropogenic effects from the late eighteenth century on. A more profound human exceptionalism (the deepest divide of nature and culture) accompanies proposals of the earliest dates, coextensive with *Homo sapiens* on the planet hunting big now-extinct prey and then inventing agriculture and domestication of animals. A compelling case for dating the Anthropocene from the multiple "great accelerations," in earth system indicators and in social change indicators, from about 1950 on, first marked by atmospheric nuclear bomb explosions, is made by Steffen et al., "The Trajectory of the Anthropocene." Zalasiewicz et al. argue that adoption of the term *Anthropocene* as a geological epoch by the relevant national and international scientific bodies will turn on stratigraphic signatures. Perhaps, but the resonances of the Anthropocene are much more disseminated than that. One of my favorite art investigations of the stigmata of the Anthropocene is Ryan Dewey's "Virtual Places: Core Logging the Anthropocene in Real-Time," in which he composes "core samples of the *ad hoc* geology of retail shelves."

43 For a powerful ethnographic encounter in the 1990s with climate-change modeling, see Tsing, *Friction*, "Natural Universals and the Global Scale," 88–112, especially "Global Climate as a Model," 101–6. Tsing asks, "What makes global knowledge possible?" She replies, "Erasing collaborations." But Tsing does not stop with this historically situated critique. Instead she, like Latour and Stengers, takes us to the really important question: "Might it be possible to attend to nature's collaborative origins without losing the advantages of its global reach?" (95). "How might scholars take on the challenge of freeing critical imaginations from the specter of neoliberal conquest—singular, universal, global? Attention to the frictions of contingent

articulation can help us describe the effectiveness, and the fragility, of emergent capitalist—and globalist—forms. In this shifting heterogeneity there are new sources of hope, and, of course, new nightmares" (77). At her first climate-modeling conference in 1995, Tsing had an epiphany: "*The global scale takes precedence—because it is the scale of the model*" (103, italics in original). But this and related properties have a particular effect: they bring negotiators to an international, heterogeneous table, maybe not heterogeneous enough, but far from full of identical units and players. "The embedding of smaller scales into the global; the enlargement of models to include everything; the policy-driven construction of the models: Together these features make it possible for the models to bring diplomats to the negotiating table" (105). That is not to be despised.

The reports of the Intergovernmental Panel on Climate Change (IPCC) are necessary documents and excellent illustrations of Tsing's accounts: *Climate Change 2014: Mitigation of Climate Change* and *Climate Change 2014: Impacts, Adaptation, and Vulnerability*.

Tsing's stakes in her intimate tracking of the relentless ethnographic specificities of far-flung chains of intimate dealings and livings are to hold in productive, nonutopian friction the scale-making power of the things climate-change models do with the life-and-death messiness of place- and travel-based worldings that always make even our best and most necessary universals very lumpy. She seeks and describes multiple situated worldings and multiple sorts of translations to engage globalism. "Attention to friction opens up the possibility of an ethnographic account of global interconnection" (6). Appreciation of what she calls "weediness" is indispensable: "To be aware of the necessity for careful coalitions with those whose knowledges and pleasures come from other sources is the beginning of nonimperialist environmentalism" (170). The *hostis* will not make an appearance in this string figuring, but mushrooms as guides for living in the ruins most certainly will. See Tsing, *The Mushroom at the End of the World*.

44 The Anthropocene Working Group, which was established in 2008 to report to the International Union of Geological Sciences and the International Commission on Stratigraphy on whether to name a new epoch in the geological timeline, aimed to issue its final report in 2016. See *Newsletter of the Anthropocene Working Group*, volumes 4 and 5.

45 For a photo gallery of fiery images of the Man burning at the end of the festival, see "Burning Man Festival 2012." Attended by tens of thousands of human people (and an unknown number of dogs), Burning Man is an annual week-long festival of art and (commercial) anarchism held in the Black Rock Desert of Nevada since 1990 and on San Francisco's Baker Beach from 1986 to 1990. The event's origins tie to San Francisco artists' celebrations of the summer solstice. "The event is described as an experiment in community, art, radical self-expression, and radical self-reliance" ("Burning Man," *Wikipedia*). The globalizing extravaganzas of the Anthropocene are not the drug- and art-laced worlding of Burning Man, but the iconography of the immense fiery "Man" ignited during the festival is irresistible. The first burning effigies on the beach in San Francisco were of a 9-foot-tall

wooden Man and a smaller wooden dog. By 1988 the Man was 40 feet tall and dogless. Relocated to a dry lakebed in Nevada, the Man topped out in 2011 at 104 feet. This is America; supersized is the name of the game, a fitting habitat for the Anthropos.

"Anthropos" (ἄνθρωπος) is an ambiguous word with contested etymologies. What Anthropos never figures is the rich generative home of a multispecies earth. The *Online Etymology Dictionary* states that it comes from the Greek *aner*, "man," "as opposed to a woman, a god, or a boy." Just what I suspected! Or, "Anthropos sometimes is explained as a compound of *aner* and *ops* (genitive *opos*) 'eye, face'; so literally 'he who has the face of a man.'" Or, sometimes, the shape of a man. Biblical scholars find it hard to make the Greek ανθρωπος include women, and it complicates translations in fascinating ways: see http://www.bible-researcher.com/anthropos.html (accessed August 7, 2015). Other sources give the meaning of the compound as "that which is below, hence earthly, human," or, the "upward looking one," and so below, lamentably on earth. Unlike the animals, man as anthropos "looks up at what he sees": http://www.science-bbs.com/114-lang/0e74f4484bff3fe0.htm (accessed August 7, 2015). The Anthropos is *not* Latour's Earthbound.

It is safe to say that Eugene Stoermer and Paul Crutzen were not much vexed by these ambiguities. Still, thank the heavens, looking up, their human eyes were firmly on the earth's atmospheric carbon burden. Or, also, swimming in too hot seas with the tentacular ones, their eyes were the optic-haptic fingery eyes of marine critters in diseased and dying coral symbioses. See Hayward, "FingeryEyes."

46 Klare, "The Third Carbon Age," writes, "According to the International Energy Agency (IEA), an inter-governmental research organization based in Paris, cumulative worldwide investment in new fossil-fuel extraction and processing will total an estimated $22.87 trillion between 2012 and 2035, while investment in renewables, hydropower, and nuclear energy will amount to only $7.32 trillion." Nuclear, after Fukushima! Not to mention that none of these calculations prioritize a much lighter, smaller, more modest human presence on earth, with all its critters. Even in its "sustainability" discourses, the Capitalocene cannot tolerate a multispecies world of the Earthbound. For the switch in Big Energy's growth strategies to nations with the weakest environmental controls, see Klare, "What's Big Energy Smoking?" See also Klare, *The Race for What's Left*.

47 Heavy tar sand pollution must break the hearts and shatter the gills of every Terran, Gaian, and Earthbound critter. The toxic lakes of wastewater from tar sand oil extraction in northern Alberta, Canada, shape a kind of new Great Lakes region, with more giant "ponds" added daily. Current area covered by these lakes is about 50 percent greater than the area covered by the world city of Vancouver. Tar sands operations return almost none of the vast quantities of water they use to natural cycles. Earthbound peoples trying to establish growing things at the edges of these alarmingly colored waters filled with extraction tailings say that successional processes for re-establishing sympoietic biodiverse ecosystems, if they prove possible at all, will be an affair of decades and centuries. See Pembina Institute, "Alberta's

Oil Sands," and Weber, "Rebuilding Land Destroyed by Oil Sands May Not Restore It." Only Venezuela and Saudi Arabia have more oil reserves than Alberta. All that said, the Earthbound, the Terrans, do not cede either the present or the future; the sky is lowering, but has not fallen, yet. Pembina Institute, "Oil Sands Solutions." First Nation, Métis, and Aboriginal peoples are crucial players in every aspect of this unfinished story. See the website for the Tar Sands Solutions Network. For melting sea ice in the Arctic, see figure 2.4, p. 48.

48 Photograph from NASA Earth Observatory, 2015 (public domain). If flame is the icon for the Anthropocene, I use the missing ice and the unblocked Northwest Passage to figure the Capitalocene. The Soufan Group provides strategic security intelligence services to governments and multinational organizations. Its report "TSG IntelBrief: Geostrategic Competition in the Arctic" includes the following quotes: "The *Guardian* estimates that the Arctic contains 30 percent of the world's undiscovered natural gas and 15 percent of its oil." "In late February, Russia announced it would form a strategic military command to protect its Arctic interests." "Russia, Canada, Norway, Denmark, and the US all make some claim to international waters and the continental shelf in the Arctic Ocean." "[A Northwest Passage] route could provide the Russians with a great deal of leverage on the international stage over China or any other nation dependent on sea commerce between Asia and Europe."

The province of Alberta in Canada ranks third in the world after Saudi Arabia and Venezuela for proven global crude reserves. Almost all of Alberta's oil is in the tar sands in the north of the province, site of the great new petrotoxic lakes of North America. See Alberta Energy, "Facts and Statistics." The Capitalocene in action! See the Indigenous Environmental Network, "Canadian Indigenous Tar Sands Campaign." Over twenty corporations operate in the tar sands in the home area of many indigenous peoples, including the First Nation Mikisew Cree, Athabasca Chipewyan, Fort McMurray, Fort McKay Cree, Beaver Lake Cree, Chipewyan Prairie, and also the Métis.

49 Klein, "How Science Is Telling Us All to Revolt"; Klein, *The Shock Doctrine*.

50 *Capitalocene* is one of those words like *sympoiesis*; if you think you invented it, just look around and notice how many other people are inventing the term at the same time. That certainly happened to me, and after I got over a small fit of individualist pique at being asked whom I got the term *Capitalocene* from—hadn't I coined the word? ("Coin"!) And why do other scholars almost always ask women which male writers their ideas are indebted to?—I recognized that not only was I part of a cat's cradle game of invention, as always, but that Jason Moore had already written compelling arguments to think with, and my interlocutor both knew Moore's work and was relaying it to me. Moore himself first heard the term *Capitalocene* in 2009 in a seminar in Lund, Sweden, when then graduate student Andreas Malm proposed it. In an urgent historical conjuncture, words-to-think-with pop out all at once from many bubbling cauldrons because we all feel the need for better netbags to collect up the stuff crying out for attention. Despite its problems, the term *Anthropocene* was and is embraced because it collects up many matters of fact, concern, and care; and I hope *Capitalocene* will roll off myriad tongues soon.

In particular, see the work of Jason Moore, a creative Marxist sociologist at Binghamton University in New York. Moore is coordinator of the World-Ecology Research Network. For his first Capitalocene argument, see Moore, "Anthropocene, Capitalocene, and the Myth of Industrialization." See Moore, *Capitalism and the Web of Life*.

51 To get over Eurocentrism while thinking about the history of pathways and centers of globalization over the last few centuries, see Flynn and Giráldez, *China and the Birth of Globalisation in the 16th Century*. For analysis attentive to the differences and frictions among colonialisms, imperialisms, globalizing trade formations, and capitalism, see Ho, "Empire through Diasporic Eyes" and *The Graves of Tarem*.

52 In "Anthropocene or Capitalocene, Part III," Jason Moore puts it this way: "This means that capital and power—and countless other strategic relations—do not act upon nature but develop through the web of life. 'Nature' is here offered as the relation of the whole. Humans live as a specifically endowed (but not special) environment-making species within Nature. Second, capitalism in 1800 was no Athena, bursting forth, fully grown and armed, from the head of a carboniferous Zeus. Civilizations do not form through Big Bang events. They emerge through cascading transformations and bifurcations of human activity in the web of life . . . [For example,] the long seventeenth century forest clearances of the Vistula Basin and Brazil's Atlantic Rainforest occurred on a scale, and at a speed, between five and ten times greater than anything seen in medieval Europe."

53 Crist, "On the Poverty of Our Nomenclature," 144. Crist does superb critique of the traps of Anthropocene discourse, as well as gives us propositions for more imaginative worlding and ways to stay with the trouble. For entangled, dissenting papers that both refuse and take up the name *Anthropocene*, see videos from the conference "Anthropocene Feminism." For rich interdisciplinary research, organized by Anna Tsing and Nils Ole Bubandt, that brings together anthropologists, biologists, and artists under the sign of the Anthropocene, see AURA: Aarhus University Research on the Anthropocene.

54 I owe the insistence on "big-enough stories" to Clifford, *Returns*: "I think of these as 'big enough' histories, able to account for a lot, but not for everything—and without guarantees of political virtue" (201). Rejecting one big synthetic account or theory, Clifford works to craft a realism that "works with open-ended (because their linear historical time is ontologically unfinished) 'big-enough stories,' sites of contact, struggle, and dialogue" (85–86).

55 Pignarre and Stengers, *La sorcellerie capitaliste*. Latour and Stengers are deeply allied in their fierce rejection of discourses of denunciation. They have both patiently taught me to understand and relearn in this matter. I love a good denunciation! It is a hard habit to unlearn.

56 It is possible to read Max Horkheimer and Theodor Adorno's *Dialectic of Enlightenment* as an allied critique of Progress and Modernization, even though their resolute secularism gets in their own way. It is very hard for a secularist to really listen to the squid, bacteria, and angry old women of Terra/Gaia. The most likely Western Marxist allies, besides Marx, for nurturing the Chthulucene in the belly

of the Capitalocene are Antonio Gramsci, *Selections from the Prison Notebooks*, and Stuart Hall. Hall's immensely generative essays extend from the 1960s through the 1990s. See, for example, Morley and Chen, *Stuart Hall*.

57 See Gilson, "Octopi Wall Street!" for the fascinating history of cephalopods figuring the depredations of Big Capital in the United States (for example, the early twentieth-century John D. Rockefeller/Standard Oil octopus strangling workers, farmers, and citizens in general with its many huge tentacles). Resignification of octopuses and squids as chthonic allies is excellent news. May they squirt inky night into the visualizing apparatuses of the technoid sky gods.

58 Hesiod's *Theogony* in achingly beautiful language tells of Gaia/Earth arising out of Chaos to be the seat of the Olympian immortals above and of Tartarus in the depths below. She/it is very old and polymorphic and exceeds Greek tellings, but just how remains controversial and speculative. At the very least, Gaia is not restricted to the job of holding up the Olympians! The important and unorthodox scholar-archaeologist Marija Gimbutas claims that Gaia as Mother Earth is a later form of a pre–Indo-European, Neolithic Great Mother. In 2004, filmmaker Donna Reed and neopagan author and activist Starhawk released a collaborative documentary film about the life and work of Gimbutas, *Signs out of Time*. See Belili Productions, "About Signs out of Time"; Gimbutas, *The Living Goddesses*.

59 To understand what is at stake in "non-Euclidean" storytelling, go to Le Guin, *Always Coming Home* and "A Non-Euclidean View of California as a Cold Place to Be."

60 "The Thousand Names of Gaia: From the Anthropocene to the Age of the Earth," International Colloquium, Rio de Janeiro, September 15–19, 2014.

61 The bee was one of Potnia Theron's emblems, and she is also called Potnia Melissa, Mistress of the Bees. Modern Wiccans re-member these chthonic beings in ritual and poetry. If fire figured the Anthropocene, and ice marked the Capitalocene, it pleases me to use red clay pottery for the Chthulucene, a time of fire, water, and earth, tuned to the touch of its critters, including its people. With her PhD writing on the riverine goddess Ratu Kidul and her dances now performed on Bali, Raissa DeSmet (Trumbull) introduced me to the web of far-traveling chthonic tentacular ones emerging from the Hindu serpentine Nagas and moving through the waters of Southeast Asia. DeSmet, *A Liquid World*.

62 Links between Potnia Theron and the Gorgon/Medusa continued in temple architecture and building adornment well after 600 BCE, giving evidence of the tenacious hold of the chthonic powers in practice, imagination, and ritual, for example, from the fifth through the third centuries BCE on the Italian peninsula. The dread-full Gorgon figure faces outward, defending against exterior dangers, and the no less awe-full Potnia Theron faces inward, nurturing the webs of living. See Busby, *The Temple Terracottas of Etruscan Orvieto*. The Christian Mary, Virgin Mother of God, who herself erupted in the Near East and Mediterranean worlds, took on attributes of these and other chthonic powers in her travels around the world. Unfortunately, Mary's iconography shows her ringed by stars and crushing the head of the snake (for example, in the Miraculous Medal dating from an early

nineteenth-century apparition of the Virgin), more than allying herself with earth powers. The "lady surrounded by stars" is a Christian scriptural apocalyptic figure for the end of time. That is a bad idea. Throughout my childhood, I wore a gold chain with the Miraculous Medal. Finally and luckily, it was her residual chthonic infections that took hold in me, turning me from both the secular and also the sacred, and toward humus and compost.

63 The Hebrew word *Deborah* means "bee," and she was the only female judge mentioned in the Bible. She was a warrior and counselor in premonarchic Israel. The *Song of Deborah* may date to the twelfth century BCE. Deborah was a military hero and ally of Jael, one of the 4Js in Joanna Russ's formative feminist science fiction novel *The Female Man*.

In April 2014, the Reverend Billy Talen and the Church of Stop Shopping exorcised the robobee from the Micro Robotics Laboratories at Harvard. The robobee is a high-tech drone bee that is intended to replace overworked and poisoned biological pollinating bees as they become more and more diseased and endangered. Honeybeealujah, old stories live! See Talen, "Beware of the Robobee," and Finnegan, "Protestors Sing Honeybeelujahs against Robobees." Or, as Brad Werner put it at the American Geophysical Union Meetings, Revolt! Do we hear the buzzing yet? It is time to sting. It is time for a chthonic swarm. It is time to take care of the bees.

64 "Erinyes 1."

65 Martha Kenney pointed out to me that the story of the Ood, in the long-running British science fiction TV series *Doctor Who*, shows how the squid-faced ones became deadly to humanity only after they were mutilated, cut off from their symchthonic hive mind, and enslaved. The humanoid empathic Ood have sinuous tentacles over the lower portion of their multifolded alien faces; and in their proper bodies they carry their hindbrains in their hands, communicating with each other telepathically through these vulnerable, living, exterior organs (organons). Humans (definitely not the Earthbound) cut off the hindbrains and replaced them with a technological communication-translator sphere, so that the isolated Ood could only communicate through their enslavers, who forced them into hostilities. I resist thinking the Ood techno-communicators are a future release of the iPhone, but it is tempting when I watch the faces of twenty-first-century humans on the streets, or even at the dinner table, apparently connected only to their devices. I am saved from this ungenerous fantasy by the SF fact that in the episode "Planet of the Ood," the tentacular ones were freed by the actions of Ood Sigma and restored to their nonsingular selves. *Doctor Who* is a much better story cycle for going-on-with than *Star Trek*.

For the importance of reworking fables in sciences and other knowledge practices, see Kenney, "Fables of Attention." Kenney explores different genres of fable, which situate what she calls unstable "wild facts" in relation to proposing and testing the strength of knowledge claims. She investigates strategies for navigating uncertain terrain, where the productive tensions between fact and fiction in actual practices are necessary.

66 "Medousa and Gorgones."

67 Suzy McKee Charnas's Holdfast Chronicles, beginning in 1974 with *Walk to the End of the World,* is great SF for thinking about feminists and their horses. The sex is exciting if very incorrect, and the politics are bracing.

68 Eva Hayward first drew my attention to the emergence of Pegasus from Medusa's body and of coral from drops of her blood. Hayward, "The Crochet Coral Reef Project," writes: "If coral teaches us about the reciprocal nature of life, then how do we stay obligated to environments—many of which we made unlivable—that now sicken us? . . . Perhaps Earth will follow Venus, becoming uninhabitable due to rampaging greenhouse effect. Or, maybe, we will rebuild reefs or construct alternate homes for the oceans' refugees. Whatever the conditions of our future, we remain obligate partners with oceans." See Wertheim and Wertheim, *Crochet Coral Reef.*

69 I am inspired by the 2014–15 Monterey Bay Aquarium exhibition *Tentacles: The Astounding Lives of Octopuses, Squids, and Cuttlefish.* See Detienne and Vernant, *Cunning Intelligence in Greek Culture and Society*, with thanks to Chris Connery for this reference in which cuttlefish, octopuses, and squid play a large role. Polymorphy, the capacity to make a net or mesh of bonds, and cunning intelligence are the traits the Greek writers foregrounded. "Cuttlefish and octopuses are pure *áporai* and the impenetrable pathless night they secrete is the most perfect image of their *metis*" (38). Chapter 5, "The Orphic Metis and the Cuttle-Fish of Thetis," is the most interesting for the Chthulucene's own themes of ongoing looping, becoming-with, and polymorphism. "The suppleness of molluscs, which appear as a mass of tentacles (*polúplokoi*), makes their bodies an interlaced network, a living knot of mobile animated bonds" (159). For Detienne and Vernant's Greeks, the polymorphic and supple cuttlefish are close to the primordial multisexual deities of the sea—ambiguous, mobile and ever changing, sinuous and undulating, presiding over coming-to-be, pulsating with waves of intense color, cryptic, secreting clouds of darkness, adept at getting out of difficulties, and having tentacles where proper men would have beards.

70 See Haraway and Kenney, "Anthropocene, Capitalocene, Chthulucene."

71 Le Guin, "'The Author of Acacia Seeds' and Other Extracts from the *Journal of the Association of Therolinguistics*," 175.

Chapter 3: Sympoiesis

This chapter is written in honor of Lynn Margulis (1938–2011) and Alison Jolly (1937–2014).

1 See *Never Alone* (Kisima Ingitchuna).

2 The large high-resolution giclée reproduction was printed on canvas with nonfading inks. Inspired by Margulis and Sagan, *Dazzle Gradually*, Dubiner's original gouache painting was 23 by 35 inches. Dubiner wrote, "The large red protozoan is *Urostyla grandis* based on a 1959 drawing by Stein in Leipzig. The purple protozoan

with 2 rows of cilia is *Didinium* . . . The blue feathered dragon-like creature at the center was inspired by a microscope image of a phospholipid cylinder by David Deamer . . . I wanted individual organisms to be accurate enough so a biologist would recognize them, but I allowed the overall painting to be a totally imaginary bioscape" (Dubiner, "New Painting"). For her blog writing on the painting, see Dubiner, "'Endosymbiosis.'" John Feldman is making a documentary film titled *Symbiotic Earth: How Lynn Margulis Rocked the Boat and Started a Scientific Revolution*. Born in 1938, Margulis died in 2011. On her UMass Amherst website, she described herself as a professor of microbial evolution and organelle heredity. See Mazur, "Intimacy of Strangers and Natural Selection"; Margulis, *Symbiotic Planet*; Margulis and Sagan, *Microcosmos*; Margulis and Sagan, *Acquiring Genomes*. See Hird, *The Origins of Sociable Life*, an important work rooted in ethnographic sociology in Margulis's laboratory.

3 In 1991, "Margulis proposed any physical association between individuals of different species for significant portions of their lifetime constitutes a 'symbiosis' and that all participants are bionts, such that the resulting association is a holobiont" (Walters, "Holobionts and the Hologenome Theory"). See Margulis, "Symbiogenesis and Symbionticism." In 1992 the term *holobiont* was used by Mindell, "Phylogenetic Consequences of Symbioses," to describe a host and its primary symbiont. In the same issue of this journal, see Margulis, "Biodiversity." Subsequently, Rohwer et al., "Diversity and Distribution of Coral-Associated Bacteria," used *holobiont* to mean the host plus all of its symbiotic microorganisms, including viruses. For an excellent recent summary of principles for holobionts and hologenomes, which nonetheless cannot evade the language of "host plus the rest," see Bordenstein and Theis, "Host Biology in Light of the Microbiome." "Safe and sound" as a meaning for *holo-* is from the *Online Etymology Dictionary*, accessed March 17, 2016.

4 Margulis, then publishing as Lynn Sagan, published her radical theory of the origin of the nucleated cell in 1967. Like many revolutionary contributions in science, such as Raymond Lindeman's paradigm-resetting "Trophic-Dynamic Aspect of Ecology," Margulis's 1967 paper was rejected many times before being accepted for publication. See Sagan, "On the Origin of Mitosing Cells"; Margulis, "Archaeal-Eubacterial Mergers in the Origin of Eukarya." For an explication of Margulis's autopoiesis and strong argument for continued use of the concept for Margulis's essential work on second-order Gaian systems theory, see Clarke, "Autopoiesis and the Planet."

5 Lovelock, "Gaia as Seen through the Atmosphere"; Lovelock and Margulis, "Atmospheric Homeostasis by and for the Biosphere."

6 Autopoietic systems theory and the figure of Gaia proved crucial to formulating the concept of the Anthropocene. Not a nurturing mother, Gaia can flip out, in system collapse after system collapse; there are limits to the power of systemic processes of homeostasis and reformulating order out of chaos at ever more complex levels. Complexity can unravel; earth can die. It matters to become response-able.

7 Dempster, "A Self-Organizing Systems Perspective on Planning for Sustainability." In 1998 Dempster thought that biology supported the conceptualization of

organisms as units, and only ecosystems and cultures are sympoietic. I argue, on biological grounds, that we can no longer think like that.

8 Margulis and Sagan, "The Beast with Five Genomes."

9 Poulsen et al., "Complementary Symbiont Contributions to Plant Decomposition in a Fungus Farming Termite." On these termite-bacterial-fungal symbioses, see the superb science writer Yong, "The Guts That Scrape the Skies."

10 For a closely argued analysis of the dead ends of competition/cooperation binaries and the relentless assumption that explanation in the last instance in biology must be competitive and individualistic, as well as for a fleshed-out description of more adequate explanatory practices, which are more and more in play among venturesome evolutionary, ecological, and behavioral biologists, see van Dooren and Despret, "Evolution."

11 Gilbert and Epel (*Ecological Developmental Biology*) document the evidence for what the authors call an "extended evolutionary synthesis," encompassing the modern synthesis, eco-devo, and eco-evo-devo.

12 Mereschkowsky, "Theorie der zwei Plasmaarten als Grundlage der Symbiogenesis." See also Anonymous, "History."

13 Gilbert, "The Adequacy of Model Systems for Evo-Devo," 57. See Black, *Models and Metaphors*; Frigg and Hartman, "Models in Science"; Haraway, *Crystals, Fabrics, and Fields*.

14 "King Lab: Choanoflagellates and the Origin of Animals."

15 Alegado and King, "Bacterial Influences on Animal Origins."

16 Choanoflagellates and their bacterial associates make an attractive model partly because sponges, long held to be the "most primitive" critters most closely related to animals, have choanoflagellate-like cells in their bodies that do things like capture their prey (bacteria and detritus). However, recent work argues that ctenophores (comb jellies) are genetically more closely related to animals than sponges are. Halanych, "The Ctenophore Lineage Is Older Than Sponges?" See Ed Yong's beautifully written science news story "Consider the Sponge." I do not know of any work exploring ctenophore-bacteria interactions, although, managing infections and responding to biofilm formations, ctenophores are tuned to bacteria and archaea, as are we all. In any case, phylogenetic relationships are not the only criteria of a good model. Up to 60 percent of the biomass of sponges is microbes. See Hill, Lopez, and Harriott, "Sponge-Specific Bacterial Symbionts in the Caribbean Sponge." What a gold mine for the study of holobionts! No wonder Nicole King started looking into all those attachment sites and signaling activities that might tie choanoflagellate-like cells in sponges to her free-living choanoflagellates, their eating, their infections, and their rosette clumping habits. If anything is, eating—not fundamentalist neo-Darwinian selfishness—is "evolutionary explanation in the last instance"; and eating is definitely both infectious and social! Biologically, eating trumps sex for innovative power; and eating is what made sex possible in the first place.

17 McGowan, "Where Animals Come From"; Yong, "Bacteria Transform the Closest Living Relatives of Animals from Single Cells into Colonies."

18 McFall-Ngai, "Divining the Essence of Symbiosis," 2. See McFall-Ngai's website from the University of Wisconsin. She has since moved to the Pacific Biosciences Research Center at the University of Hawaii. Other emerging model systems for symbiosis tuned to EcoEvoDevo include mouse gut development with bacterial symbionts (Jeffrey Gordon's lab at Washington University in St. Louis) and mouse brain development as well as immune system development tuned to signals from specific gut bacteria (Sarkis Mazmanian's lab at CalTech). See also EcoEvoDevo research with spadefoot toads (David Pfennig's lab at UNC Chapel Hill). Working on pea aphid symbiosis with *Buchnera*, Nancy Moran's lab at the University of Texas has done wonderful work on the coevolution of aphids and symbionts, but they have not emphasized development. Thanks to Scott Gilbert, personal communication, June 10, 2015.

The inaugural meeting of the Pan-American Society for Evolutionary Developmental Biology was held August 5–9, 2015, on the campus of University of California, Berkeley. Out of three hundred EvoDevo scientists who indicated an interest in attending, the ten organizers invited twenty-five attendees from a broad range of scientific backgrounds and approaches and set up an online portal for other EvoDevo participants. The European Society for Evolutionary Developmental Biology was founded in Prague in 2006. The international research community in EcoDevo and EvoDevo, as well as EcoEvoDevo, is both sizable and growing. Rudolf Raff edits the journal *Evolution and Development*, founded in 2011. See Abouheif et al., "Eco-Evo-Devo." A strong Russian tradition established by late nineteenth- and early twentieth-century workers contributed prominently to the conceptual formation of what became EvoDevo and EcoDevo. See Olsson, Levit, and Hossfeld, "Evolutionary Developmental Biology." See also Tauber, "Reframing Developmental Biology and Building Evolutionary Theory's New Synthesis."

19 McFall-Ngai, "Divining the Essence of Symbiosis."

20 Moran, "Research in the Moran Lab," website for "Nancy Moran's Lab."

21 See Gilbert, Sapp, and Tauber, "A Symbiotic View of Life"; McFall-Ngai et al., "Animals in a Bacterial World." This multiauthored paper is the result of a workshop supported by the National Evolutionary Synthesis Center (NESC). Michael Hadfield first introduced me to Margaret McFall-Ngai in Hawaii in 2010, and their collaborative thinking and publishing deeply informs mine. Asking Sapp (a historian of biology who writes on evolutionary biology beyond the neo-Darwinian framework) and Tauber (biochemist, philosopher, and historian of science who writes on immunology) to join him, Gilbert (developmental biologist and historian of biology) cowrote a separate paper because of unresolved disagreement at the NESC workshop over the extent of deviation from neo-Darwinian evolutionary theory ("competition in the last instance" and the power of cheaters in evolutionary game theory) that is in Gilbert's theory of the holobiont as a unit of selection. Gilbert thinks immune systems are very good at managing a dialogue with (not exterminating) cooperation-destroying cheaters in holobionts. Gilbert et al., "Symbiosis as a Source of Selectable Epigenetic Variation." Gilbert emphasizes that we have always been lichens. See also Guerrero, Margulis, and Berlanga, "Symbiogenesis."

22 McFall-Ngai et al., "Animals in a Bacterial World," 3229.

23 At Hadfield and McFall-Ngai's request, I provided minor help in revising the introduction and conclusion to McFall-Ngai et al., "Animals in a Bacterial World." Hadfield began teaching me about invertebrate marine developmental and ecological biology in the early 1970s when we were in a commune together in Honolulu. Gilbert and I have been close friends and colleagues exchanging papers and ideas since he was a PhD biology student at the Johns Hopkins University, and I was an assistant professor in the History of Science Department and Gilbert's advisor for his simultaneous MA in history of science.

24 Wertheim, *A Field Guide to Hyperbolic Space*. Hyperbolic space might be defined as "an excess of surface," the title of the first section of Wertheim's book. The very existence of such a thing seemed frankly pathological to Euclidean thinkers until the curves of worlding became undeniable to mathematicians. Such crenellated realities had long been in the repertoire of other critters, including a proud woman of the silk-weaving families in nineteenth-century Spitalfields, as she crocheted a nice ruffle onto the edges of a milk jug cover while she listened to Darwin talking about fancy racing pigeons with her husband and sons.

25 Hustak and Myers, "Involutionary Momentum," 79, 97, 106.

26 Hustak and Myers, "Involutionary Momentum," 77.

27 xkcd, Bee Orchid, https://xkcd.com/1259/, accessed August 10, 2015. Although gone from everywhere but one region, the not-quite-extinct solitary bee is from the genus *Eucera*. The orchid is *Ophrys apifera*. See "Bee Orchid."

28 On resurgence, see Tsing, "A Threat to Holocene Resurgence Is a Threat to Livability." Tsing argues that the Holocene was, and still is in some places, the long period when refugia, places of refuge, still existed, even abounded, to sustain reworlding in rich cultural and biological diversity after tremendous disturbance. Perhaps the outrage meriting a name like Anthropocene is about the destruction of places and times of refuge for people and other critters. My Chthulucene, even burdened with its problematic Greek-ish tendrils, entangles myriad temporalities and spatialities and diverse intra-active entities-in-assemblages—including the more-than-human, other-than-human, inhuman, and human-as-humus. The symchthonic ones are not extinct, but they are mortal. One way to live and die well as mortal critters in the Chthulucene is to join forces to reconstitute refuges, to make possible partial and robust biological-cultural-political-technological recuperation and recomposition, which must include mourning irreversible losses.

29 Meaning "real or genuine person," *Inupiaq* refers both to the person and to the language, which is closely related to Canadian Inuit and Greenlandic dialects and is distinct from the Yupik of western Alaska. Referring to the people collectively, *Inupiat* is the plural of *Inupiaq*. See University of Alaska Fairbanks, "Alaska Native Language Center."

30 "Crochet Coral Reef"; "Ako Project"; "Never Alone"; "Black Mesa Water Coalition"; "Black Mesa Trust" (founded by Hopi activists); "Black Mesa Weavers for Life and Land"; "Navajo Sheep Project"; "Diné be'iiná/The Navajo Lifeway"; "Black Mesa Indigenous Support."

31 Hustak and Myers, "Involutionary Momentum," 77.

32 In "Welcome to a New Planet," Michael Klare cites the figure from the World Wildlife Fund report in September 2015 of 850 million people depending on coral reef ecologies for food security. The same report notes that 85 percent of the reefs are listed officially as "threatened" in the so-called coral triangle, which encompasses the waters of Indonesia, Malaysia, the Philippines, Papua New Guinea, Solomon Islands, and Timor Leste, including Raja Ampat off the coast of West Papua, considered the global epicenter of marine biodiversity. Irreversible failure of the reefs, a real possibility by as early as 2050, could set off human misery and mass migrations of unprecedented proportions, not to mention nonhuman misery and double death. Climate justice and environmental justice are truly multispecies affairs. Raja Ampat is also the epicenter of ongoing innovative coalitional work for resurgence. See World Wildlife Fund, "Living Blue Planet."

33 The deepwater coral refugia hypothesis is difficult to test, but see Greenwood, "Hope from the Deep."

34 "The regrowing forest is an example of what I am calling *resurgence*. The cross-species relations that make forests possible are renewed in the regrowing forest. Resurgence is the work of many organisms, negotiating across differences, to forge assemblages of multispecies livability in the midst of disturbance. Humans cannot continue their livelihoods without it" (Tsing, "A Threat to Holocene Resurgence Is a Threat to Livability"). Not all reforestation is equal, not everything that grows on disturbed land constitutes resurgence. Reforestation in Madagascar with native species is very difficult because the soils of deforested areas are severely damaged. Reforestation with exotic species, some of which become invasive, is practiced with eucalyptus, pine, silver wattle (an acacia), silky oak, and paperbark. See "Deforestation in Madagascar." Plantation system "reforestation," for example, with oil palms, has not been common until recently in Madagascar.

35 For an example of Navajo, Hopi, and settler environmental alliance, see "Sierra Club Sponsors 'Water Is Life' Forum with Tribal Partners." The Sierra Club was a major ally with Black Mesa Navajo and Hopi activists who shut down the Mohave Generating Station and the Black Mesa mine in 2005. See Francis, *Voices from Dzil'íjiin (Black Mesa)*. The Sierra Club was founded in the late nineteenth century as a white settler colony institution joining the category of nature to conservation, eugenics, and native exclusion from land. The Sierra Club's current efforts to learn to be a decolonial ally with indigenous peoples is heartening.

36 Lustgarten, "End of the Miracle Machines." Lustgarten's twelve-part series from ProPublica, "Killing the Colorado," is indispensable reading for thinking about how to nurture the Chthulucene in the midst of the Anthropocene's practices of fossil making by ceaseless fossil burning.

37 The website of Peabody Energy insists on a very different picture, one filled with restored native plants, productive revitalized grasslands, prize-winning safety records, economic benefits for everybody, and happy people. In 2006, "Peabody's environmental and community practices on Black Mesa were recognized as a world model for sustainability at the Energy Globe Awards in Brussels, Belgium"

(Peabody Energy, "Powder River Basin and Southwest"); see also Peabody Energy, "Factsheet: Kayenta."

In the early 1990s Fred Palmer, Peabody's main lobbyist in 2015 for government affairs, founded the Greening Earth Society, which actively promoted the idea that climate change is beneficial to plants and public health. Peabody Energy led the charge against President Obama's efforts at the end of his second term to regulate coal emissions more forcefully through the EPA. In the 2000s, as its chief environmental officer, Peabody hired Craig Idso, cofounder and former president of the Center for the Study of Carbon Dioxide and Global Change, a think tank dedicated to attacking mainstream climate science. Greg Boyce, Peabody's CEO in 2015, regularly criticizes "flawed computer models" as the basis of "climate theory." See Goldenberg, "The Truth behind Peabody Energy's Campaign to Rebrand Coal as a Poverty Cure." Critical to the industry's attempt to rebrand coal-generated electricity as the solution for the world's poor, Peabody is a major force behind Advanced Energy for Life. "Advanced Energy for Life" produces a slick procoal website that argues that more not less investment in coal, coupled to ever more elaborate and expensive technologies, is critical to global well-being. Peabody Energy is the only non-China-based partner in the Shenhua Coal Group. See Peabody Energy, "Peabody in China." Nonetheless, Peabody is facing serious economic losses as the global coal industry becomes less and less sustainable in the face of competition from fracking-related natural gas abundance. Including the People's Climate Movement and the Indigenous Environmental Movement, among others, the growing global movement to leave fossil fuels in the ground could have profound effects. "Leave It in the Ground," http://leave-it-in-the-ground.org/, accessed March 17, 2016. Peabody Energy declared bankruptcy in 2016.

38 For pictures of the Navajo Generating Station and much more, see Friberg, "Picturing the Drought." For the Black Mesa mine site, see the photos by Minkler in "Paatuaqatsi/Water Is Life."

39 For Navajo-Hopi-Peabody issues on Black Mesa, see Nies, "The Black Mesa Syndrome: Indian Lands, Black Gold." This is my source for the $2.7 million payout to Boyden. See Nies, *Unreal City*; Ali, *Mining, the Environment, and Indigenous Development Conflicts*, 77–85. For Navajo voices, see Benally, *Bitter Water*.

40 See the Academy Award–winning documentary film (1986) by Floria and Mudd, *Broken Rainbow*, on the scandal of coal extraction and the relocation of Navajo from Black Mesa, beginning in 1864, to aid mining speculation.

41 I am indebted to many activist sources for my condensed synopsis of Black Mesa issues: Lacerenza, "An Historical Overview of the Navajo Relocation"; "Short History of Big Mountain–Black Mesa"; Begaye, "The Black Mesa Controversy"; Rowe, "Coal Mining on Navajo Nation in Arizona Takes Heavy Toll"; Black Mesa Water Coalition, "Our Work."

42 Wertheim, *A Field Guide to Hyperbolic Space*, 35.

43 Wertheim and Wertheim, *Crochet Coral Reef*, 17. This is a two-hundred-plus-page book of sumptuous photographs and astute essays, plus the names of everybody who crocheted for this experimental art science model ecosystem.

44 New Zealand TV series, 1995–2001. "Dreamworker" was an episode of series 1, in September 1995, when I imagine Christine and Margaret glued to the screen, concocting their own material dream passage. "Gabrielle is kidnapped to become the bride of Morpheus, the god of dream, so Xena has to go through her Dreamscape Passage to save her friend." See *Xena Warrior Princess*, "Dreamworker."

45 On the crochet coral reefs as experimental life-forms, in some ways analogous to ALife worlds, but with very different narrative, material, political, and human and nonhuman social ecologies, see Roosth, "Evolutionary Yarns in Seahorse Valley."

46 Wertheim and Wertheim, *Crochet Coral Reef*, 21.

47 See Hayward, "The Crochet Coral Reef Project."

48 Wertheim and Wertheim, *Crochet Coral Reef*, 23.

49 Wertheim and Wertheim, *Crochet Coral Reef*, 17.

50 Wertheim and Wertheim, *Crochet Coral Reef*, 202.

51 Margaret Wertheim, "The Beautiful Math of Coral."

52 Christine Wertheim, "CalArts Faculty Staff Directory."

53 Metcalf, "Intimacy without Proximity."

54 See the Australian Earth Laws Alliance website. This uncredited photo appears on http://www.earthlaws.org.au/wp-content/uploads/2014/09/turtle-and-reef.jpg, as well as many other places on the Internet, always uncredited. Accessed August 11, 2015. The geo-eco-techno materiality of visual cultures matters to holding open space for critters in place.

55 National Oceanic and Atmospheric Administration, "Green Turtles."

56 See "Ako Project: The Books," written by Alison Jolly, illustrated by Deborah Ross, Malagasy text by Hantanirina Rasamimanana, 2005–12. The Lemur Conservation Foundation publishes the books in the United States and Canada. The books are published by UNICEF in Madagascar (fifteen thousand of each book and six thousand of each poster). Outside Madagascar, unilingual books are available in English and Chinese, with more translations planned. Each book features a different lemur species in a different kind of habitat, including the aye-aye, ring-tailed lemurs, sifaka, indri, red ruffed lemurs, and mouse lemurs.

57 See Jolly, *Thank You, Madagascar*, for a funny, astute, quirky, informed, gorgeously written, often tragic account of major tangles in the history of Malagasy-Western conservation encounters and projects over the late twentieth and early twenty-first centuries, all of which Jolly participated in. I am grateful to Jolly's daughter Margaretta Jolly for copies of documents and correspondence on the Ako Project.

58 Patricia Wright, a friend and colleague of Alison's, must also be foregrounded for her extraordinary knowledge and work; Wright plays a large role in *Thank You, Madagascar*. Without her, Ranomafana National Park, with its projects for Malagasy and foreign scientists, wildlife, and local people, would not exist. See "Centre ValBio: Ranomafana National Park"; "Patricia Wright"; and Wright and Andriamihaja, "Making a Rain Forest National Park Work in Madagascar." None of that stops me—or Jolly or Wright—from recording that many local people around the park consider that their land, including graves of their ancestors, was illegitimately taken from them to make the park and its kind of boundaries in ongoing scientific

and state colonial practices. Similarly, none of that stops informed players in this region from judging that the trees and critters of this specific area would now be gone if the park had failed; there is no innocent or simple way to stay with all the faces of the trouble; that is precisely why we must do so. See Jolly, *Thank You, Madagascar*, 214–28.

Shifting cultivators like the Malagasy, who clear small hillside plots and also use irrigated paddies for rice, are conventionally accused of destroying land and its future productivity; the truth has often been the opposite. The issue is controversial, but see Survival International, "Shifting Cultivation," and Cairns, *Shifting Cultivation and Environmental Change*. Kull, *Isle of Fire*, is the harshest critic of the history of conservation through fire suppression in Madagascar. He argues for community-based fire management rather than ongoing—and ineffective—criminalization of burning. Regeneration from plots used by shifting cultivators and then left fallow has been critical to forest species diversity and abundance in most tropical areas for a long time—unless fallow times are too short and pressure for new croplands too heavy. Private property regimes and their state apparatuses have a hard time with shifting cultivators (and with pastoralists called nomads). To put it mildly, the state wants people to settle down with definite property boundaries.

In solidarity with other pastoral/mobile peoples pressed by centralizing, resource-extracting, national governments, on July 13, 2015, the Black Mesa Water Coalition (BMWC) posted on its Facebook page a *New York Times* article on the current Chinese government's efforts to settle, forcibly if necessary, the "nomads" of its tribal western regions. The relation of the stepped-up efforts of sedentarization are hardly independent of intensified coal and other energy and mineral extraction in western China, and similar forces on Black Mesa on Navajo and Hopi lands since the mid-nineteenth century. See Jacobs, "China Fences in Its Nomads." The poster for BMWC commented, "This story sounds very familiar doesn't it? It's like what BIA has been doing to Diné people and it continues to happen today on NPL and HPL." NPL, Navajo Partition Lands; HPL, Hopi Partition Lands. https://www.facebook.com/blackmesawc?fref=ts. Accessed August 9, 2015. See the last section of this chapter, "Navajo Weaving."

A recent study in Madagascar attempted to quantify whether and how much fallow times have decreased in shifting cultivation/tavy land use in one rain forest corridor region of eastern Madagascar. The study claims to have instigated methods to ensure taking the knowledge systems and statements of both agricultural experts and local farmers equally seriously. See the conclusion in Styger et al., "Influence of Slash-and-Burn Farming Practices on Fallow Succession and Land Degradation in the Rainforest Region of Madagascar," 257: "Over the last 30 years, fallow periods decreased from 8–15 years to 3–5 years. Hence, fallow vegetation is changing within 5–7 fallow/cropping cycles after deforestation from tree (*Trema orientalis*) to shrub (*Psiadia altissima, Rubus moluccanus, Lantana camara*) to herbaceous fallows (*Imperata cylindrical* and ferns) and grasslands (*Aristida* sp.), when land falls out of crop production. This sequence is 5–12 times faster than previously reported. The frequent use of fire is replacing native species with exotic, aggressive ones and favors grasses over woody species, creating treeless landscapes that are of minimal productive and ecological value." The study highlights that the local people, the Betsimisaraka, have rich knowledge of fallowing and regeneration but are pressed by multiple forces in a spiraling process of land degradation. Ecological, ethnic, social hierarchal, populational, regional, national, international, and economic pressures tangle to strangle biodiversity and diverse livelihoods for people and other critters.

Typically, shifting cultivators have not traditionally wanted large families and have used many means to limit births. Why demographic spirals and land pressure have developed as they have since the mid-twentieth century in the upland rice plots and forests of Madagascar is not simple; but private property, nation state, and colonial apparatuses bear much of the responsibility, but not all the responsibility; the heavy toll of human numbers on today's earth cannot be addressed by laying all the responsibility on someone else's plate (or womb). Population in Madagascar is hard to estimate in part because no census has been taken since 1993; the first census was taken in 1975. The "method" for the following statements is inference: "According to the 2010 revision of the World Population Prospects the total population was 20,714,000 in 2010, compared to only 4,084,000 in 1950 . . . UN projections give about 50 million by 2050. Birthrates have fallen in both urban and rural areas, more in urban. 70 percent of population is rural/subsistence farming" (United Nations, "World Population Prospects").

59 Deborah Ross is a book artist with work in major magazines and in zoos and botanical gardens. She has run watercolor workshops for Walt Disney Studios, DreamWorks, Pixar, and Cal Arts. Most significant for the Ako Project have been her rural art workshops for the Malagasy villagers of Kirindy and Tampolo. See Ross, "Deborah Ross Arts." With degrees in scientific illustration and ecology and environment, Janet Mary Robinson is the poster artist for the Ako Project. Thanks to Margaretta Jolly for information on the origin of the project; e-mail from Jolly to Haraway, June 28, 2015.

60 Jolly, "Alison Jolly and Hantanirina Rasamimanana," 45. For the story of her first meeting and then working with Rasamimanana, see Jolly, *Lords and Lemurs*. For a taste of Rasamimanana and Jolly thinking with each other as scientists, see Jolly et al., "Territory as Bet-Hedging."

61 Without giving up, Jolly laments that even Rasamimanana's promotion of teaching and research related to the Ako books has not yet been able to overcome the reticence of many teachers to use such unorthodox materials. Jolly, *Thank You, Madagascar*, 51. In "Conservation Education in Madagascar," Dolins et al. argue "that while nongovernmental organizational efforts are and will be very important, the Ministry of Education urgently needs to incorporate biodiversity education in the curriculum at all levels, from primary school to university" (abstract).

62 Fifth International Prosimian Congress, website; Durrell Wildlife Conservation, "World Primate Experts Focus on Madagascar." For a list of related publications, see ValBio, "ICTE-Centre ValBio Publications."

63 Jolly, *Thank You, Madagascar*, 362.

64 The quotation is the English subtitle on a screen shot in *Never Alone*, Announcement Trailer, showing Nuna, the arctic fox, and a spirit helper. For extracts from an interview with Amy Fredeen of the Cook Inlet Tribal Council and Sean Vesce of E-Line Media for National Public Radio, see Demby, "Updating Centuries-Old Folktales with Puzzles and Power-Ups." From the interview: "The last living person to tell the story was a master storyteller named Robert Cleveland. Amy and her team did an amazing job, and they located the oldest living offspring of Robert—a woman named Minnie Gray, who is in her 80s, I believe. They discovered that Minnie lived just a few blocks from the Cook Inlet Tribal Council headquarters. And we brought her in and we were able to start a series of conversations with her. We introduced her to the team and what we wanted to do. And we were delighted when she really was encouraging us not only to use her father's story for inspiration, but to adapt it and evolve it for the game context. One of the things she taught us was that storytelling is not a fixed act." The process of making the game is described in detail. One consequential decision: "We made the creative decision to keep the only [spoken] audio in the game in Inupiaq, and it's presented in 10 languages for subtitles. What we were looking to do was re-create the experience of being told a story by an elder in their own language. It's hard to describe that sense, but we wanted to try to re-create that for players so they got a sense of how powerful it would have been to hear one of these stories back then." In the interview, Amy Fredeen quotes Daniel Starkey, an American Indian game reviewer on *Eurogamer.net*, who writes, "*Never Alone* (*Kisima Ingitchuna* in Inupiaq) is different. Its very existence challenges me. Instead of eliciting self-pity, it stands in absolute defiance of everything that I've grown to be, not only telling me to be better, but showing me how" (Starkey, "*Never Alone* Review").

About forty members of the Inupiat community supported the project in various ways throughout its making, and many more at key points. Making sure the game was grounded in Inupiat environmental conditions, experience, and ideas was a central concern of the indigenous collaborators, including the kids who helped by playing early versions. The NPR interview described the kids who were very engaged in deciding which animal—fox? owl? wolf?—should be the human girl Nuna's companion character.

65 *Never Alone* tells a story about an endless storm that threatens the people. Contemporary arctic peoples have well-developed accounts of climate alterations and of the changes in their environments, but the relevant idiom is not the Anthropocene. For example, see the website for the ISUMA film, *Inuit Knowledge and Climate Change*, which states, "Nunavut-based director Zacharias Kunuk (*Atanarjuat The Fast Runner*) and researcher and filmmaker Dr. Ian Mauro (*Seeds of Change*) have teamed up with Inuit communities to document their knowledge and experience regarding climate change. This new documentary, the world's first Inuktitut language film on the topic, takes the viewer 'on the land' with elders and hunters to explore the social and ecological impacts of a warming Arctic. This unforgettable film helps us to appreciate Inuit culture and expertise regarding environmental change and indigenous ways of adapting to it."

See Callison, *How Climate Change Comes to Matter*, for fieldwork exploring the vernaculars in which one group of Alaskan Inuit address climate change.

66 Situated in complex histories and politics, there are many formats besides computer games to consider when thinking about indigenous digital cultures. There are also computer games designed with indigenous cultural material, but not like *Never Alone*. See Ginsberg, "Rethinking the Digital Age"; Ginsberg, Abu-Lughod, and Larkin, *Media Worlds*; Lewis, *Navajo Talking Picture*.

67 The notion of "Sila" is explained in a "Cultural Insight" that has to be earned by players of *Never Alone*. I always die before I get that far in the game, but it is possible to cheat on YouTube. See "*Never Alone* Cultural Insights—Sila Has a Soul," in which Fannie Kuutuuq and others discuss Sila. A pan-Inuit term, *Sila* means something like "the weather" to Anglophone southerners, but only if the weather means the sky and the air, breath-soul, the element that enfolds the world and invests beings with life, as well as the environment from the earth to the moon, with its dynamic changes and powers. See Merker, "Breath Soul and Wind Owner." The concept of climate change will not engulf Sila, nor vice versa; but these ideas/work objects have met each other. There will be consequences for what counts as agencies, temporalities, and response-abilities. It matters what thoughts think thoughts, what stories tell stories, what knowledges know knowledges.

68 I am in agreement with William Elliott on these cautions and with his engagements with native stories and thinkers, including new approaches to located animisms. He generously shared two manuscripts with me: Elliott, "Ravens' World: Environmental Elegy and Beyond in a Changing North," and Elliott, "*Never Alone:* Alaska Native Storytelling, Digital Media, and Premodern Posthumanisms." In an NPR interview, speaking of the collaboration with E-Line Media, Cook Inlet Tribal Council member Amy Fredeen noted, "I know this may come across a little strong, but when we talked about creating the first indigenous video game company, we wanted to set the bar high. And we wanted to own this space around telling traditional cultural stories through video games" (Demby, "Updating Centuries-Old Folktales"). Fredeen is clear that sharing indigenous stories outside the usual terms of colonizing appropriation depends on owning the stories and the storytelling apparatus.

69 Quotation from *Never Alone (Kisima Ingitchuna)* website.

70 Takahashi, "After *Never Alone*, E-Line Media and Alaska Native Group See Big Opportunity in 'World Games.'" Takahashi continues: "[The game] got more than 700 reviews in a wide array of publications (including *GamesBeat*), and it has been discussed around the world. It was on more than fifty 'best of 2014' awards. YouTube and Twitch player videos have drawn millions of views." Thanks to Marco Harding for the reference and for teaching me how to play the game.

71 Eduardo Viveiros de Castro, personal communication, October 2, 2014.

72 That is one of the reasons that "belief" has nothing to do with the practices of the sciences. Sciences are sensible practices, in all their material semiotic workings, including mathematics and physics. Isabelle Stengers has been relentlessly cogent on this point; her love for Galileo's inclined planes depends on understanding

that science is sensible. Asking if one "believes" in evolution or climate change is a Christian question, in both religious and secular formats, for which only a confessional reply is accepted. In such worlds, Science and Religion reign, and there it is impossible to play *Never Alone*. Harding, "Secular Trouble," is my guide to the history of the category of belief, especially in Protestant colonizing cultures. See Harvey, *The Handbook of Contemporary Animism*.

73 "Dzit Yíjiin bikáa'gi iiná náánásdláadóó ha'níigo biniiyé da'jitt'ó," translated by Mae Washington. See Black Mesa Weavers for Life and Land, "Black Mesa Weavers and Wool."

74 Continuous weaving is a material-semiotic practice in my idiom. The Black Mesa Water Coalition posts pictures of wonderful contemporary blankets for sale and of the weavers, including children learning the skill, as well as pictures of blankets in process, on its Facebook page.

Black Mesa Weavers for Life and Land commissioned three limited editions of Black Mesa Blankets. A sympoietic work, these blankets were designed by Diné shepherds and weavers, spun from Navajo-Churro fleece, produced in collaboration with the Black Mesa Weavers for Life and Land, the San Jose Museum of Quilts and Textiles, the Christensen Fund, and the Pendleton Woolen Mills. For a picture of the Black Mesa Blanket, see San Jose Museum of Quilts and Textiles, "Black Mesa Blanket."

A late nineteenth-century giant in wool blankets targeting the Indian trade, Pendleton Woolen Mills played a large role in the history of harsh conditions for Navajo weaving. The "major blanket manufacturers usurped the Native American market and also appropriated a large portion of the Anglo market" (M'Closkey, *Swept under the Rug*, 87). Today, Pendleton's American Indian College Fund blankets provide scholarships for indigenous students, and Navajo families often treasure their own Pendleton blankets as well as Navajo weaving.

75 "Thinker/maker" is a way of designating those engaged in the inextricable thinking/making practices called art that I learned from Loveless, "Acts of Pedagogy."

76 Willink and Zolbrod, *Weaving a World*, 8. This volume is based on extensive discussions in the 1990s of weavings and weaving with more than sixty elders from the eastern part of the Navajo Nation in and around Crownpoint, New Mexico. The beginnings of the Crownpoint Navajo Rug Auction in 1968 and the founding of the Crownpoint Rug Weavers Association, composed of Navajo weavers throughout the Southwest, mark critical junctions in strengthening weavers' well-being and control of their markets, designs, and stories. Buyers from all over the world purchase directly from the weavers, who run the auction. See "Crownpoint Navajo Rug Auction" and Iverson, *Diné*, 268. However, most weavers still get much too little for the work, much less everything else that goes into making a blanket. The Crownpoint Auction experienced lethal financial problems in 2014 and reorganized as the new Crownpoint Rug Auction.

By 1996 Willink and Zolbrod had worked together for more than twenty-five years. A member of the faculty at the University of New Mexico, Roseann Willink is a member of the Mexican Clan and was born for the Towering House Clan.

Arguing that Navajo poetics and stories are intimately enfolded into the conduct of daily life, tying together relations among persons of the community with the cosmos, Paul Zolbrod published *Diné bahane': The Navajo Creation Story*, the most complete version in English. See Denetdale, *Reclaiming Diné History*, 23–26.

For Navajo artisans' stories of economic and cultural survival through the art of weaving, see the film written and directed by Bennie Klain, *Weaving Worlds*.

77 M'Closkey, *Swept under the Rug*, 17–23, 205–52, argues persuasively for Navajo weaving as cosmological performance. She draws from her own work with weavers, as well as from previous scholarship, especially Witherspoon and Peterson, *Dynamic Symmetry and Holistic Asymmetry*, as well as Willink and Zolbrod, *Weaving a World*. Because of the long history of Navajo weavers selling their blankets under highly unequal terms to traders, as well as working with patterns, fibers, and dyes dictated by the art and tourist markets, most scholars and museologists have treated Navajo weaving as a commodity or an art product, but not as indigenous cosmological performance essential to maintaining *hózhó*. Among other things, that has meant no legal copyright protection for Navajo patterns and reproduction of cheap knock-offs in places like Oaxaca and Pakistan under all-too-imaginable conditions of labor. See M'Closkey and Halberstadt, "The Fleecing of Navajo Weavers."

Both the number and quality of contemporary Navajo weavings are extraordinary; and contemporary weavers have to compete in a market crowded both by copies produced abroad and by older exquisite and authentic Navajo blankets, which were sold from the late nineteenth century until the 1960s *by the pound* at reservation trading posts to get credit to buy necessities. These heritage blankets sometimes sell for hundreds of thousands of dollars today in an art market that returns none of that money to the families of the original weavers, while blankets of similar or better quality, using both old and new designs, sell at auction to individual buyers as well as to craft market buyers at better prices than in the past, but still at rates that cannot sustain most of the weavers and their families. See M'Closkey, *Swept under the Rug*, for detailed analysis of the exploitation of Navajo weavers and weavings. Much of M'Closkey's information comes from archives of the Hubbell Trading Post, which became a National Historic Site in 1967. See Hubbell Trading Post, "History and Culture."

Chapter 1, "Playing String Figures with Companion Species," argued that Navajo *na'atl'o'*, string figure games, are tied to the creation stories and performances of Spider Woman and the Holy Twins. *Na'atl'o'* is also called "continuous weaving."

78 See Begay, "Shi'Sha'Hane (My Story)," 13–27. For an exhibit of the innovative fourth-generation weaver D. Y. Begay's tapestries at UC Davis's Gorman Museum in 2013, see Dave Jones, "Navajo Tapestries Capture the Soul of Her Land." See also the "Weaving in Beauty" website and Monument Valley High School, "Ndahoo'aah Relearning/New Learning Navajo Crafts/Computer Design." Ndahoo'aah is a summer program in design, computer programming, mathematics, and traditional Navajo crafts held every summer at Monument Valley High School. The website states, "Ndahoo'aah teaches some of the Navajo crafts that are still practiced on

the Reservation . . . At the same time, Ndahoo'aah teaches LOGO graphics programming, focusing on mathematics (especially geometry). Graphics tools are then used to produce traditional designs and colorations." For stories by weavers and other thinkers/makers, click on the "Stories" button of the Ndahoo'aah website. See also Eglash, "Native American Cybernetics." Learning with young people from Black Mesa as well as the Diné College, and with weavers who helped the visitors understand their algorithms, Eglash and his coworkers link the robustness of such knowledge worlds to what he and his collaborators call "generative justice." The point of such an approach is *not* mixing Indigenous and Western knowledge practices and stirring, but rather exploring the fraught possibility of generative contact zones without denying long histories of violence. E-mail from Ron Eglash to Donna Haraway, March 2, 2016.

79 "Inherent in the beauty of rug weaving is the artistic achievement reflecting every weaver's frame of mind, creating designs of continual mathematical movement and regeneration of symbolism acknowledged for the Diné" (Clinton, "The Corn Pollen Path of Diné Rug Weaving"). In her book-in-progress, "Attaching, for Climate Change: A Sympoiesis of Media," Katie King proposes *khipu*, Inca knotted strings, as a model for, but also a performance of, complex systems, media, and geo-change as "transcontextual transdisciplinary tangles." King writes, "Fiber arts and ethnomathematics are necessarily transdisciplinary resources for the kind of scholarship being continually re-created [in khipu worlds] in a zone of unusual cultural continuity even after conquest. The Andes are a multitemporal geo-political zone for caring for object/ecology" (draft book proposal, 2015). See King, "In Knots."

Coral reefs, forests in Madagascar, the Inuit Arctic, and the Navajo-Hopi Black Mesa are also "multi-temporal geo-political zones for caring for object/ecology even after conquest" (King, draft proposal, 2015). In particular, continuous weaving, cosmological performance, world games, and "writing without words" in khipu, *Never Alone*, Navajo weaving, and the Crochet Coral Reef form complex string figures of thinking/making/acting. Again in King's borrowings, these are "reciprocities made visible." Boone and Mignolo, *Writing without Words*. For "reciprocities made visible," see Salomon, *The Cord Keepers*, 279. Thanks to Katie King for these references.

80 When I was researching Spanish rough sheep (Churro sheep) in the U.S. desert Southwest, I stumbled onto one of my favorite indigenous media projects—a Shoshoni Claymation video. The Gosiute of eastern Nevada and western Utah are Shoshone peoples. Like all peoples of the U.S. Southwest, the Gosiute are embroiled in the ecologies, economies, and politics of nuclear mining, war, waste processing, and storage. Their relatives have lived in these deserts for more than a thousand years; and, living and dead, they are indigenous to the ongoing Chthulucene, tangled in the grip of the colonial and imperial Anthropocene and Capitalocene. A sort of audio collage, the sound track for the Claymation video *Frog Races Coyote/Itsappeh wa'ai Wako* was constructed from the archives of several Shoshone-language storytellers by the Gosiute/Shoshoni Project of the University

of Utah. Frogs think-with frogs; frog wins the race against coyote around the lake. Collective action can defeat the wiliest opponent.

The Frog and Coyote story is taught today in the Utah Indian Curriculum Guide, "The Goshutes." Listening to and learning a Shoshoni language today in public schools and on the Internet is part of indigenous America *not* disappearing, but traveling in tongues to unexpected places to reopen questions for ongoingness, accountability, and lived storying.

On the importance of fostering actual indigenous language use, in all its "emergent vitalities," among the young who are no longer fluent native speakers, see Perley, "Zombie *Linguistics*."

81 See Denetdale, *Reclaiming Diné History*, 62–86; Johnson, *Navajo Stories of the Long Walk Period*; Morrison, *Paradise*. The twentieth-century Navajo-Hopi reservation land partition laws and forced removal of thousands of Navajo from Black Mesa/Big Mountain/Dzil ni Staa to make way for industrial coal mining are sometimes called the second great Hwéeldi. Beginning in 1977, Pauline Whitesinger, her clan allies, and other Diné elders started a resistance that has not ceased. "By 1980, Big Mountain Dineh resisters and their few but growing non-Native allies began network strategies that reached as far as Washington State, Southern California, and the east coast. Non-Native support collectives began bringing themselves and logistics out to the now restricted zones. Both the indigenous community and non-Natives shared the need to document the deliberate violations of human rights, to stop forcible occupation to extract fossil fuel, to halt the desecration of human religions, and to let the world know that the U.S. is committing genocide" (NaBahe [Bahe] Keediniihii [Katenay],"The Big Mountain Dineh Resistance"). Efforts to relocate sheep-herding Navajo and their animals intensified again in 2014, with strong BIA and tribal police efforts to break the bond between Diné and non-Native allies. See the Black Mesa Indigenous Support website. Black Mesa Indigenous Support runs Big Mountain spring training camps for activists. I owe "originally trauma" to Kami Chisholm.

82 For a rich argument about Native Americans making kin with each other and with plants and animals—processes disrupted by forced commodity relations and Christian kinship systems—see Kim TallBear, "Failed Settler Kinship, Truth and Reconciliation, and Science." TallBear drew from Dakota histories for this blog post. TallBear is a leader in thinking about the "making of love and relations beyond settler sexualities."

83 Several sources inform my sketch of the near extermination of Navajo-Churro sheep in the 1930s, but especially the thoroughly researched book by Weisiger, *Dreaming of Sheep in Navajo Country*. See also Weisiger, "Gendered Injustice"; the website of the Navajo Sheep Project; White, *The Roots of Dependency*; Johnson and Roessel, *Navajo Livestock Reduction*; and McPherson, "Navajo Livestock Reduction in Southeastern Utah, 1933–46." In *A Plague of Sheep*, Melville argues that Spanish sheep were devastatingly effective colonizers, creatures of empire, which forever altered the ecology and native society of highland central Mexico in favor of the conquerors. The same could be said of sheep in the U.S. Southwest. True enough,

but origins are not closed destinies; and sheep and indigenous and allied peoples of these lands have forged remarkably durable multispecies ways of living and dying with each other on the Colorado Plateau, in complex resistance to ongoing colonial practices.

84 Horoshko, "Rare Breed," and Navajo Sheep Project, "History."

85 See Black Mesa Weavers for Life and Land, "Diné Navajo Weavers and Wool," and Halberstadt, "Black Mesa Weavers for Life and Land."

86 Diné be'iína/The Navajo Lifeway, "Dibé be'iína/Sheep Is Life."

87 Strawn and Littrel, "Returning Navajo-Churro Sheep for Weaving."

88 For a story about Roy Kady, one of the Navajo Nation's best-known male weavers, who has dedicated his life to the well-being of Navajo-Churro sheep, see Kiefel, "Heifer Helps Navajos Bolster Sheep Herd." Blystone and Chanler, *A Gift from Talking God*, is a moving DVD produced in 2009, narrated by Jack Loeffler, featuring Roy Kady, Jay Begay, Lyle McNeal, and Gary Paul Nabhan. See also Kraker, "The Real Sheep," and Cannon, "Sacred Sheep Revive Navajo Tradition, for Now."

89. "Behavioural studies in our laboratory using choice mazes and operant discrimination tasks have revealed quite remarkable face-recognition abilities in sheep, similar to those found in humans . . . These experiments showed that sheep could discriminate between sheep and human faces, between different breeds of sheep and between sexes in the same breed" (Tate et al., "Behavioural and Neurophysiological Evidence for Face Identity and Face Emotion Processing in Animals," 2155).

90 Says cofounder Peter Hagerty, a sheep and horse farmer who bought wool in 1985 from the Soviet Union to somehow unknot the Cold War, "I used to describe Peace Fleece as an international yarn company doing business with historic enemies like Palestinians and Israelis and Russians and Americans. Today that description still holds true but recently I have grown to see it more as a place where very normal people come together on a regular basis to help each other get through the day" (Peace Fleece, "The Story"). See Peace Fleece, "Irene Bennalley." On Irene Bennalley, see Benanav, "The Sheep Are Like Our Parents."

91 Black Mesa Water Coalition, "About." On the founding and goals of BMWC, see Paget-Clarke, "An Interview with Wahleah Johns and Lilian Hill." Johns is Diné from Forest Lake, a community on Black Mesa (Johns, website). Working out of the San Francisco Bay Area beginning in 2013, Johns was the Solar Project Coordinator of the BMWC. Hill is from Kykotsmovi, Tobacco clan. Living in Kykotsmovi, she is a Certified Permaculture Designer and Natural Builder (Hill, "Hopi Tutskwa Permaculture"). BMWC activists were active at COP21 in Paris in 2015 in the Peoples Climate Justice Summit/Indigenous Rising. Executive Directior of BMWC Jihan Gearon testified on September 23, 2015, at the Peoples Tribunal. For an audio recording, go to Gearon, Peoples Tribunal.

92 Haraway and Tsing, "Tunneling in the Chthulucene." For thinking with another people of the Southwest, see Basso, *Wisdom Sits in Places*.

93 BMWC, "Our Work"; Communities United for a Just Transition, "Our Power Convening." See BMWC, "10th Anniversary Video," narrated by Executive Director Jihan Gearon. For a video of codirectors of BMWC in 2009, see Johns and Begay,

Speech at Power Shift '09. See also BMWC, "Green Economy Project." For a powerful reflection in 2015 on how to continue working together across time and difference, see Gearon, "Strategies for Healing Our Movements." Tódích'ii'nii (Bitter Water) clan and African American, Gearon earned a BS from Stanford in earth systems, focusing on energy science and technology. See Afro-Native Narratives, "Jihan Gearon, Indigenous People's Rights Advocate." In her generation, the urgent conversation between needed concepts and practices is more possible—politically, culturally, spiritually, scientifically. Gearon made *Grist* magazine's "Grist 50: The 50 People You'll Be Talking About in 2016," https://grist.org/grist-50/profile/jihan-gearon/, accessed March 17, 2016.

94 Giovanna Di Chiro, professor of environmental studies at Swarthmore College, has for many years been my guide to bringing together the feminist movement, multiethnic and antiracist environmental justice, critters of the seas and inland waters, urban antitoxic coalitions, and action research. We are also linked by research on symbiosis and evolutionary relationships. String figures linking women—and men—through friendship, mentoring, and research projects in all these worlds have shaped the pattern. See DiChiro, "Cosmopolitics of a Seaweed Sisterhood," "A New Spelling of Sustainability," "Acting Globally," and "Beyond Ecoliberal 'Common Futures.'" Part of a sympoietic seaweed sisterhood that shaped her life, in 1979 Giovanna was an undergraduate at UC Santa Cruz working with phycologist Linda Goff. Giovanna was a collector for the team at the Oahu, Hawaii, coral reef research site off Coconut Island that characterized *Prochloron didemni*, the cyanobacterial symbiont living in the gut of its ascidian partner. The molecular and ultrastructural analysis provided evidence of an evolutionary relationship between *Prochloron* and eukaryotic chloroplasts of green plants. See Giddings, Withers, and Staehlin, "Supramolecular Structure of Stacked and Unstacked Regions of the Photosynthetic Membranes of *Prochloron*." A few years before, teaching biology and the history of science at the University of Hawaii on Oahu, I had written chapters of my PhD thesis on organismic metaphors that shape embryos in developmental biology while living on Coconut Island in a commune that included Michael Hadfield, an important marine developmental and ecological biologist in the current flowering of EcoEvoDevo, discussed in the first part of this chapter. See Haraway, *Crystals, Fabrics, and Fields*. I was Giovanna's PhD adviser in History of Consciousness at UCSC; her degree was awarded in 1995. Cat's cradle, indeed.

95 Stengers, *Cosmopolitics I* and *Cosmopolitics II*; Stengers, "The Cosmopolitical Proposal."

Chapter 4: Making Kin

1 Intra-action is a concept given us by Karen Barad, *Meeting the Universe Halfway*. I keep using interaction too in order to remain legible to audiences who do not yet understand the radical change Barad's analysis demands, but probably out of my linguistically promiscuous habits, as well.

2 Tsing, "Feral Biologies."

3 Moore, *Capitalism in the Web of Life*.

4 I owe Scott Gilbert for pointing out, during the *Ethnos* conversation and other interactions at Aarhus University in October 2014, that the Anthropocene (and Plantationocene) should be considered a boundary event like the K-Pg boundary, not an epoch.

5 In a recorded conversation for *Ethnos* at the University of Aarhus in October 2014, the participants collectively generated the name Plantationocene for the devastating transformation of diverse kinds of human-tended farms, pastures, and forests into extractive and enclosed plantations, relying on slave labor and other forms of exploited, alienated, and usually spatially transported labor. See Tsing et al., "Anthropologists Are Talking about the Anthropocene." See AURA, website. Scholars have long understood that the slave plantation system was the model and motor for the carbon-greedy machine-based factory system that is often cited as an inflection point for the Anthropocene. Nurtured in even the harshest circumstances, slave gardens not only provided crucial human food, but also refuges for biodiverse plants, animals, fungi, and soils. Slave gardens are an underexplored world, especially compared to imperial botanical gardens, for the travels and propagations of myriad critters. Moving material semiotic generativity around the world for capital accumulation and profit—the rapid displacement and reformulation of germ plasm, genomes, cuttings, and all other names and forms of part organisms and of deracinated plants, animals, and people—is one defining operation of the Plantationocene, Capitalocene, and Anthropocene taken together. The Plantationocene continues with ever greater ferocity in globalized factory meat production, monocrop agribusiness, and immense substitutions of crops like oil palm for multispecies forests and their products that sustain human and nonhuman critters alike. The participants in the *Ethnos* conversation included Noboru Ishikawa, Anthropology, Center for South East Asian Studies, Kyoto University; Anna Tsing, Anthropology, University of California at Santa Cruz; Donna Haraway, History of Consciousness, University of California at Santa Cruz; Scott F. Gilbert, Biology, Swarthmore; Nils Bubandt, Department of Culture and Society, Aarhus University; and Kenneth Olwig, Landscape Architecture, Swedish University of Agricultural Sciences. Gilbert adopted the term *Plantationocene* for key arguments in the coda to the second edition of his widely used textbook, Gilbert and Epel, *Ecological Developmental Biology*.

6 According to personal e-mail communications from both Jason Moore and Alf Hornborg in late 2014, Malm proposed the term *Capitalocene* in a seminar in Lund, Sweden, in 2009, when he was still a graduate student. I first used the term independently in public lectures starting in 2012. Moore edited a book titled *Anthropocene or Capitalocene?* (PM Press, 2016), which has essays by Moore, myself, and others. Our collaborative webs thicken.

7 The suffix *-cene* proliferates! I risk this overabundance because I am in the thrall of the root meanings of *-cene/kainos*, namely, the temporality of the thick, fibrous, and lumpy "now," which is ancient and not.

8 Os Mil Nomes de Gaia/The Thousand Names of Gaia was the generative international conference organized by Eduardo Viveiros de Castro, Déborah Danowski,

and their collaborators in September 2014 in Rio de Janeiro. See The Thousand Names of Gaia, "Videos," and Haraway, "Entrevista."

9 Clifford, *Returns*, 8, 64, 201, 212.

10 Van Dooren, *Flight Ways*; Despret, "Ceux qui insistent." For a wealth of Despret's important essays translated into English, see Buchanan, Bussolini, and Chrulew, "Philosophical Ethology II: Vinciane Despret."

11 Card, *Speaker for the Dead*.

12 Making kin must be done with respect for historically situated, diverse kinships that should not be either generalized or appropriated in the interest of a too-quick common humanity, multispecies collective, or similar category. Kinships exclude as well as include, and they should do that. Alliances must be attentive to that matter. The sorry spectacle of many white liberals in the U.S., in the wake of African American and allied organizing against police murders of Black people and other outrages, resisting #BlackLivesMatter by insisting that #AllLivesMatter is instructive. Making alliances requires recognizing specificities, priorities, and urgencies. Alicia Garza, who created #Black Lives Matter with Patrisse Cullors and Opal Tometi as a call to action, wrote a powerful history of the hashtag and subsequent movement and efforts to delegitimize it with a false universal kinship rather than accountable alliances in the celebration and humanization of Black lives. See Garza, "A Herstory of the #BlackLivesMatter Movement." As she argues, when Black people get free, everybody gets free, but that requires central focus on Black lives because their ongoing degradation is fundamental to U.S. society.

Similar issues attach to the core relation of BlackLivesMatter and environmental justice, a topic explored in a series of astute posts on *Grist* by the justice editor, Brentin Mock, https://grist.org/author/brentin-mock/, accessed March 17, 2016. On many levels, making kin is not separate from these topics.

Similar questions also apply to the too-easy term *reconciliation*, which is used as a nation- and kin-making term. Intending to make kin while not seeing both past and ongoing colonial and other policies for extermination and/or assimilation augurs for very dysfunctional "families," to say the least. Kim TallBear and Erica Lee are doing fundamental work in this area, within a generative explosion of feminist indigenous public thinking, acting, and scholarship. Here is where common worlds—cosmopolitics—to go on with might have a chance to be constructed. See TallBear, "Failed Settler Kinship, Truth and Reconciliation, and Science," and Lee, "Reconciling in the Apocalypse." I am informed by TallBear's critique of settler sexualities and arguments for both inherited and yet-to-be-invented practices of kinship, infused with situated, historically attentive, ongoing, and experimental indigenous worldings. Listen to TallBear, "Making Love and Relations Beyond Settler Sexualities." IdleNoMore, like BlackLivesMatter, taps the roots of any possible multicritter multipeople mycorrhizal "holoent" on a damaged planet.

13 Strathern, *The Gender of the Gift*.

14 Latour, "Facing Gaïa."

15 Robinson, *2312*. This extraordinary SF narrative won the Nebula Award for best novel.

16 Strathern, "Shifting Relations." Making kin is a surging popular practice, and new names are also proliferating. See Skurnick, *That Should Be a Word*, for *kinnovator*, a person who makes family in nonconventional ways, to which I add *kinnovation*. Skurnick also proposes *clanarchist*. These are not just words; they are clues and prods to earthquakes in kin making that is not limited to Western family apparatuses, heteronormative or not. I think babies should be rare, nurtured, and precious; and kin should be abundant, unexpected, enduring, and precious.

17 *Gens* is another word, patriarchal by origin, with which feminists are playing. Origins and ends do not determine each other. Kin and gens are littermates in the history of Indo-European languages. In hopeful intra-actional communist moments, see Bear et al., "Gens." The writing is perhaps too dry (although the summary bullet points help), and there are no juicy examples to make this Manifesto seduce the spoiled reader; but the references give huge resources to do all that, most the fruit of long-term, intimately engaged, deeply theorized ethnographies. See especially Tsing, *The Mushroom at the End of the World*. The precision of the methodological approach in "Gens" is in its address to those would-be Marxists or other theorists who resist feminism, and who therefore don't engage the heterogeneity of real life worlds but stay with categories like Markets, the Economy, and Financialization (or, I would add, Reproduction, Production, and Population—in short, the supposedly adequate categories of standard liberal and nonfeminist socialist political economy). Go, Honolulu's Revolution Books and all your kin!

18 My experience is that those I hold dear as "our people," on the Left or whatever name we can still use without apoplexy, hear neo-imperialism, neoliberalism, misogyny, and racism (who can blame them?) in the "Not Babies" part of "Make Kin Not Babies." We imagine that the "Make Kin" part is easier and ethically and politically on firmer ground. Not true! "Make Kin" and "Not Babies" are both hard; they both demand our best emotional, intellectual, artistic, and political creativity, individually and collectively, across ideological and regional differences, among other differences. My sense is that "our people" can be partially compared to some Christian climate-change deniers: beliefs and commitments are too deep to allow rethinking and refeeling. For our people to revisit what has been owned by the Right and by development professionals as the "population explosion" can feel like going over to the dark side.

But denial will not serve us. I know "population" is a state-making category, the sort of "abstraction" and "discourse" that remake reality for everybody, but not for everybody's benefit. I also think that evidence of many kinds, epistemologically and affectively comparable to the varied evidence for rapid anthropogenic climate change, shows that 7–11 billion human beings make demands that cannot be borne without immense damage to human and nonhuman beings across the earth. This is not a simple causal affair; ecojustice has no allowable one-variable approach to the cascading exterminations, immiserations, and extinctions on today's earth. But blaming Capitalism, Imperialism, Neoliberalism, Modernization, or some other "not us" for ongoing destruction webbed with human numbers will not work either. These issues demand difficult, unrelenting work; but they also demand joy, play, and response-ability to engage with unexpected others. All parts of these issues are much too important for Terra to hand them over to the Right or to development professionals or to anybody else in the business-as-usual camps. Here's to Oddkin—non-natalist and off-category!

We must find ways to celebrate low birth rates and personal, intimate decisions to make flourishing and generous lives (including innovating enduring kin—kinnovating) without making more babies—urgently and especially, but not only, in wealthy high-consumption and misery-exporting regions, nations, communities, families, and social classes. We need to encourage population and other policies that engage scary demographic issues by proliferating other-than-natal kin—including nonracist immigration, environmental and social support policies for new comers and "native-born" alike (education, housing, health, gender and sexual creativity, agriculture, pedagogies for nurturing other-than-human critters, technologies and social innovations to keep older people healthy and productive, etc., etc.).

The inalienable personal "right" (what a word for such a mindful bodily matter!) to birth or not to birth a new baby is not in question for me; coercion is wrong at every imaginable level in this matter, and it tends to backfire in any case, even if one can stomach coercive law or custom (I cannot). On the other hand, what if the new normal were to become a cultural expectation that every new child have at least three lifetime committed parents (who are not necessarily each other's lovers and who would birth no more new babies after that, although they might live in multichild, multigenerational households)? What if serious adoption practices for and by the elderly became common? What if nations that are worried about low birth rates (Denmark, Germany, Japan, Russia, Singapore, Taiwan, white America, more) acknowledged that fear of immigrants is a big problem and that racial purity projects and fantasies drive resurgent pronatalism? What if people everywhere looked for non-natalist kinnovations to individuals and collectives in queer, decolonial, and indigenous worlds, instead of to European, Euro-American, Chinese, or Indian rich and wealth-extracting sectors?

As a reminder that racial purity fantasies and refusal to accept immigrants as full citizens actually drive policy now in the "progressive" "developed" world, see Hakim, "Sex Education in Europe Turns to Urging More Births." In response to this piece, science writer Rusten Hogness posted on his Facebook page on April 9, 2015, "What is wrong with our imaginations and with our ability to look out for one another (human and non-human alike) if we can't find ways to address issues raised by changing age distributions without making ever more human babies? We need to find ways to celebrate young folks who decide not to have kids, not add nationalism to the already potent mix of pro-natalist pressures on them"(https://www.facebook.com/rusten.hogness?fref=ts, accessed March 17, 2016).

Pronatalism in all its powerful guises ought to be in question almost everywhere. I keep "almost" as a reminder about the consequences of genocide and displacement for peoples—an ongoing scandal. The "almost" is also a prod to remember contemporary sterilization abuse, shockingly inappropriate and unusable means of contraception, reduction of women and men to ciphers in old and new population control policies, and other misogynist, patriarchal, and ethnicist/racist practices built into business as usual around the world. For example, see Wilson, "The 'New' Global Population Control Policies."

For an indispensable critical analysis of geopolitics and global intellectual history of population control discourse, see Bashford, *Global Population*. For a focused critical study of the oppressive social life of numbers in Guatemala, see Nelson, *Who Counts?* Such studies show why reemphasizing the burden of growing human numbers, especially as a global demographic abstraction, can be so dangerous. Thanks to Michelle Murphy for the references and the resistance to my arguments, no matter how well intended. I still think they are necessary. See Murphy, "Thinking against Population and with Distributed Futures."

We need each others' risk-taking support, in conflict and collaboration, big time on all these matters.

Chapter 5: Awash in Urine

1 Cyborgs might be considered to be "holoents" in the sense developed in chapter 3.

2 "Diethylstilbesterol," *Wikipedia*. This *Wikipedia* article is a first stop with a useful bibliography, but is worse than worthless for tracking feminist and women's health activist connections to DES. See Bell, *DES Daughters, Embodied Knowledge, and the Transformation of Women's Health Politics in the Late Twentieth Century*. Barbara Seaman, who died in 2008, is one of my heroes in this story; her work was crucial to persuading the U.S. federal government to convene a task force on DES. In 1975, she cofounded the National Women's Health Network. For a bit of the history and tributes to Seaman, see Editors, "A Tribute to Barbara Seaman," and Seaman, "Health Activism, American Feminist." Jewish women have been central in the history of feminist women's health activism. Another hero who recently died, Pat Cody, also worked effectively to change a personal tragedy caused by DES into a global feminist health movement. See Rosen, "Pat Cody." For innovative and standard-setting feminist science studies, see Oodshourn, *Beyond the Natural Body*.

First-generation users of DES have increased incidence of and mortality rates from breast cancer. Second-generation offspring, "DES daughters," develop dangerous vaginal and breast cancers, as well as other problems like infertility and "abnormal pregnancy outcomes." That means damaged or dead children. DES is the only transplacental carcinogen known in our species. What a distinction! It is also a teratogen; see "abnormal pregnancy outcome." DES sons have nasty effects too.

3 From Forney, "Diethylstilbesterol for Veterinary Use": "The most-serious side-effect of estrogen therapy [in dogs] is bone-marrow suppression and toxicity that may progress to a fatal aplastic-anemia . . . Side effects are more common in older animals."

4 See Brooks, "Diethylstilbesterol": "As the uses of DES dwindled to a few veterinary uses, its manufacturer found it unprofitable to continue production and DES

went off the market in the late 1990s. Fortunately for the numerous incontinent female dogs hoping to lead indoor lives, the human carcinogenicity issues have not crossed over into the canine health arena. The low doses and infrequent dosing schedule has [*sic*] positioned DES as a medication of unparalleled safety and convenience in the treatment of canine incontinence. Compounding pharmacies now make this medication readily available to patients who need it on a prescription basis." An important player in companion animal consumer marketing, Foster and Smith, "Diethylstilbesterol," tells us that prolonged use in pets can cause ovarian cancer.

5 Raun and Preston, "History of Diethylstilbestrol Use in Cattle."

6 Popular sites on estrogens in and for women include "Estrogen," HealthyWomen.org, and "Estrogen," Midlife-Passages.com. Premarin® is a mix of conjugated estrogens branded with a registered trademark. These equine estrogens are chemically different from those made in the human body; i.e., they are not bioidentical, but they are bioactive across species. Ethinyl estradiol is the artificial estrogen commonly found in contemporary birth control pills and, like DES, is manufactured in the laboratory. Nomenclatures for hormones around terms like *natural*, *synthetic*, *biomimetic*, *bioidentical*, and *artificial* can be confusing, biologically and politically. For example, soy-derived estrogens often are called natural, but not because they are chemically identical to naturally occurring human estrogens. "Natural" is about branded biovalue in many contested senses. Cenestin is a conjugated estrogen marketed by Duramed Pharmaceuticals. Made from plant sources, Cenestin is called "natural," but it is a conjugated mix that is chemically a copy of Premarin, so neither is bioidentical to human estrogens. Because it is derived from horses, Premarin gets called synthetic, although it is heavily processed but not synthesized in the laboratory. See Petras, "Making Sense of HRT." Substituting a plant source for horse urine, Cenestin is a the kind of work-around that becomes cost effective when the biopolitical/bioethical cost of a technoscientific product gets too high in a particular naturalcultural ecology. Duramed calls its conjugated estrogens "synthetic" and an "advanced form of Premarin," emphasizing that "Cenestin does not contain any hormone synthesized by the horse." So Cenestin is at once natural, synthetic, mimetic, and advanced, while Premarin's relation to horses forbids the label *natural*. Keeping corporate lawyers busy, the patenting fur has flown thickly over the naming rights for these drugs.

7 Hormone Health Network, "Emminen." The first hormonal treatment for symptoms of human menopause in the United States, in 1929, was with a derivative of calf amniotic fluid. Emmenin, from Canadian pregnant women's urine, was first marketed in the United States in 1933. In 1939 DES was marketed as a more potent estrogen than Emmenin. An obvious cyborg cocktail, the mix of historically situated organic and technological species, both human and not-human, could hardly be missed.

8 Women's Health Initiative, "Risks and Benefits of Estrogen Plus Progestin in Healthy Postmenopausal Women"; Vance, "Premarin." See also Wilks, "The Comparative Potencies of Birth Control and Menopausal Hormone Drug Use."

9 North American Equine Ranching Information Council, "About the Equine Ranching Industry."

10 North American Equine Ranching Information Council, "Equine Veterinarians' Consensus Report on the Care of Horses on PMU Ranches."

11 HorseAid, "What Are the Living Conditions of the Mares?"

12 "Premarin Controversy," *Wikipedia*; Horse Fund, "Fact Sheet"; HorseAid website; Hall, "The Bright Side of PMU."

13 HorseAid Report, "PREgnant MARes' urINe, Curse or Cure?"

14 Horse Fund, "Fact Sheet." The horsefund.org fact sheet paints a very different picture than the NAERIC website. HorseAid has been more careful in its descriptions and dates than the International Fund for Horses, which continued to publish as if they were current, as late as 2011, conditions originally described in the 1980s and 1990s, but also printed an interesting and damning timeline on Premarin and its checkered corporate and medical history on http://www.horsefund.org/premarin-timeline.php, online in November 2011, but unavailable in August 2015. For data on Premarin sales and PMU farm numbers from 1965 to 2010, see Allin, "Wyeth Wins, Horses Lose." In 2010 Allin was a research analyst with the International Fund for Horses. Pfizer will increase PMU collection in 2016–17.

15 Working in Manitoba for about nine years, Equine Angels Rescue Sanctuary narrated a story from October 2011 about helping PMU farmers get out of the breeding business. The plan attended to the needs of farmers, foals, mares, and stallions. The note on the website on February 3, 2015, updated progress on the project. See Weller, *Equine Angels*. There were several other PMU horse rescue sites active in 2011. The last time I checked the unfinished story was August 13, 2015; see Equine Advocates, "PMU Industry." Reflecting the sharp decline in farm numbers since 2002, rescue operations now attend more to out-of-work PMU horses than to overproduced foals, but pregnant mare's urine is still produced and still used in Premarin-containing products.

Chapter 6: Sowing Worlds

1 In everything I write about companion species, I am instructed by Anna Tsing's "Unruly Edges: Mushrooms as Companion Species." Without the deceptive comforts of human exceptionalism, Tsing succeeds both in telling world history from the point of view of fungal associates and also in rewriting Engels's *The Origin of the Family, Private Property, and the State*. Tsing's is a tale of speculative fabulation, an SF genre crucial to feminist theory. She and I are in a relation of reciprocal induction, that fundamental evolutionary ecological developmental worlding process that is basic to all becoming-with. See Gilbert and Epel, *Ecological Developmental Biology*.

2 Rose, *Reports from a Wild Country*, taught me that recuperation, not reconciliation or restoration, is what is needed and maybe just possible. I find many of the words that begin with *re-* useful, including *resurgence* and *resilience*. *Post-* is more of a problem.

3 For Le Guin, see especially *The Word for World Is Forest* and "'The Author of Acacia Seeds' and Other Extracts from the *Journal of the Association of Therolinguistics*." "The Author of Acacia Seeds" was first published in 1974 in *Fellowship of the Stars*. For Butler, see especially *Parable of the Sower* and *Parable of the Talents*. Butler inspired a new generation of "stories from social justice movements." See Brown and Imarisha, *Octavia's Brood*. Le Guin's work also pervades much writing for environmental justice and environmental resurgence.

4 Le Guin, "The Carrier Bag Theory," shaped my thinking about narrative in evolutionary theory and of the figure of woman the gatherer for *Primate Visions*. Le Guin learned about the Carrier Bag Theory of Evolution from Fisher, *Women's Creation*, in that period of large, brave, speculative, worldly stories that burned in feminist theory in the 1970s and 1980s. Like speculative fabulation, speculative feminism was, and is, an SF practice.

5 Le Guin, "The Carrier Bag Theory," 166.

6 Le Guin, "The Carrier Bag Theory," 169.

7 Le Guin, "The Carrier Bag Theory," 169.

8 My guide with and through SF, my "mystra," here is LaBare, "Farfetchings." (LaBare's term *mystra* begins to accrue meanings on p. 17.) LaBare argues that SF is not fundamentally a genre, even in the extended sense that includes film, comics, and much else besides the printed book or magazine. The SF mode is, rather, a mode of attention, a theory of history, and a practice of worlding. He writes, "What I call the 'sf mode' offers one way of focusing that attention, of imagining and designing alternatives to the world that is, alas, the case" (1). LaBare suggests that the SF mode pays attention to the "*conceivable, possible, inexorable, plausible*, and *logical*" (italics in original, 27). One of his principal mystras is Le Guin, especially in the lure of her understanding of "talking backwards" in the postapocalyptic Northern Californian SF novel *Always Coming Home*. Reading *Parable of the Sower* together with *Always Coming Home* is a good way for coastal travelers to fill the carrier bag for recuperative terraforming *before* the apocalypse instead of just afterward. Instructed in this SF mode, perhaps human people and earth others can avert inexorable disaster and plant the conceivable germ of possibility for multispecies, multiplacetime recuperation before it is too late.

9 *Myrmex* is the Greek word for ant, and one story has it that an Attic maiden named Myrmex annoyed Athena by claiming the invention of the plow as the maiden's own and so was turned into an ant by the goddess. Me, judging from the tunnels ants dig all over the world and comparing that to Athena's more sky-looking and heady credentials, I think Myrmex probably had the stronger claim to having authored the plow. Breaking out of daddy's brain is really not the same as tunneling and runneling in the earth, whether one is goddess, woman, or ant. For actual ants, one could not do better than Deborah Gordon, *Ants at Work*; *Ant Encounters*; and "The Ecology of Collective Behavior." For contrasting approaches to explanation, see Hölldobler and Wilson, *The Superorganism*, and Hölldobler and Wilson, *The Ants*. Based on her studies of developing behavior in harvester ant colonies in the Arizona desert and evidence that individual ants switch tasks over their lifetimes,

Gordon has been a critic of E. O. Wilson's emphasis on rigidly determined ant behavior. For me, Wilson is the heroic Athena to Gordon's inventive Attic maid Myrmex with a seedbag and a digging tool. To get started with acacias, go to *Wikipedia*'s "Acacia" entry, and then to "Biology of Acacia," a special issue of *Australian Systematic Botany* (2003). Lest one think all the world-building action is an ant story, check out Mann, "Termites Help Build Savannah Societies."

10 Le Guin, "'The Author of Acacia Seeds,'" 174.

11 Le Guin, "'The Author of Acacia Seeds,'" 174.

12 See, for example, Gilbert and Epel, *Ecological Developmental Biology*; Gilbert et al., "Symbiosis as a Source of Selectable Epigenetic Variation"; McFall-Ngai, "The Development of Cooperative Associations between Animals and Bacteria"; McFall-Ngai, "Unseen Forces"; and Hird, *The Origins of Sociable Life*. On symbiogenesis as the driver of evolutionary change, see Margulis and Sagan, *Acquiring Genomes*.

13 See the website of the Global Invasive Species Database for information about troublesome Australian acacias in South America and South Africa. See also "Pacific Islands Ecosystems at Risk" for information about *Acacia mearnisii* (black wattle). Several acacia species, especially the coastal wattle *Acacia cyclops*, worry conservationists in California. All of these disputed travelers teach us to stay with the multispecies trouble that motivates most of my work and play these days.

14 See Bonfante and Anca, "Plants, Mycorrhizal Fungi, and Bacteria." This article draws our attention to the many-faceted practices of communication among members of multispecies consortia. As the abstract summarizes, "Release of active molecules, including volatiles, and physical contact among the partners seem important for the establishment of the bacteria/mycorrhizal fungus/plant network. The potential involvement of quorum sensing and Type III secretion systems is discussed, even if the exact nature of the complex interspecies/interphylum interactions remains unclear."

15 "Acacia," *Wikipedia*; Heil et al., "Evolutionary Change from Induced to Constitutive Expression of an Indirect Plant Resistance."

16 Attenborough, "Intimate Relations"; A. Ross, "Devilish Ants Control the Garden."

17 My debts to Deborah Bird Rose are obvious here and throughout this essay. See especially her development of the idea of double death in "What If the Angel of History Were a Dog?" Double death signifies the killing of ongoingness and the blasting of generations. Rose teaches me about Aboriginal ways of crafting responsibility, inhabiting time, and the need for recuperation in her *Reports from a Wild Country*. See also the important open-access journal *Environmental Humanities*, now under the protection of Duke University Press.

18 Le Guin, "'The Author of Acacia Seeds,'" 175.

Chapter 7: A Curious Practice

Epigraph 1: Vinciane Despret, personal communication.

Epigraph 2: Arendt, *Lectures on Kant's Political Philosophy*, 43.

1 Despret, "'Sheep Do Have Opinions,'" 360.
2 Despret, "Domesticating Practices," 24.
3 Despret, "Domesticating Practices," 36.
4 Despret, "Domesticating Practices," 31.
5 Despret, "The Becoming of Subjectivity in Animal Worlds," 124. With courage and precision, Porcher has also studied horrific industrial pig facilities that can never be called farms. For her vision of what is possible, see Porcher, *Vivre avec les animaux*.
6 Despret, "The Becoming of Subjectivity in Animal Worlds," 133.
7 Despret, "The Becoming of Subjectivity in Animal Worlds," 135.
8 Despret, "The Body We Care For."
9 Stengers and Despret, *Women Who Make a Fuss*.
10 Stengers and Despret, *Women Who Make a Fuss*, 46.
11 Stengers and Despret, *Women Who Make a Fuss*, 47.
12 Stengers and Despret, *Women Who Make a Fuss*, 159.
13 Stengers and Despret, *Women Who Make a Fuss*, 162–63.
14 Quotations from draft sent to me in 2013, no pagination; subsequently published as Despret, "Why 'I Had Not Read Derrida.'"
15 For example, see the film written and directed by Davaa and Falorni, *The Story of the Weeping Camel*.
16 See Despret, *Au bonheur des morts*.
17 Subsequently revised for chapter 1 of *Staying with the Trouble*, the Cerisy paper was published in French as Haraway, "Jeux de ficelles avec les espèces compagnes."
18 See Crasset, "Capsule," and figure 1.5, p. 26.
19 Despret, "Ceux qui insistent."

Chapter 8: The Camille Stories

1 Zoutini, Strivay, and Terranova, "Les enfants du compost, les enfants des monarques." Inspired by the unruly litter of the children of compost, Terranova made a full length narrative film portrait, *Donna Haraway: Story Telling for Earthly Survival*.
2 "Anthropocene: Arts of Living on a Damaged Planet," Santa Cruz, CA, May 8–10, 2014. Videos of presentations from the conference are available on the website. Anna Tsing and colleagues organized this conference. See also Tsing et al., *Arts of Living on a Damaged Planet*.
3 The "Make Kin Not Babies" and the "Children of Compost" projects will have a collective digital world for story posting and gaming. Fragments, polished stories, plot sketches, scientific speculations, drawings, plausible biological and technolog-

ical mechanisms for the syms' ongoing transformations, designs, images, animations, games, characters, pamphlets, manifestos, histories and critiques, bestiaries, field guides, blogposts, slogans: all are welcome. Sym fiction can change plots, introduce new characters and stories, play with media, argue, draw, speculate, more. Stay tuned for the websites and blogs to come. See also the website for Stories for Change, an "online meeting place for digital story telling facilitators and activists."

4 This slogan joins a litter of symbiogenetic and sympoietic provocations that lure my writing. In the 1980s, Elizabeth Bird, then a graduate student in History of Consciousness, gave me "Cyborgs for Earthly Survival." More recently, over breakfast at home Rusten Hogness gave me "Not Posthuman but Compost!" as well as humusities rather than humanities. Camille gives us "Make Kin Not Babies." Breaking the "necessity" of the tie between kin and reproduction is a crucial task for feminists now. It is past time to make a fuss. Disloyal to patriarchal genealogy, we have helped disable the sense of natural necessity of the ties of race and nation, although that work is never done; and we have unraveled the bonds of sex and gender, although we are not finished there either. Feminists have been powerful players disabling the pretensions of human exceptionalism too. No wonder that there is much more collaborative work to do strengthening webs, cutting some ties and knotting others, to live and die well in a habitable world. Adele Clarke and I organized a panel at the Denver meetings in November 2015 of the Society for Social Studies of Science titled "Make Kin Not Babies," arguing for innovative antiracist, prowomen, prochildren, proindigenous, nonsettler, non-natalist approaches. Besides Clarke and myself, panelists were Alondra Nelson, Michelle Murphy, Kim TallBear, and Chia-Ling Wu.

Both inheriting and also reweaving ongoing webs of affective and material relationships are the stakes; such webs are necessary for staying with the trouble. In scholarly circles, ethnographers have understood best that making kin involves all sorts of categories of players—including gods, technologies, critters, expected and unexpected "relatives," and more—and diverse processes, which taken together make the characterization of "kinship" as relations formed solely by genealogical descent and reproduction, or alliance and lineage, unsustainable. There is an immense literature on kinship and on the making of relationships that should not be called genealogical and processes that should not be called reproduction. The Camille Stories rely especially on ethnographies like those of Strathern (*The Gender of the Gift* and *The Relation*), Goslinga ("Embodiment and the Metaphysics of Virgin Birth in South India"), and Ramberg ("Troubling Kinship" and *Given to the Goddess*). The word *kin* is too important to let the critics have it, and *family* cannot do the work of *kin* and its routes/roots to *kind*, with all the multivalences of that term. My sense of kin making requires not just situated deities and spirits—still an unnerving act for so-called moderns—but also heterogeneous critters of biological persuasions. In Camille's world, kin making must be *sensible* in both animist and biological registers. The important words must be *resignified*, *repopulated*, and *reinhabited*. "Making Kin Not Babies" is about making oddkin, including reimagining worlding with the help of Daniel Heath Justice's *Kynship Chronicles* and website, "Justice, Imagine Otherwise." SF is indispensable.

Marilyn Strathern, immersed in Melanesian worlds, proposed that "a person is the form relationships take, a composite of relations rather than a proprietary individual" (Ramberg, "Troubling Kinship," 666; Strathern, *The Gender of the Gift*). The Communities of Compost rely on this string figural approach to making more-than-human persons.

5 *Motley* as a noun is a multihued fabric. *Motely* as an adjective is about incongruous diversity. I need both tones.

6 Some of the other settlements and migrations of the Communities of Compost focused on the consequences of mining include: (1) In China, on ruined lands and villages near several coal-mining sites, led by older rural women and Chinese Green movement activists. See "China and Coal." Thanks to my colleagues Chris Connery and Lisa Rofel and the invaluable Chinese Marxist feminist activist and cultural critic Dai Jinhua for help. (2) In Alberta and Nunavut in Canada, linked to resistance to tar sands extraction and other fossil fuel projects by indigenous and southern coalitions. (3) In the Galilee Basin in Australia, in solidarity with Wangan and Jagalingou peoples' resistance to a Carmichael coal mine. See Palese, "It's Not Just Indigenous Australians v. Adani." (4) In the Navajo and Hopi nations on Black Mesa, in alliance with Anglo, Latino, and native activists. See chapter 3. (5) In Peru and Bolivia in solidarity with antimining movements. See de la Cadena, "Indigenous Cosmopolitics in the Andes." (6) In the Putumayo region of Colombia, in resistance to its designation for mining and agro-industrial extraction. See K. Lyons, "Soils and Peace" and "Soil Science, Development, and the 'Elusive Nature' of Colombia's Amazonian Plains"; and Forest Peoples Program, "Indigenous Peoples of Putumayo Say No to Mining in Their Territories."

7 What constitutes reproductive freedom remained a vexed question in the gatherings of the Children of Compost, especially in the first few generations, for which the consequences of vast human numbers on earth were still intense. The Great Acceleration of human numbers began after about 1950 and topped out by the end of the twenty-first century, but the process of rebalancing and human reduction without exacerbating deep inequalities was difficult in every imaginable way across the earth. Some of the communities worked to adapt rights approaches, locating decision making in the last instance in persons conceived as individuals. But other communities inherited and invented very different ways of thinking about and practicing making new persons and the ramifying obligations and powers of other persons who were involved. Coercion to make or not to make a new child was considered a crime and could result in banishment from the community, but sometimes violent conflict over bringing new babies into existence or over determining who and what were kin did occur. The Communities of Compost insisted that kin did not mean undifferentiated universal relatedness; exclusions and inclusions remained the name of the game of kin, as it always had. Which exclusions and inclusions, and which expansions and contractions, underwent radical, often bumpy change. The Children of Compost understood recrafting definitions and practices

of reproductive freedom in symbiotic terms to be one of their chief obligations, but it never stopped being necessary to oppose oppressive and totalitarian forces in the intentional communities, just as it never ceased to be necessary to interrogate the discursive categories marked by "bio-" that informed symbiogenesis.

8 *Symbiont* and *symbiote* are synonyms; both refer to an organism living in a state of symbiosis, whether beneficial to one or both (or more) or not. Thus, in our stories, both the human and the nonhuman partner should be called a symbiont or symbiote. Symbiogenesis refers to the cobbling together of living entities to make something new in the biological, rather than digital or some other, mode. Symbiogenesis results in novel sorts of organization, not just novel critters. Symbiogenesis opens up the palette (and palate) of possible collaborative living. Many Communities of Compost decided to nurture symbiogenetic transformations with fungal or plant symbionts as the principal partners for the human babies and fetuses, and all the symbioses necessarily entailed intimate assemblages of bacteria, archaea, protists, viruses, and fungi. Camille's community was more tightly webbed with others whose children were bonded with animals as their principal partners, but these distinctions weakened over the generations as multispecies socialities of many previously unimagined kinds emerged.

9 For provocative thinking about corridors, see Hannibal, *The Spine of the Continent*; Soulé and Terborgh, *Continental Conservation*; Hilty, Lidicker, and Merelender, *Corridor Ecology*; and Meloy, *Eating Stone*. The Yellowstone to Yukon Conservation Initiative (see website) is inspiring and critical. I love these sciences and these writings, even as I long for vigorous encounters with sophisticated, situated, multinatural, multilingual, multicultural scholarship—indeed, in a nonromantic, nondismissive, practical mode, with indigenous cosmopolitics and its diverse peoples. For a positive example, see Koelle, "Rights of Way." The ongoing separations and misunderstandings between decolonial thinking and projects and biodiversity thinking and projects is a tragedy for people, peoples, and other critters alike. Practicing corridor thinking, the Communities of Compost do everything they can to bring these disparate worlds into contact.

10 Ecological Evolutionary Developmental Biology, or EcoEvoDevo, is one of the most important knowledge practices to reshape the sciences in the late twentieth and early twenty-first centuries CE. See Gilbert and Epel, *Ecological Developmental Biology*, especially the coda by Gilbert.

11 See "Mountaintop Removal Mining"; Stephens and Sprinkle, *Goodbye Gauley Mountain*. I am placing Camille's community in the area along the Kanawha River, where artist Beth Stephens grew up and to which she and her spouse Annie Sprinkle returned to make their film. Making kin, Annie and Beth married the mountain, one of the many earthy spouses they have knitted together in their traveling ecosexual practices. The history and ongoing vitality of working-class culture and politics rooted in coal mining are fundamental to what Camille and per's community must keep learning. Historically, West Virginia was home and hunting territory to Shawnee, Cherokee, Delaware, Seneca, Wyandot, Ottawa, Tuscarora, Susquehannock, Huron, Sioux, Mingo, Iroquois, and other native peoples. As Scottish and

Irish settlers pushed west from Virginia in the eighteenth century, many Native Americans sought refuge in the Blue Ridge Mountains. The divergent racial class histories of (1) the white settler and African American coal-mining families and communities devastated by twentieth- and twenty-first-century mountaintop-removal mining and by economic restructuring of the coal industry in West Virginia (King Coal), and (2) the adjacent Indian/Black Monacan people of Amherst, Virginia, and elsewhere in the Southeast, who also crafted resurgent politics and identities in the twentieth and twenty-first centuries, were important for Camille's community to know if they were to ally effectively with local people for multispecies futures. See Cook, *Monacans and Miners*. For her account of growing up and returning to West Virginia, see Stephens, "Goodbye Gauley Mountain." The violence wreaked on land and people by open pit mining is impossible to exaggerate. To compare Appalachia (coal) and Peru (copper), see Gallagher, "Peru," and "Mountain Justice Summer Convergence." Surface mining is far from the whole story, a fact made plain by the rapid increase in fracking for natural gas, including in Appalachia; see Cocklin, "Southwestern Plans to Step on the Gas Pedal in Appalachia Next Year."

12 Oberhauser and Solensky, *The Monarch Butterfly*; Rea, Oberhauser, and Quinn, *Milkweed, Monarchs and More*; Pyle, *Chasing Monarchs*; Kingsolver, *Flight Behavior*. For controversy based on recent summer counts about whether the numbers of North American monarchs are declining, or whether the migration, but not population sizes, is declining, see Burnett, "Monarch Migration Rebounds" and Kaplan, "Are Monarch Butterflies Really Being Massacred?" See also websites for "Monarch Butterfly"; "Monarch Butterfly Conservation in California"; "Monarch Butterfly Biosphere Reserve"; "West Virginia State Butterfly." For a good map of the migrations, see "Flight of the Butterflies."

Western monarch butterflies overwinter in California, including my town of Santa Cruz, where we avidly seek them out each year in eucalyptus and Monterey pine and cypress groves at Natural Bridges State Park and Lighthouse Park. Monarchs in Santa Cruz numbered about 120,000 in 1997 but had plummeted to 1,300 by 2009, a few dozen in 2014, and maybe a couple hundred in winter 2015. See Jepsen et al., "Western Monarchs at Risk."

13 There are approximately 110 milkweed species in North American and about 3,000 species worldwide.

14 Tucker, "Community Institutions and Forest Management in Mexico's Monarch Butterfly Reserve"; Farfán et al., "Mazahua Ethnobotany and Subsistence in the Monarch Butterfly Biosphere Reserve, Mexico"; Zebich-Knos, "A Good Neighbor Policy?"; Vidal, López-Garcia, and Rendón-Salinas, "Trends in Deforestation and Forest Degradation in the Monarch Butterfly Biosphere Reserve in Mexico"; Rendón-Salinas and Tavera-Alonso, "Forest Surface Occupied by Monarch Butterfly Hibernation Colonies in December 2013."

See also "Mazahua People." Arauho et al., "Zapatista Army of Mazahua Women in Defense of Water," write: "The Mazahuas are an indigenous people of Mexico, inhabiting the north-western portion of the State of Mexico and north-eastern area of Michoacán, with a presence also in the Federal District owing to recent migration. The largest concentration of Mazahua is found in the municipalities of San Felipe del Progreso and San José del Rincón, both in Mexico state (*Estado de México*), near Toluca. According to the 1990 Mexican census Mazahua speakers numbered 127,826; the last census counted some 350,000 Mazahua. The word *Mazahua* is of Nahuatl origin, meaning 'the owners of deer,' probably referring to the rich fauna of the mountainous region inhabited by the Mazahua. However, they refer to themselves as *Hñatho*."

15 It was possible that no pregnant person in the following generations would choose to continue the symbiotic bond of a new child with monarch butterflies, but instead would let this symbiosis end in favor of initiating another. Such decisions occurred, but most birth parents of the following generations felt intensely in both flesh and spirit that the five-generational reproductive symbioses were vital for them as well.

16 Born in 1930, Oren R. Lyons Jr., Turtle Clan, Seneca Nation, Haudenosaunee Confederacy, wrote: "We are looking ahead, as is one of the first mandates given us as chiefs, to make sure and to make every decision that we make relate to the welfare and well-being of the seventh generation to come" and "What about the seventh generation? Where are you taking them? What will they have?" See O. Lyons, "An Iroquois Perspective," 173, 174. See also O. Lyons et al., *Exiled in the Land of the Free*.

"The Great Binding Law" reads: "In all of your deliberations in the Confederate Council, in your efforts at law making, in all your official acts, self-interest shall be cast into oblivion. Cast not over your shoulder behind you the warnings of the nephews and nieces should they chide you for any error or wrong you may do, but return to the way of the Great Law, which is just and right. Look and listen for the welfare of the whole people and have always in view not only the present but also the coming generations, even those whose faces are yet beneath the surface of the ground—the unborn of the future Nation" (*Constitution of the Iroquois Nations*).

See also Barker, *Native Acts* and "Indigenous Feminisms." For an introduction to extraordinary thinking and research on indigeneity and technoscience, see Kim TallBear website.

17 United Nations, "World Population Prospects." Professional demographers, who formulated their thinking within the category of "populations," understood that even extreme war and pandemic disease in the mid-twenty-first century would not much affect the ultimate global burden of 11 billion or so by century's end. Only a radical one-child policy enforced across the earth could have done that. Aside from being impossible to effect, the unequal, coercive, misogynist, racist, and political implications of that approach were clear even to the most abstract thinkers. The infectious rather than coercive practices of the Communities of Compost were much more radical than a one-child policy, leading to a three-parent norm taking hold rapidly, in a matrix of both traditional and newly invented nonbiogenetic multichild and multigenerational living practices. The history of one-child policies in the twentieth century showed them to be punitive and rooted in coercive and unequal sacrifice. The three-parent approach, which had seemed merely utopian

at first, proved otherwise, especially in concert with exuberant decolonial practices nurtured by the Children of Compost. Three-or-more parent practices that were developed over several decades by the Communities of Compost proved intensely prochild, proparent, profriendship, and procommunity, for human people and other critters alike. Historians look to this inventive period as a time of proliferation of rituals, ceremonies, and celebrations for making kin sympoietically rather than biogenetically. One of the strongest results of this inventiveness was a resurgence of practices of both child and adult friendship across the earth.

See Murphy, "Thinking against Population and with Distributed Futures"; Nelson, *Who Counts?*; Bashford, *Global Population*; and Crist, "Choosing a Planet of Life."

In "A Manifesto for Abundant Futures," Collard, Dempsey, and Sundberg write: "By *abundance* we mean more diverse and autonomous forms of life and ways of living together. In considering how to enact multispecies worlds, we take inspiration from Indigenous and peasant movements across the globe as well as decolonial and postcolonial scholars" (322).

18 The four major genders available in the Western world by the early twenty-first century were cis-female, cis-male, trans-female, and trans-male; but most knowledgeable people considered this list to be misleading, geographically and historically restricted, and impoverished. The beauty of this naming system, however, was its links with stereo-isomerism in chemistry and spatial sensibilities in taxonomizing across domains. Questions like how to be a good cis lover to a trans partner were widely discussed in this early period of gender reformatting. Several members of Camille 1's settlement had been active in trans circles before joining a Community of Compost. The historically anomalous extreme gender binarism of the so-called Modern Period in the West continued to plague both perceptions and naming practices well into the Late Dithering, by which time the Communities of Compost were beginning to make a serious difference for multigender as well as multispecies flourishing. See Weaver, "Trans Species" and "Becoming in Kind."

19 Robinson, *2312*. The dates of the Great Dithering in this SF novel are 2005–60, followed by the Years of Crisis with multiple system failures, and then the balkanization of earth into space, with earth left as a necessary but hopeless cesspool of multispecies immiseration and relentless ineffective human action.

20 *Oddkin* was the colloquial term for other-than-conventional biogenetic relatives.

21 A note on English pronouns: *per* was the gender-neutral pronoun applying to every person. Marge Piercy in *Woman on the Edge of Time* suggested this choice as early as 1976. Whether or not a person decided in the course of a life to develop the mindful bodily habitus of one or more genders, *per* remained the usual pronoun, although some persons preferred a gender descriptive pronoun. Having the advantage of intelligibility in many languages, *sym* was frequently used as the pronoun for the human and animal partners in a symbiosis or other kinds of radical sympoiesis.

22 In 2012, more than five hundred rare, threatened, or endangered critters (insects, arachnids, molluscs, fish, reptiles, amphibians, birds, mammals) were officially

registered with the West Virginia Natural Heritage Program. See West Virginia Department of Natural Resources, "Rare, Threatened, and Endangered Animals."

The Appalachian region is a salamander biodiversity hotspot unique on the earth, and mountaintop removal is a major threat through habitat destruction and water pollution. Migratory salamanders that move between dry land and their breeding pools are especially threatened by habitat fragmentation. Higher temperatures from global warming will have major effects on salamander habitat, and protecting corridors for their movement to cool enough habitats is crucial. See Lanno, *Amphibian Declines*; "Appalachian Salamanders"; "Biodiversity of the Southern Appalachians"; Conservation and Research Center of the Smithsonian National Zoological Park, "Proceedings of the Appalachian Salamander Conservation Workshop, May 30–31, 2008"; IUCN Red List of Threatened Species, "*Ambystoma barbouri*."

For these wonderful catadromous fish that undergo myriad metamorphoses, there is no more moving and informative account than Prosek, *Eels*. For official dithering over their protection, see the U.S. Fish and Wildlife Service, "The American Eel."

American kestrels were not officially threatened as a species in Camille 1's time, but they had declined precipitously in many parts of their range as agribusiness progressively wiped out habitat they depended on. Earlier, kestrels had benefited from less toxic and nonmonocrop agricultural practices, including pasturing animals outdoors, because these raptors prefer open fields, meadows, pastures, and roadside verges to forests for hunting. Also, reforestation in abandoned agricultural fields in the U.S. Northeast is a problem for kestrels. These lovely small falcons prey on diminutive mammals such as voles and mice, small birds and reptiles, and insects like grasshoppers, cicadas, beetles, dragonflies, butterflies, and moths. Kestrel counts made at the Hawk Mountain raptor sanctuary in Pennsylvania showed that these winged predators increased from the 1930s through mid-1970s, decreased in late 1970s and early 1980s, were relatively stable in late 1980s and early 1990s, and became uncommon by the early 2000s. West Nile virus might have been a significant factor; 95 percent of tested birds had antibodies to this virus by 2015. By the time Camille 1's cohort was born in 2025, despite their long history of coadapting well with people, American kestrels were in deep trouble in many places.

Some kestrel populations stay in place year round; others migrate long distances north and south. Unique to the Americas, these kestrels make their living from Tierra del Fuego to the boreal forests of Alaska and Canada. In New Gauley, the best time to look for kestrels is mid-September through mid-October during their migration. See Hawk Mountain, "American Kestrel," and Hawk Mountain, "Long-Term Study of American Kestrel Reproductive Ecology."

In 2015, the U.S. Fish and Wildlife Service petitioned for official Endangered Species Act (ESA) status for two crayfish living in West Virginia. In both cases, mountaintop-removal coal mining had devastated their waterways. These were the first species of many, whose futures were blasted by mountaintop-removal mining, for which ESA protection was sought. See the April 6, 2015, press release of the Center for Biological Diversity, "Two Crayfishes Threatened by Mountain-Top Removal Mining in West Virginia, Kentucky, Virginia Proposed for Endangered Species Act Protection," which states that crayfish "are considered to be a keystone animal because the holes they dig create habitat used by other species including fish. Crayfish keep streams cleaner by eating decaying plants and animals, and they are eaten, in turn, by fish, birds, reptiles, amphibians, and mammals, making them an important link in the food web. The Big Sandy and Guyandotte River crayfishes are sensitive to water pollution, making them indicator species of water quality."

23 Scientists called such requirements philopatry; noncompostist Anglophones spoke of patriotism; compostists spoke of love and need for home, *this* home, not some supposed equivalent. They learned to think that way from the Little Penguins of Sydney Harbor, as they told their stories through van Dooren, *Flight Ways*.

24 For research into the ecology and evolution of parasites and their hosts as well as the genetics of monarch butterfly migration, see the de Roode Lab website.

25 I learned to think and rethink with the locution "in conflict and collaboration" from the University of California Santa Cruz Research Cluster of Women of Color in Conflict and Collaboration, which was established during Angela Davis's term as UC Presidential Chair from 1995 to 1998.

26 An early twenty-first-century conference in Denmark that examined questions in close relation to the ecological urgencies of the Anthropocene proved especially useful to compostists in New Gauley. From the conference announcement of AURA, "Postcolonial Natures": "Three of the proposed start dates for the Anthropocene directly link planetary change to colonial processes: The Columbian Exchange, a product of Portuguese and Spanish Imperialism; nineteenth century industrialism, an offspring of British colonial efforts; and the 'Great Acceleration' of the 1950s, profoundly tied to American imperialism and the forms of consumer capitalism it brought into being . . . Centering questions of power, colonialism, and capitalist relations, the conference aims to probe how histories of inequality and oppression haunt landscapes and shape multispecies relations." Note that the "Great Acceleration" is precisely the time human numbers on the earth started their devastating rate increases. The connections of human numbers with the questions of this conference are multiple and close.

27 See the websites for "Nausicaä: Character" and "Nausicaä of the Valley of the Wind." The Japanese anime film was released in 1984. Thanks to Anna Tsing for drawing my attention back to this amazing story.

28 In his *Insectopedia* (166–67) Hugh Raffles refers to the translation of the twelfth-century Japanese story as "The Lady Who Loved Worms," namely, the gorgeous caterpillars of butterflies and moths. The story, "The Lady Who Loved Insects," is collected among the ten short stories of the *Tsutsumi Chūnagon Monogatari*, authorship unknown.

29 In an interview in 1995, "The Finale of Nausicaä," Miyazaki commented that he wanted to create a heroine who was "not consummately normal." In this same in-

terview, Miyazaki described the Ohmu as larval throughout their lives, the young as well as the adults. No wonder Camille 1 was charmed!

See Mirasol, "Commentary on *Nausicaä of the Valley of the Wind*," a YouTube video by the Filipino film critic Michael Mirasol.

30 Scott Gilbert is my guide to imagining the developmental biological mechanisms to give Camille 2 a beard of butterfly antennae (personal e-mail to the author, April 7, 2015): "There could be several ways of grafting butterfly antennae to the human jaw: One is to induce central tolerance. If the community knew that Camille would become affiliated with the monarch, they could inject [the newborn] with the butterfly antennae extracts (presumably grown in culture or from dead butterflies) as soon as [the child is] born. The immune system is still developing and would be trained to recognize these substances as 'self.' If the decision comes later in life, one could try inducing clonal anergy with the injection of butterfly material in the absence of certain T-cells. This hasn't been a great success story, as peripheral tolerance hasn't exactly cured our allergies yet. So, to the future . . . One interesting way to do it would be to induce the neighboring tissue to become placental! The placenta makes an environment that prevents the mother's immune system from destroying the fetus. Placentally derived factors appear to make T-regulatory cells and limit T-helper formation (see Svenson-Arveland et al., 'The Human Fetal Placenta Promotes Tolerance against the Semiallogenic Fetus by Producing Regulatory T Cells and Homeostatic M_2 Macrophages' for a recent article). Another SF way of doing it might be to have symbiotic bacteria expressing the butterfly antigens induce tolerance. This is being explored as a route to prevent or ameliorate peanut allergies (Ren et al., 'Modulation of Peanut-Induced Allergic Immune Responses by Oral Lactic Acid Bacteria-Based Vaccines in Mice') and atopic dermatitis (Farid et al., 'Effect of a New Synbiotic Mixture on Atopic Dermatitis in Children'). Prolonged contact may induce the antigen-specific T-reg cells. That would be interesting because it would involve a symbiosis to produce the tolerance."

So there were many options for New Gauley scientists to discuss with the fifteen-year-old Camille 2. Camille 2 first tried the placentally derived factors, but the initial efforts failed. Success was finally achieved by giving Camille 2 symbiotic bacteria that expressed butterfly antennal antigens and induced tolerance for the implants in the human host.

From "Butterfly Anatomy": "The antennae of Monarchs *Danaus plexippus* are covered in over 16,000 olfactory (scent detecting) sensors—some scale-like, others in the form of hairs or olfactory pits. The scale-like sensors, which number about 13,700, are sensitive to sexual pheromones, and to the honey odour, which enables them to locate sources of nectar. Butterfly antennae, like those of ants and bees, may also be used to communicate physically—e.g., it is common to see male Small Tortoiseshells, *Aglais urticae*, drumming their antennae on the hindwings of females during courtship, possibly to 'taste' pheromones on the female's wings. . . . Butterflies are often observed 'antenna dipping'—dabbing the antennal tips onto soil or leaves. In this case they are sampling the substrate to detect its chemical qualities. They do this to establish whether soil contains essential nutrients. Male butterflies often drink mineralised moisture to obtain sodium, which they pass to the females during copulation."

31 In Appalachia coal was king, and the busted unions, blasted towns, ravaged human lungs, resilient people, and disappeared mountains, waters, and critters were at the heart of Camille 1's heritage. Other key regions and peoples for New Gauley and the Camilles were the many indigenous peoples of the Canadian tar sands and the Diné and Hopi of the Black Mesa plateau in struggle with Peabody Energy's coal operations. See chapter 3. The Diné (Navajo) hold that the only true evil in the world is greed.

The destruction of aquifers, creeks, lakes, seas, wetlands, and rivers through fossil fuel extraction linked the Camilles to the peoples and other critters of the transvolcanic belt in Mexico, winter home of monarch butterflies, where water was diverted in vast transbasin projects across the mountains to Mexico City.

32 Decolonial approaches to multispecies worlding and rehabilitation that helped Camille 2 prepare for per's sojourn in Mexico: Basso, *Wisdom Sits in Places*; Danowski and Viveiros de Castro, "L'Arrêt du Monde," 221–339; de la Cadena, *Earth Beings*; Escobar, *Territories of Difference*; Green, *Contested Ecologies*; Hogan, *Power*; Kaiser, "Who Is Marching for Pachamama?"; Kohn, *How Forests Think*; Laduke, *All Our Relations*; Tsing, *Friction*; Weisiger, *Dreaming of Sheep in Navajo Country*.

Working for many years in the Putumayo region of Colombia, in "Can There Be Peace with Poison?," Kristina Lyons argues, "Rural communities articulate increasingly ecological conceptions of territoriality in their struggles to defend not only the possibility for human life, but also the relational existence of a continuum of beings and elements (soils, forests, rivers, insects, animals, food crops, medicinal plants, and humans) sharing the contingencies of life and labor under military duress." Lyons proposes the term *selva* as both a theoretical term and situated naturalcultural places. *Selva* does not do the same colonial work as *nature*. The compostists remember their heritage from the British anthropologist Marilyn Strathern, who insisted it matters which ideas think other ideas. See also K. Lyons, *Fresh Leaves*, and de la Cadena, "Uncommoning Nature."

For a taste of the ambiguous, controversial, sometimes vital, sometimes shocking work done by environmental organizations on many sides of "green" questions, see the recent essay by a founder of feminist political ecology, Dianne Rocheleau, "Networked, Rooted and Territorial: Green Grabbing and Resistance in Chiapas." Birthplace of the Zapatista movements, Chiapas is part of the region Camille 2 came to know well. See Harcourt and Nelson, *Practicing Feminist Political Ecologies*. A geographer, Rocheleau changed my world from the moment Anna Tsing first placed her writing in my hands in the first of the three graduate seminars we cotaught at UCSC on geofeminisms (in 2002, 2007, and 2010): Rocheleau and Edmunds, "Women, Men and Trees."

33 For context for Mazahuan formations since 1968 in the setting of Mexican and international indigenous movements, see Gallegos-Ruiz and Larsen, "Universidad Intercultural."

34 Gómez Fuentes, Tire, and Kloster, "The Fight for the Right to Water."

35 Molina, "Zapatistas' First School Opens for Session." The uprising of the Ejército Zapatista de Liberación Nacional began in Chiapas in 1994.

36 For an eloquent story of Amazonian butterflies drinking turtle tears, see Main, "Must See."

37 Leaving out powerful lines, I take the epigraph above from sections of Gallegos-Ruiz and Larsen, "Universidad Intercultural," 24–25. In 1980 a small girl read the poem in Mazahua and Spanish to the presidential candidate Miguel de la Madrid. For the Spanish poem, see Garduño Cervantes, "Soy mazahua." For a primary school girl reciting the poem in Spanish in 2011, see Guadalupe, "Soy mazahua." On the danger of the extinction of the Mazahuan language, see Domínguez, "De la extinción de su lengua mazahua." I have not found a written or oral version of the poem in Mazahua.

38 Remember, sciences are not Modern in the sense meant here; they are not Science. The poverty of language defeats even avid compostists.

39 Camille had read about the extraordinarily destructive transbasin and panregional water transfers of Colorado River water in the U.S. West in the twentieth and twenty-first centuries. A borehole was drilled through the San Jacinto Mountains in Arizona to deliver water to Southern California, and the river was pumped from west to east through a vast trans-Rockies system of tunnels and aqueducts to deliver water to Denver. Lustgarten, "End of the Miracle Machines." See chapter 3 of *Staying with the Trouble*. For a pro-engineering approach, see the website of the Colorado Water Users Association. Alliances of indigenous and other environmental groups were critical to shutting down these practices through sustained politics and ecologies organized around environmental justice and just transition. See the websites of the Just Transition Alliance and Indigenous Environmental Network.

40 Enciso L., "Mexico"; Geo-Mexico, "Where Does Mexico City Get Its Water?"; "Water Management in Greater Mexico City," *Wikipedia*.

In September 2004, Mazahuan women took decisive action. In "The Fight for the Right to Water," Gómez Fuentes, Tire, and Kloster write: "That was when we said to ourselves: 'They're pulling the men's legs and playing with them, because we're not seeing any action.' 'That was when we decided to gather our courage and lead the struggle,' says Rosalva Crisóstomo, from San Isidro . . . The affected communities have finally managed to receive some compensation for their flooded fields, the return of lands that were expropriated for the Cutzamala System but never used, and drinking water for many communities, although some have been excluded from these benefits. As part of their plan for the sustainable development of the region, the communities themselves have begun the process of reforestation and wetland restoration. They have organized micro-enterprises and cooperatives for the production and trade of agricultural products, which helps prevent out-migration. But above and beyond all of this, the Mazahua communities have reaffirmed their culture, their identity, and their customs and traditions as a people, as well as recapturing a pride in their language and clothing, especially among the women."

In 2006 at the Fourth World Water Forum, including the Latin American Water Tribunal, there were many alternative forums and demonstrations. The Mazahua movement was strongly represented. See Trujillo, "The World Water Forum." In September 2010, Agustina Araujo, Guadalupe Acevedo, Ofelia Lorenzo, and Irma Romero, *comandantas* of the Zapatista Army of Mazahua Women, gave testimony before the First Symposium of Indigenous Women, who were leading the struggle for resources and territorial control. See "Zapatista Army of Mazahua Women in Defence of Water in the Cutzamala Region," and Wickstrom, "Cultural Politics and the Essence of Life."

41 A Community of Compost in Amazonian Peru was especially attuned to the local butterflies, who sip the tears of turtles for vital minerals. In this community, human babies were joined in symbiosis with vulnerable turtles and butterflies. From "Mariposas que beben lágrimas de tortuga": "Phil Torres, un científico del Centro de Investigación de la Reserva Natural de Tambopata, en Perú, explicó recientemente en la revista LiveScience que estos bonitos insectos alados absorben sodio y otros nutrientes que necesitan de las saladas lágrimas de las tortugas que viven en el Amazonas." Thanks for this site, and for much else, to Marisol de la Cadena, who teaches me about indigenous cosmopolitics.

42 Russ, *Adventures of Alyx*; Pullman, *His Dark Materials*; Czerneda, *Beholder's Eye*.

43 Monarchs, their protozoan parasites (*Ophryocystis elektroscirrha*), milkweeds, and arbuscular mycorrhizal fungi in the soil associated with the roots of milkweeds constitute a string figural holobiome. The fungi can determine how much of the toxins critical for monarchs to keep their parasites in check will be made by their milkweed food plants. The effects will differ among the many milkweed species and their situated ecologies. Diseases in the soil biota, such as my fabulated twenty-fourth-century runaway viral infection of mycorrhizal fungi important across the taxon of Danainae, can profoundly affect the interactions of other members of the holobiome, such as monarchs and protozoans, resulting in such diseased developmental phenomena as mass failed emergence of adults from the chrysalis. That would be a short-term boon for such critters as paper wasps, which select distressed emergent butterflies as sites to lay their eggs, so that the butterfly corpse can provide food for their offspring. However, large-scale failure of fungal, protozoan, plant, and butterfly holobiomes make wasp eggs fail too, in the end. See Tao et al., "Disease Ecology across Soil Boundaries."

44 Camille 4 drew insight from Dovey, *Only the Animals*. In a book review in *The Guardian*, Romy Ash wrote, "*Only the Animals* is a story told by the souls of ten dead animals. Each animal is caught up in a human conflict over the [twentieth] century. They tell the stories of their deaths. A mussel speaks of its death in the Pearl Harbour bombings, an elephant from the 1987 Mozambique Civil War, a bear of its death during the Bosnia-Herzegovina conflict of 1992."

45 The Communities of Compost were avid readers of SF, and they drew from Orson Scott Card's *Speaker for the Dead* (1986), the sequel to *Ender's Game*, winner of both the Hugo and Nebula Awards. The story was rooted in Ender Wiggin's unknowing but culpable militaristic, exterminating acts as a young boy against the

Hive Queen and her species. Wiggin undertook lifelong atonement as Speaker for the Dead, traveling to those unreconciled with each other and with the deceased, collecting up and decomposing the trouble, and recomposing peace for the living and the dead. The species of the Hive Queen was not totally extinct; Ender held open space for resurgence of these sentient insectoid others that remained utterly different from human beings. Communication was never simple, without ruptures and failures; and it was therefore generative of possibilities.

Another approach to speakers for the dead is practiced in a contemporary documentary about the deliberately plowed-under nineteenth-century tombstones of the forcibly removed and actively forgotten original black settlers of Princeville, Ontario. See The Original People, "Speakers for the Dead."

46 Harding, personal communication, October 7, 2014. See Tagaq, "Animism," for links to Tagaq's work and for the tentacular design of the young woman entwined with the two-headed wolf. For the music, go to Tagaq, "Polaris Prize Performance and Introduction." See also the website for the *Animism* album trailer.

47 Tagaq, "Tagaq Brings Animism to Studio Q."

48 Eduardo Viveiros de Castro, personal communication, October 2, 2014. See Harvey, *Handbook of Contemporary Animism.*

49 This hymn of the Communities of Compost was sung in ceremonies for resurgence attuned to the wounded earth. This song was inherited from the twentieth- and twenty-first-century neopagan witch Starhawk (*Truth or Dare*, 30–31). Starhawk's song insists on the importance of feeling pain as an active historical sensibility, as a practice in what Stengers calls an ecology of practices. Quoted also in Stengers, in *Hypnose entre magie et science*. Precious as a word and a process, *resurgence* is a gift from Tsing, "A Threat to Holocene Resurgence Is a Threat to Livability."

Bibliography

Abouheif, E., M. J. Favé, A. S. Ibarrarán-Viniegra, M. P. Lesoway, A. M. Rafiqi, and R. Rajakumar. "Eco-Evo-Devo: The Time Has Come." *Advances in Experimental Medicine and Biology* 781 (2014): 107–25. doi: 10.1007/978-94-007-7347-9_6. Accessed August 2, 2015.

"Acacia." *Wikipedia*. http://en.wikipedia.org/wiki/Acacia. Accessed August 21, 2015.

"Advanced Energy for Life." https://www.advancedenergyforlife.com/. Accessed August 10, 2015.

Afro-Native Narratives. "Jihan Gearon, Indigenous People's Rights Advocate." http://iloveancestry.com/americatoday/afro-native-truth/item/261-jihan-gearon-indigenous-peoples-rights-advocate-black-mesa-water-coalition. Accessed August 12, 2015.

"Ako Project." http://www.lemurreserve.org/akoproject.html. Accessed August 11, 2015.

"Ako Project: The Books." http://www.lemurreserve.org/akobooks.html. Accessed August 11, 2015.

Alberta Energy. "Facts and Statistics." http://www.energy.alberta.ca/oilsands/791.asp. Accessed August 8, 2015.

Alegado, Rosanna, and Nicole King. "Bacterial Influences on Animal Origins." *Cold Spring Harbor Perspectives in Biology* 6 (2014): a016162. http://cshperspectives.cshlp.org/content/6/11/a016162.full.pdf+html. Accessed June 8, 2015.

Ali, Saleem H. *Mining, the Environment, and Indigenous Development Conflicts*. Tucson: University of Arizona Press, 2003.

Allen, Robert, James DeLabar, and Claudia Drossel. "Mirror Use in Pigeons." http://psychology.lafayette.edu/mirror-use-in-pigeons/. Accessed August 3, 2015.

Allin, Jane. "Wyeth Wins, Horses Lose in the Premarin® Drug Sales Sweepstakes." *Tuesday's Horse*. April 7, 2010. http://tuesdayshorse.wordpress.com/2010/04/07/wyeth-wins-horses-lose-in-the-premarin%C2%AE-drug-salessweepstakes/.

Anderson, Virginia DeJohn. *Creatures of Empire: How Domestic Animals Transformed Early America*. New York: Oxford University Press, 2004.

Anidjar, Gil. *Blood: A Critique of Christianity*. New York: Columbia University Press, 2014.

Anonymous. "History: The Formation of the Endosymbiotic Hypothesis." https://endosymbiotichypothesis.wordpress.com/history-the-formation-of-the-endosymbiotic-hypothesis/. Accessed August 9, 2015.

"Anthropocene: Arts of Living on a Damaged Planet." Conference sponsored by the Anthropology Department, University of California at Santa Cruz, and Aarhus University Research on the Anthropocene (AURA), Aarhus, Denmark. Santa Cruz, CA, May 8–10, 2014. http://anthropocene.au.dk/arts-of-living-on-a-damaged-planet/. Accessed August 31, 2015.

"Anthropocene Feminism." Conference videos. University of Wisconsin–Milwaukee. April 10–12, 2014. http://c21uwm.com/anthropocene/conference-videos/. Accessed August 8, 2015.

Anthropocene Working Group. *Newsletter of the Anthropocene Working Group* 4 (June 2013): 1–17; 5 (September 2014): 1–19. http://quaternary.stratigraphy.org/workinggroups/anthropo/anthropoceneNl4a.pdf and http://quaternary.stratigraphy.org/workinggroups/anthropo/anthropoceneworkinggroupnewslettervo15.pdf. Accessed August 7, 2015.

"Appalachian Salamanders." Smithsonian Conservation Biology Institute. http://nationalzoo.si.edu/SCBI/SpeciesSurvival/AmphibianConservation/salamander/. Accessed September 1, 2015.

Arauho, Augustina, Guadalupe Acevedo, Ofelia Lorenzo, and Irma Romero. "Zapatista Army of Mazahua Women in Defense of Water." In *Dialogue: Indigenous Women in Defense of Life and Land*, edited by Marisa Belausteguigoitia Rius, Mariana Gómez Alvarez Icaza, and Iván González Márquez. *Development* 54, no. 4 (2011): 470–72. http://www.palgrave-journals.com/development/journal/v54/n4/pdf/dev201192a.pdf. Accessed September 1, 2015.

Arendt, Hannah. *Eichmann in Jerusalem: A Report on the Banality of Evil*. New York: Penguin, 1964.

———. *Lectures on Kant's Political Philosophy*. Brighton, UK: Harvester Press, 1982.

———. "Truth and Politics." In *Between Past and Future: Eight Exercises in Political Thought*, 227–64. New York: Penguin, 1977.

Ash, Romy. "*Only the Animals* by Ceridwen Dovey: A Book Review." *Guardian*, May 16, 2014. http://www.theguardian.com/books/australia-culture-blog/2014/may/16/only-the-animals-by-ceridwen-dovey-book-review. Accessed March 19, 2016.

Attenborough, David. "Intimate Relations." *Life in the Undergrowth*. http://www.bbc.co.uk/sn/tvradio/programmes/lifeintheundergrowth/prog_summary.shtml#4. Accessed August 21, 2015.

AURA (Aarhus University Research on the Anthropocene). http://anthropocene.au.dk/. Accessed August 8, 2015.

———. "Postcolonial Natures: Landscapes of Violence and Erasure." Conference at the University of Aarhus. June 17, 2015. http://anthropocene.au.dk/currently/events/show/artikel/conference-postcolonial-natures-landscapes-of-violence-and-erasure/. Accessed September 1, 2015.

Australian Earth Laws Alliance. http://www.earthlaws.org.au/rights-of-nature-tribunal/. Accessed March 19, 2016.

Barad, Karen. *Meeting the Universe Halfway*. Durham, NC: Duke University Press, 2007.

Barash, David P. *Buddhist Biology: Ancient Eastern Wisdom Meets Modern Western Science*. New York: Oxford University Press, 2013.

———. *Natural Selections: Selfish Altruists, Honest Liars and Other Realities of Evolution*. New York: Bellevue Literary Press, 2007.

Barker, Joanne. "Indigenous Feminisms." In *The Oxford Handbook of Indigenous People's Politics*, edited by Jose Antonio Lucero, Dale Turner, and Donna Lee VanCott (online January 2015). doi: 10.1093/oxfordhb/9780195386653.013.007. Accessed September 24, 2015.

———. *Native Acts*. Durham, NC: Duke University Press, 2011.

Bashford, Alison. *Global Population: History, Geopolitics, and Life on Earth*. New York: Columbia University Press, 2013.

Basso, Keith. *Wisdom Sits in Places: Landscape and Language among the Western Apache*. Albuquerque: University of New Mexico Press, 1996.

"Batman Park." *Wikipedia*. http://en.wikipedia.org/wiki/Batman_Park. Modified January 13, 2015. Accessed August 3, 2015.

"Batman's Treaty." *Wikipedia*. http://en.wikipedia.org/wiki/Batman's_Treaty. Modified July 5, 2015. Accessed August 3, 2015.

Bear, Laura, Karen Ho, Anna Tsing, and Sylvia Yanagisako. "Gens, a Feminist Manifesto for the Study of Capitalism." *Cultural Anthropology Online*. March 30, 2015. http://culanth.org/fieldsights/652-gens-a-feminist-manifesto-for-the-study-of-capitalism. Accessed August 12, 2015.

"Bee Orchid." http://www.explainxkcd.com/wiki/index.php/1259:_Bee_Orchid; https://xkcd.com/1259/. Accessed August 10, 2015.

Begay, D. Y. "Shi'Sha'Hane (My Story)." In *Woven by the Grandmothers*, edited by Eulalie Bonar, 13–27. Washington, DC: Smithsonian Institution Press, 1996.

Begaye, Enei. "The Black Mesa Controversy." *Cultural Survival Quarterly* 29, no. 4 (winter 2005). http://www.culturalsurvival.org/publications/cultural-survival-quarterly/united-states/black-mesa-controversy. Accessed August 10, 2015.

Belili Productions. "About Signs out of Time." http://www.belili.org/marija/aboutSIGNS.html. Accessed August 8, 2015.

Bell, Susan. *DES Daughters, Embodied Knowledge, and the Transformation of Women's Health Politics in the Late Twentieth Century*. Philadelphia: Temple University Press, 2009.

Benally, Malcolm D., ed. and trans. *Bitter Water: Diné Oral Histories of the Navajo-Hopi Land Dispute*. Tucson: University of Arizona Press, 2011.

Benanav, Michael. "The Sheep Are Like Our Parents." *New York Times*, July 27, 2012. http://www.nytimes.com/2012/07/29/travel/following-a-navajo-sheep-herder.html?pagewanted=all&_r=1. Accessed August 12, 2015.

Berokoff, Tanya. "Attachment," "Love," and "Let's Hear." *Racing Pigeon Posts*. http://www.articles.racing-pigeon-post.org/Attachment.html; http://www.articles.racing-pigeon-post.org/Love.html; http://www.articles.racing-pigeon-post.org/Lets_hear.html. Accessed February 17, 2012; not available on August 3, 2015.

"Biodiversity of the Southern Appalachians." Highlands Biological Station, Western Carolina University. http://highlandsbiological.org/nature-center/biodiversity-of-the-southern-appalachians/. Accessed September 1, 2015.

"Biology of Acacia." Advances in Legume Systematics Series Part 11, special issue of *Australian Systematic Botany* 16, no. 1 (2003). http://www.publish.csiro.au/issue/650.htm. Accessed August 21, 2015.

Black, Max. *Models and Metaphors: Studies in Language and Philosophy*. Ithaca, NY: Cornell University Press, 1962.

"Black Mesa Indigenous Support." http://supportblackmesa.org/about/mission/. Accessed August 10, 2015.

"Black Mesa Trust." http://www.blackmesatrust.org. Accessed August 10, 2015.

Black Mesa Weavers for Life and Land. http://www.culturalsurvival.org/ourpublications/csq/article/black-mesa-weavers-life-and-land. Accessed August 10, 2015.

———. "Diné Navajo Weavers and Wool." http://www.migrations.com/blackmesa/blackmesa.html. Accessed August 11, 2015.

Blystone, Peter, and Margaret Chanler. *A Gift from Talking God: The Story of the Navajo Churro*. Documentary film. Blyportfolio, 2009.

BMWC (Black Mesa Water Coalition). http://www.blackmesawatercoalition.org/. Accessed August 10, 2015.

———. "About." http://www.blackmesawatercoalition.org/about.html. Accessed August 10, 2015.

———. "Green Economy Project." http://gardenwarriorsgoodseeds.com/2014/10/04/black-mesa-water-coalition-green-economy-project-pinon-az/. Accessed August 12, 2015.

———. "Our Work." http://www.blackmesawatercoalition.org/ourwork.html. Accessed August 10, 2015.

———. "Photos." https://www.facebook.com/blackmesawc/photos_stream. Accessed August 11, 2015.

———. "Tenth Anniversary Video." Narrated by Jihan Gearon. Paper Rocket Productions, 2011. http://www.blackmesawatercoalition.org/index.html. Accessed August 12, 2015.

Bonfante, Paola, and Iulia-Andra Anca. "Plants, Mycorrhizal Fungi, and Bacteria: A Network of Interactions." *Annual Review of Microbiology* 63 (2009): 363–83.

Boone, E. H., and W. Mignolo, eds. *Writing without Words: Alternative Literacies in Mesoamerica and the Andes*. Durham, NC: Duke University Press, 1994.

Bordenstein, S. R., and K. R. Theis. "Host Biology in Light of the Microbiome: Ten

Principles of Holobionts and Hologenomes." *PLoS Biol* 13, no. 8 (2015): e1002226. doi: 10.1371/journal.pbio.1002226. Accessed September 24, 2015.

Brooks, Wendy. "Diethylstilbesterol." *The Pet Pharmacy*. http://www.veterinarypartner.com/Content.plx?P=A&C=31&A=487&S=0. Accessed August 13, 2015.

Brown, Adrienne Maree, and Walidah Imarisha, eds. *Octavia's Brood: Science Fiction Stories from Social Justice Movements*. Oakland, CA: AK Press, 2015.

Buchanan, Brett, Jeffrey Bussolini, and Matthew Chrulew, eds. "Philosophical Ethology II: Vinciane Despret." Special issue. *Angelaki* 20, no. 2 (2015). doi: 10.1080/0969725X.2015.1039819.

Burnett, Victoria. "Monarch Migration Rebounds, Easing Some Fears." *New York Times*, February 27, 2016, A11. http://www.nytimes.com/2016/02/28/world/americas/monarch-butterfly-migration-rebounds-easing-some-fears.html. Accessed February 27, 2016.

"Burning Man." *Wikipedia*. https://en.wikipedia.org/wiki/Burning_Man. Accessed August 7, 2015.

"Burning Man Festival 2012: A Celebration of Art, Music, and Fire." *New York Daily News*, September 3, 2012. http://www.nydailynews.com/news/burning-man-festival-2012-celebration-art-music-fire-gallery-1.1150830. Accessed August 9, 2015.

Busby, Kimberly Sue. "The Temple Terracottas of Etruscan Orvieto: A Vision of the Underworld in the Art and Cult of Ancient Volsinii." PhD diss., University of Illinois, 2007.

Butler, Octavia E. *Parable of the Sower*. New York: Four Walls Eight Windows Press, 1993.

———. *Parable of the Talents*. New York: Seven Stories Press, 1998.

"Butterfly Anatomy." *Butterflies: Complete Guide to the World of Butterflies and Moths*. http://www.learnaboutbutterflies.com/Anatomy.htm. Accessed September 1, 2015.

Cairns, Malcolm F., ed. *Shifting Cultivation and Environmental Change: Indigenous People, Agriculture, and Forest Conservation*. New York: Routledge, 2014.

Callison, Candis. *How Climate Change Comes to Matter: The Communal Life of Facts*. Durham, NC: Duke University Press, 2014.

Cameron, James. *Avatar*. Film. U.S. release, 2009.

Cannon, Hal. "Sacred Sheep Revive Navajo Tradition, for Now." NPR, June 13, 2010. http://www.npr.org/templates/story/story.php?storyId=127797442. Accessed August 12, 2015.

Card, Orson Scott. *Ender's Game*. New York: Tor Books, 1985.

———. *The Speaker for the Dead*. New York: Tor Books, 1986.

Ceballos, Geraldo, Paul Ehrlich, Anthony Barnosky, Andres Garcia, Robert Pringle, and Todd Palmer. "Accelerated Modern Human-Induced Species Losses: Entering the Sixth Mass Extinction." *Science Advances* 1, no. 5 (June 19, 2015): e1400253. Accessed August 7, 2015.

Cenestin®. Duramed. http://www.cenestin.net/. Accessed November 15, 2011.

Center for Biological Diversity. "Two Crayfishes Threatened by Mountain-Top Removal Mining in West Virginia, Kentucky, Virginia Proposed for Endangered Species Act Protection." Press release, April 6, 2015. http://www.biologicaldiversity.org/news/press_releases/2015/crayfish-04-06-2015.html. Acessed September 1, 2015.

"Centre ValBio: Ranomafana National Park." http://www.stonybrook.edu/commcms/centre-valbio/about_us/ranomafana.html. Accessed August 11, 2015.

Charnas, Suzy McKee. *Walk to the End of the World*. New York: Ballantine, 1974.

"China and Coal." SourceWatch: The Center for Media and Democracy. http://www.sourcewatch.org/index.php/China_and_coal#Opposition_to_coal_and_government_repression. Accessed August 31, 2015.

Chisholm, Kami. "The Transmission of Trauma." PhD diss., University of California, Santa Cruz, 2007.

Clarke, Bruce. "Autopoiesis and the Planet." In *Impasses of the Post-Global: Theory in the Era of Climate Change*, volume 2, edited by Henry Sussman, 60–77. Ann Arbor: Michigan University Library/Open Humanities Press, 2012. http://quod.lib.umich.edu/o/ohp/10803281.0001.001. Accessed March 20, 2016.

Clifford, James. *Returns: Becoming Indigenous in the Twenty-First Century*. Cambridge, MA: Harvard University Press, 2013.

———. *Routes: Travel and Translation in the Late Twentieth Century*. Cambridge, MA: Harvard University Press, 1997.

Clinton, Verna. "The Corn Pollen Path of Diné Rug Weaving." 2006, http://www.migrations.com/blackmesa/weavingsforsale.html. Accessed August 11, 2015.

Cocklin, Jamison. "Southwestern Plans to Step on the Gas Pedal in Appalachia Next Year." *NGI's Shale Daily*, December 14, 2014. http://www.naturalgasintel.com/articles/100875-southwestern-plans-to-step-on-the-gas-pedal-in-appalachia-next-year. Accessed September 1, 2015.

Collard, Rosemary-Claire, Jessica Dempsey, and Juanita Sundberg. "A Manifesto for Abundant Futures." *Annals of the Association of American Geographers* 105, no. 2 (2015): 322–30.

Colorado Water Users Association. http://www.crwua.org/colorado-river/uses/urban-uses. Accessed September 2, 2015.

Communities United for a Just Transition. "Our Power Convening." http://www.ourpowercampaign.org/convenings/our-power-convening/. Accessed August 12, 2015.

Conservation and Research Center of the Smithsonian National Zoological Park. "Proceedings of the Appalachian Salamander Conservation Workshop, May 30–31, 2008." http://nationalzoo.si.edu/SCBI/SpeciesSurvival/AmphibianConservation/AppalachianSalamanderReport.pdf. Accessed September 1, 2015.

Constitution of the Iroquois Nations. "The Great Binding Law." Gayanashogowa. http://www.indigenouspeople.net/iroqcon.htm. Accessed September 24, 2015.

Convention on Biological Diversity. *Global Biodiversity Outlook*, 1–4. https://www.cbd.int/gbo/. Accessed August 7, 2015.

Cook, Samuel R. *Monacans and Miners: Native American and Coal Mining Communities in Appalachia*. Lincoln: University of Nebraska Press, 2000.

Crasset, Matali. "Capsule." *Artconnexion*, November 2003. http://www.artconnexion.org/espace-public-public-realm/37-matali-crasset-capsule. Accessed August 3, 2015.

Crist, Eileen. "Choosing a Planet of Life." In *Overpopulation, Overdevelopment, Overshoot*, edited by Tom Butler. San Francisco: Foundation for Deep Ecology and Goff Books, 2015.

———. "On the Poverty of Our Nomenclature." *Environmental Humanities* 3 (2013): 129–47.

"Crochet Coral Reef." http://crochetcoralreef.org/. Accessed August 10, 2015.

"Crownpoint Navajo Rug Auction." http://www.crownpointrugauction.com/. Accessed August 11, 2015.

Crutzen, Paul. "Geology of Mankind." *Nature* 415 (2002): 23.

Crutzen, Paul, and Eugene Stoermer. "The 'Anthropocene.'" *Global Change Newsletter*, International Geosphere-Biosphere Program Newsletter, no. 41 (May 2000): 17–18. http://www.igbp.net/download/18.316f18321323470177580001401/NL41.pdf. Accessed August 7, 2015.

Czerneda, Julie E. *Beholder's Eye*. Web Shifters No. 1. New York: Daw Books, 1998.

da Costa, Beatriz. "Beatriz da Costa's Blog and Project Hub." http://nideffer.net/shaniweb/pigeonblog.php. Accessed August 3, 2015.

———. *Dying for the Other*, selections, 2011. https://vimeo.com/33170755. Accessed August 3, 2015.

———. "PigeonBlog." In "Interspecies." Special issue. *Antennae*, no. 13 (summer 2010): 31–48. http://www.antennae.org.uk/back-issues-2010/4583475279. Accessed February 17, 2012.

da Costa, Beatriz, with Cina Hazegh and Kevin Ponto. "Interspecies Coproduction in the Pursuit of Resistant Action." N.d. http://nideffer.net/shaniweb/files/pigeonstatement.pdf. Accessed August 3, 2015.

da Costa, Beatriz, and Kavita Philips, ed. *Tactical Biopolitics: Art, Activism, and Technoscience*. Cambridge, MA: MIT Press, 2008.

Danowski, Déborah, and Eduardo Viveiros de Castro. "L'Arrêt du Monde." In *De l'univers clos au monde infini*, 221–339. Paris: F. Dehors, 2014.

Davaa, Byambasuren, and Luigi Falorni, writers and directors. *The Story of the Weeping Camel*. Mongolkina Production Company, 2003.

Davidson, Daniel Sutherland. "Aboriginal Australian String Figures." *Proceedings of the American Philosophical Society* 84, no. 6 (August 26, 1941): 763–901. http://www.jstor.org/stable/984876. Accessed August 3, 2015.

Dawkins, Richard. *The Selfish Gene*. 2nd ed. Oxford: Oxford University Press, [1976] 1990.

"Deforestation in Madagascar." https://en.wikipedia.org/wiki/Deforestation_in_Madagascar. Accessed August 10, 2015.

de la Cadena, Marisol. *Earth Beings*. Durham, NC: Duke University Press, 2015.

———. "Indigenous Cosmopolitics in the Andes: Conceptual Reflections Beyond 'Politics.'" *Cultural Anthropology* 25, no. 2 (2010): 334–70. http://dx.doi.org/10.14506/. Accessed August 31, 2015.

———. "Uncommoning Nature." e-flux journal 56th Venice Biennale, August 22, 2015.

http://supercommunity.e-flux.com/texts/uncommoning-nature/. Accessed August 23, 2015.

Demby, Gene. "Updating Centuries-Old Folktales with Puzzles and Power-Ups." National Public Radio. November 30, 2014. http://www.npr.org/sections/codeswitch/2014/11/21/365791351/updating-centuries-old-folklore-with-puzzles-and-power-ups. Accessed August 11, 2015.

Dempster, M. Beth. "A Self-Organizing Systems Perspective on Planning for Sustainability." MA thesis, Environmental Studies, University of Waterloo, 1998. http://www.bethd.ca/pubs/mesthe.pdf. Accessed August 6, 2015.

Denega, Danielle M. *The Cold War Pigeon Patrols: And Other Animal Spies*. New York: Children's Press/Scholastic, 2007.

Denetdale, Jennifer Nez. *Reclaiming Diné History: The Legacies of Navajo Chief Manuelito and Juanita*. Tucson: University of Arizona Press, 2007.

de Roode, Jaap. "De Roode Lab." Emory University. http://www.biology.emory.edu/research/deRoode/publications.html. Accessed March 19, 2016.

DeSmet, Raissa Trumbull. "A Liquid World: Figuring Coloniality in the Indies." PhD diss., History of Consciousness Department, University of California at Santa Cruz, 2013.

Despret, Vinciane. *Au bonheur des morts: Récits de ceux qui restent*. Paris: La Découverte, 2015.

———. "The Becoming of Subjectivity in Animal Worlds." *Subjectivity* 23 (2008): 123–39.

———."The Body We Care For: Figures of Anthropo-zoo-genesis." *Body and Society* 10, nos. 2–3 (2004): 111–34.

———. "Ceux qui insistent: Les nouveaux commanditaires." In *Faire art comme on fait société*, edited by Didier Debaise, X. Douroux, C. Joschke, A. Pontégine, and K. Solhdju. Part I, chapter 7. Dijon: Les Presses du Réel, 2013.

———. "Domesticating Practices: The Case of Arabian Babblers." In *Routledge Handbook of Human-Animal Studies*, edited by Garry Marvin and Susan McHugh, 23–38. New York: Routledge, 2014.

———. "'Sheep Do Have Opinions.'" In *Making Things Public*, edited by Bruno Latour and Peter Weibel, 360–68. Cambridge, MA: MIT Press, 2005.

———. "Why 'I Had Not Read Derrida': Often Too Close, Always Too Far Away." Translated by Greta D'Amico. In *French Thinking about Animals*, edited by Louisa Mackenzie and Stephanie Posthumus, 91–104. East Lansing: Michigan State University Press, 2015.

Detienne, Marcel, and Jean-Pierre Vernant. *Cunning Intelligence in Greek Culture and Society*. Translated from the French by Janet Lloyd. Brighton, UK: Harvester Press, 1978.

DeVries, Karen. "Prodigal Knowledge: Queer Journeys in Religious and Secular Borderlands." PhD diss., History of Consciousness Department, University of California at Santa Cruz, 2014.

Dewey, Ryan. "Virtual Places: Core Logging the Anthropocene in Real-Time." November 13, 2014. http://www.ryandewey.org/blog/2014/11/13/virtual-places-core-logging-the-anthropocene-in-real-time. Accessed March 16, 2016.

DiChiro, Giovanna. "Acting Globally: Cultivating a Thousand Community Solutions for Climate Justice." *Development* 54, no. 2 (2011): 232–36. doi: 10.1057/dev.2011.5. Accessed August 12, 2015.

———. "Beyond Ecoliberal 'Common Futures': Toxic Touring, Environmental Justice, and a Transcommunal Politics of Place." In *Race, Nature, and the Politics of Difference*, edited by Donald Moore, Jake Kosek, and Anand Pandian, 204–32. Durham, NC: Duke University Press, 2003.

———. "Cosmopolitics of a Seaweed Sisterhood." Paper for the American Society for Literature and the Environment, Moscow, Idaho, June 26, 2015. In *Humanities for the Environment: Integrated Knowledges and New Constellations of Practice*, edited by Joni Adamson, Michael Davis, and Hsinya Huang. New York: Routledge, forthcoming.

———. "A New Spelling of Sustainability: Engaging Feminist-Environmental Justice Theory and Practice." In *Practicing Feminist Political Ecologies: Moving Beyond the 'Green Economy,'* edited by Wendy Harcourt and Ingrid Nelson. London: Zed Books, 2015.

"Diethylstilbesterol." *Wikipedia*. http://en.wikipedia.org/wiki/Diethylstilbestrol. Accessed August 13, 2015.

Diné be'iiná/The Navajo Lifeway. http://www.navajolifeway.org/. Accessed August 10, 2015.

———. "Dibé be'iína/Sheep Is Life." http://www.navajolifeway.org/. Accessed August 12, 2015.

"Diné String Games." http://dine.sanjuan.k12.ut.us/string_games/games/opening_a/coyotes_opposite.html. Accessed August 3, 2015.

Dobzhansky, Theodosius. *Genetics and the Origin of Species*. Columbia Classics in Evolution. New York: Columbia University Press, [1937] 1982.

Dolins, Francine, Alison Jolly, Hantanirina Rasamimanana, Jonah Ratsimbazafy, Anna T. C. Feistner, and Florent Ravoavy. "Conservation Education in Madagascar: Three Case Studies." *American Journal of Primatology* 72 (2010): 391–406.

Domínguez, María Albina. Grupo Amanecer del Llano. "De la extinción de su lengua mazahua" (palabras en mazahua y en español). Omáwari—Conferencia de prensa. Teatro Experimental—Centro Cultural Paso del Norte. Ciudad Juárez, Chih. México. September 23, 2011. Producción Revista Rancho Las Voces. Posted September 25, 2011. 6:07 minutes. https://www.youtube.com/watch?v=oOYqJAkFaVo. Accessed September 2, 2015.

Dovey, Ceridwen. *Only the Animals*. Melbourne: Penguin, 2014.

Downing, Samantha. "Wild Harvest—Bird Poo." Pitchfork Projects. December 16, 2010. http://pitchforkdesign.blogspot.com/. Accessed August 3, 2015.

Dubiner, Shoshanah. "'Endosymbiosis': Homage to Lynn Margulis." February 3, 2012. http://www.cybermuse.com/blog/2012/2/13/endosymbiosis-homage-to-lynn-margulis.html. Accessed August 9, 2015.

———. "New Painting in Honor of Lynn Margulis." *Science in Service to Society*, issue 3, October 2012. College of Natural Sciences, UMass Amherst. https://www.cns.umass.edu/about/newsletter/october-2012/memorial-painting-in-honor-of-lynn-margulis. Accessed August 9, 2015.

Durrell Wildlife Conservation. "World Primate Experts Focus on Madagascar." August 12, 2013. http://www.durrell.org/latest/news/world-primate-experts-focus-on-madagascar/. Accessed August 24, 2015.

Eglash, Ron. "Native American Cybernetics: Indigenous Knowledge Resources in Information Technology." http://homepages.rpi.edu/~eglash/eglash.dir/nacyb.htm. Accessed March 17, 2016.

Elliott, William. "*Never Alone*: Alaska Native Storytelling, Digital Media, and Premodern Posthumanisms." Paper for the American Society for Literature and the Environment, Moscow, Idaho, June 24, 2015.

———. "'Ravens' World: Environmental Elegy and Beyond in a Changing North." In *Critical Norths: Space, Nature, Theory*, edited by Sarah Ray and Kevin Maier. Fairbanks: University of Alaska Press, forthcoming.

Enciso L., Angélica. "Mexico: Warning against Practice of Inter-Basin Water Transfers." Translated by Louise McDonnell. *La Journada*, May 9, 2015. http://mexicovoices.blogspot.com/2015/05/mexico-warning-against-practice-of.html. Accessed September 3, 2015.

Engels, Frederick. *The Origin of the Family, Private Property, and the State*, edited by Eleanor Burke Leacock. New York: International Publishers, 1972.

Environmental Humanities. http://environmentalhumanities.org/. Accessed August 21, 2015.

"Environmental Humanities South." http://www.envhumsouth.uct.ac.za/why-environmental-humanities. Accessed August 6, 2015.

Epstein, R., R. P. Lanza, and B. F. Skinner. "'Self-awareness' in the Pigeon." *Science* 212 (1981): 695–96.

Equine Advocates. "PMU Industry." http://www.equineadvocates.org/issueDetail.php?recordID=5. Accessed August 13, 2015.

Equine Angels Rescue Sanctuary. http://www.foalrescue.com/. Accessed August 13, 2015.

"Erinyes 1." *Theoi Greek Mythology*. http://www.theoi.com/Khthonios/Erinyes.html. Accessed August 8, 2015.

Escobar, Arturo. *Territories of Difference*. Durham, NC: Duke University Press, 2008.

"Estrogen." Healthy Women.org. http://www.healthywomen.org/condition/estrogen. Accessed August 13, 2015.

"Estrogen." Midlife-Passages.com. http://www.midlife-passages.com/estrogen.html. Accessed August 13, 2015.

Extinction Studies Working Group. http://extinctionstudies.org/. Accessed August 6, 2015.

Farfán, Berenice, Alejandro Casas, Guillermo Ibarra-Manríquez, and Edgar Pérez-Negrón. "Mazahua Ethnobotany and Subsistence in the Monarch Butterfly Biosphere Reserve, Mexico." *Economic Botany* 61, no. 2 (2007): 173–91.

Farid, R., H. Ahanchian, F. Jabbari, and T. Moghiman. "Effect of a New Synbiotic Mixture on Atopic Dermatitis in Children." *Iranian Journal of Pediatrics* 21, no. 2 (2011): 225–30. http://www.ncbi.nlm.nih.gov/pubmed/23056792. Accessed September 1, 2015.

Feldman, John. *Symbiotic Earth: How Lynn Margulis Rocked the Boat and Started a Scientific Revolution*. Documentary film, in production. http://hummingbirdfilms.com/margulis-revolution/. Accessed August 9, 2015.

Fifth International Prosimian Congress. https://lemurconservationfoundation.wordpress.com/2013/09/18/5th-prosimian-congress/. Accessed August 24, 2015.

Finnegan, John P. "Protestors Sing Honeybeelujahs against Robobees." *Harvard Crimson*, April 23, 2014. http://www.thecrimson.com/article/2014/4/23/protesters-sing-honeybeelujahs-robobees/. Accessed August 8, 2015.

Fisher, Elizabeth. *Women's Creation*. New York: McGraw-Hill, 1975.

"Flight of the Butterflies." http://www.flightofthebutterflies.com/epic-migrations/. Accessed September 1, 2015.

Floria, Maria, and Victoria Mudd. *Broken Rainbow*. Documentary film. 1986. http://www.earthworksfilms.com/templates/ewf_br.html. Accessed August 10, 2015.

Flynn, Dennis O., and Arturo Giráldez. *China and the Birth of Globalisation in the 16th Century*. Farnum, UK: Ashgate Variorium, 2012.

Forest Peoples Program. "Indigenous Peoples of Putumayo Say No to Mining in Their Territories." March 23, 2015. http://www.forestpeoples.org/topics/rights-land-natural-resources/news/2015/03/indigenous-peoples-putumayo-say-no-mining-their-te. Accessed September 1, 2015.

Forney, Barbara. "Diethylstilbesterol for Veterinary Use." http://www.wedgewoodpetrx.com/learning-center/professional-monographs/diethylstilbestrol-for-veterinary-use.html. Accessed August 13, 2015.

Foster and Smith, Doctors. "Diethylstilbesterol." http://www.peteducation.com/article.cfm?c=0+1303+1470&aid=3241. Accessed August 13, 2015.

Francis, Cherrylee, dir. *Voices from Dzil'ijiin (Black Mesa)*. Black Mesa United, Inc. October 15, 2011. http://empowerblackmesa.org/documentary.htm. Accessed August 10, 2015.

Friberg, Michael. "Picturing the Drought." Photo Essay. Special to ProPublica. July 7, 2015. https://projects.propublica.org/killing-the-colorado/story/michael-friberg-colorado-water-photo-essay. Accessed August 10, 2015.

Frigg, Roman, and Stephen Hartman. "Models in Science." *Stanford Encyclopedia of Philosophy*. 2012. http://plato.stanford.edu/entries/models-science/. Accessed August 9, 2015.

"Gaea, the Mad Titan." A Tribute to John Varley's Gaean Trilogy. http://ammonra.org/gaea/. Accessed August 6, 2015.

Gallagher, Erin. "Peru: Ongoing Protests and Strikes against Tia Maria Mining Project." *Revolution News*, May 14, 2015. http://revolution-news.com/peru-ongoing-protests-strikes-against-tia-maria-mining-project/. Accessed September 1, 2015.

Gallegos-Ruiz, M. Antonieta, and Robin Larsen. "Universidad Intercultural: Mexico's New Model University for Indigenous Peoples." *International Perspectives, Jour-*

nal of the CSUSB International Institute, Focus on the Study of the Americas 3 (fall 2006): 18–31. http://internationacsusb.edu/download/journa106.pdf#page=24. Accessed September 2, 2015.

Garduño Cervantes, Julio. "Soy Mazahua!!!" Posted on "Ixtlahuaca, mi tiera." April 2, 2011. https://suarezixtlamas.wordpress.com/2011/04/02/%C2%A1%C2%A1%C2%A1soy-mazahua/. Accessed September 2, 2015.

———. "Soy mazahua. Un poeta mazahua canta a su tierra." In *Colección Cuandernos Regionales*. Serie Cuandernos del Estado de México, 2 (1982): 30.

Garza, Alicia. "A Herstory of the #BlackLivesMatter Movement." *The Feminist Wire*, October 7, 2014. http://www.thefeministwire.com/2014/10/blacklivesmatter-2/. Accessed March 17, 2016.

Gearon, Jihan R. "Strategies for Healing Our Movements." *Last Real Indians*. February 28, 2015. http://lastrealindians.com/strategies-for-healing-our-movements-by-jihan-r-gearon/. Accessed August 12, 2015.

Gearon, Jihan R. Peoples Climate Justice Summit, People's Tribunal, September 23, 2015. Posted October 8, 2014 by *Indigenous Rising*. 7:05 minutes. http://indigenousrising.org/jihan-gearon-of-black-mesa-water-coalition-shares-testimony-at-the-peoples-climate-justice-summit-indigenous-rising/. Accessed March 17, 2016.

Geo-Mexico, the Geography and Dynamics of Modern Mexico. "Where Does Mexico City Get Its Water?" http://geo-mexico.com/?p=9043. Accessed September 3, 2015.

Giddings, Thomas, Nancy Withers, and Andrew Staehlin. "Supramolecular Structure of Stacked and Unstacked Regions of the Photosynthetic Membranes of *Prochloron*, sp, a Prokaryote." *Proceedings of the National Academy of Science* 77, no. 1 (1980): 352–56. http://www.pnas.org/content/77/1/352.full.pdf. Accessed August 12, 2015.

Gilbert, Scott F. "The Adequacy of Model Systems for Evo-Devo." In *Mapping the Future of Biology: Evolving Concepts and Theories*, edited by A. Barberousse, T. Pradeu, M. Morange, 57–68. New York: Springer, 2009.

———. "We Are All Lichens Now." http://cstms.berkeley.edu/baysts/ai1ec_event/we-are-all-lichens-now-scott-gilbert-philosophy-colloquium/?instance_id. Accessed August 6, 2015.

Gilbert, Scott F., and David Epel. *Ecological Developmental Biology: The Environmental Regulation of Development, Health, and Evolution*. 2nd ed. Sunderland, MA: Sinauer Associates, 2015.

Gilbert, Scott F., Emily McDonald, Nicole Boyle, Nicholas Buttino, Lin Gyi, Mark Mai, Neelakantan Prakash, and James Robinson. "Symbiosis as a Source of Selectable Epigenetic Variation: Taking the Heat for the Big Guy." *Philosophical Transactions of the Royal Society B* 365 (2010): 671–78.

Gilbert, Scott F., Jan Sapp, and Alfred I. Tauber. "A Symbiotic View of Life: We Have Never Been Individuals." *Quarterly Review of Biology* 87, no. 4 (December 2012): 325–41.

Gilson, Dave. "Octopi Wall Street!" *Mother Jones*, October 6, 2011. http://www

.motherjones.com/mixed-media/2011/10/occupy-wall-street-octopus-vampire-squid. Accessed August 8, 2015.

Gimbutas, Marija. *The Living Goddesses*, edited by Miriam Robbins Dexter. Berkeley: University of California Press, 1999.

Ginsberg, Faye. "Rethinking the Digital Age." In *Global Indigenous Media: Cultures, Poetics, and Politics*, edited by Pamela Wilson and Michelle Stewart, 287–306. Durham, NC: Duke University Press, 2008.

Ginsberg, Faye, Lila Abu-Lughod, and Brian Larkin, eds. *Media Worlds: Anthropology on New Terrain*. Berkeley: University of California Press, 2002.

Global Invasive Species Database. http://www.issg.org/database/species/ecology.asp?si=51. Accessed August 21, 2015.

Goldenberg, Suzanne. "The Truth behind Peabody Energy's Campaign to Rebrand Coal as a Poverty Cure." *Guardian*, May 19, 2015. http://www.theguardian.com/environment/2015/may/19/the-truth-behind-peabodys-campaign-to-rebrand-coal-as-a-poverty-cure. Accessed August 10, 2015.

Goldenthal, Baila. "Painting/Cats Cradle." http://www.bailagoldenthal.com/painting/cats_cradle/cats_cradle.html. Accessed August 6, 2015.

———. "Resume." http://www.bailagoldenthal.com/resume.html. Accessed August 6, 2015.

Gómez Fuentes, Anahí Copitzy, Magali Iris Tire, and Karina Kloster. "The Fight for the Right to Water: The Case of the Mazahuan Women of Mexico." *Aqua Rios y Pueblos*, December 21, 2009. http://www.aguariosypueblos.org/en/mazahuan-women-mexico/. Accessed September 3, 2015.

Gordon, Deborah M. *Ant Encounters: Interaction Networks and Colony Behavior*. Princeton, NJ: Princeton University Press, 2010.

———. *Ants at Work: How an Insect Society Is Organized*. New York: W. W. Norton, 2000.

———. "The Ecology of Collective Behavior." *PloS Biology* 12, no. 3 (2014): e1001805. doi: 10.1371/journal.pbio.1001805.

Gordon, Jeffrey. "Gordon Lab." Washington University in St. Louis. https://gordonlab.wustl.edu/. Accessed August 9, 2015.

Gosiute/Shoshoni Project of the University of Utah. *Frog Races Coyote/Itsappeh wa'ai Wako*. Claymation video. English subtitles. 7 minutes. http://stream.utah.edu/m/dp/frame.php?f=72b1a0fc6341cb41542. Accessed August 11, 2015.

Goslinga, Gillian. "Embodiment and the Metaphysics of Virgin Birth in South India: A Case Study." In *Summoning the Spirits: Possession and Invocation in Contemporary Religion*, edited by Andrew Dawson, 109–23. London: I. B. Tauris, 2011.

Goushegir, Aladin. *Le combat du colombophile: Jeu aux pigeons et stigmatisation sociale (Kashâ yâ nabard-e kabutarbâz)*. Tehran: Institut français des etudes iraniennes, 1997, Bibliothèque Iranienne, No. 47.

Gramsci, Antonio. *Selections from the Prison Notebooks*, edited by Quintin Hoare and G. N. Smith. New York: International, 1971.

Grebowicz, Margaret, and Helen Merrick. *Beyond the Cyborg*. New York: Columbia University Press, 2013.

Green, Lesley, ed. *Contested Ecologies: Dialogues in the South on Nature and Knowledge*. Cape Town: HSRC Press, 2013.

Greenwood, Veronique. "Hope from the Deep." *Nova Next*. March 4, 2015. http://www.pbs.org/wgbh/nova/next/earth/deep-coral-refugia/. Accessed August 10, 2015.

Guadalupe, Carlos. "Soy mazahua." Posted April 12, 2011. 1:45 minutes. https://www.youtube.com/watch?v=FQkrAWixzJA. Accessed March 20, 2016.

Guerrero, R., L. Margulis, and M. Berlanga. "Symbiogenesis: The Holobiont as a Unit of Evolution." *International Microbiology* 16, no. 3 (2013): 133–43.

Hakim, Danny. "Sex Education in Europe Turns to Urging More Births." *New York Times*, April 9, 2015. http://www.nytimes.com/2015/04/09/business/international/sex-education-in-europe-turns-to-urging-more-births.html?_r=0. Accessed August 12, 2015.

Halanych, K. M. "The Ctenophore Lineage Is Older Than Sponges? That Can't Be Right! Or Can It?" *Journal of Experimental Biology* 218 (2015): 592–97.

Halberstadt, Carol Snyder. "Black Mesa Weavers for Life and Land." *Cultural Survival Quarterly* 25, no. 4 (2001). http://www.culturalsurvival.org/ourpublications/csq/article/black-mesa-weavers-life-and-land. Accessed August 12, 2015.

Hall, Leslie. "The Bright Side of PMU." *Apples 'n Oats* (winter 2006). http://www.applesnoats.com/html/olddefault.html. Accessed November 15, 2011. Not online August 13, 2015.

Hambling, David. "Spy Pigeons Circle the World." *Wired*, October 25, 2008. http://www.wired.com/dangerroom/2008/10/stop-that-spy-p/. Accessed August 3, 2015.

Hannibal, Mary Ellen. *The Spine of the Continent*. Guilford, CT: Lyons, 2012.

Haraway, Donna J. *Crystals, Fabrics, and Fields: Metaphors that Shape Embryos*. Berkeley, CA: North Atlantic Books, 2004. First published by Yale University Press, 1976.

———. "Entrevista com Donna Haraway feita em 21/08/2014 por Juliana Fausto, Eduardo Viveiros de Castro e Déborah Danowski e exibida no Colóquio Internacional Os Mil Nomes de Gaia: do Antropoceno à Idade da Terra no dia 18/09/2014." Posted September 24, 2014. In English. 36:24 minutes. https://www.youtube.com/watch?v=1x0oxUHOlA8. Accessed March 20, 2016.

———. "Jeux de ficelles avec les espèces compagnes: Rester avec le trouble." Translated by Vinciane Despret and Raphael Larrière. In *Les Animaux: Deux ou trois choses que nous savons d'eux*, edited by Vinciane Despret and Raphael Larrière, 23–59. Paris: Hermann, 2014.

———. *Primate Visions: Gender, Race, and Nature in the World of Modern Science*. New York: Routledge, 1989.

———. "SF: Science Fiction, Speculative Fabulation, String Figures, So Far." Acceptance Speech for the Pilgrim Award of the Science Fiction Research Association, July 2011. 29:04 minutes. https://vimeo.com/28892350. Accessed August 3, 2015.

———. *SF: Speculative Fabulation and String Figures/SF: Spekulative Fabulation und String-Figuren*. No. 33 in "100 Notes/Notizen, 100 Thoughts/Gedanken," dOCUMENTA (13). Ostfildern, Germany: Hatje Cantz Verlag, 2011.

———. "Situated Knowledges: The Science Question in Feminism as a Site of Dis-

course on the Privilege of Partial Perspective." *Feminist Studies* 14, no. 3 (1988): 575–99.
———. *When Species Meet*. Minneapolis: University of Minnesota Press, 2008.
Haraway, Donna, and Martha Kenney. "Anthropocene, Capitalocene, Chthulucene." Interview for *Art in the Anthropocene: Encounters among Aesthetics, Politics, Environment, and Epistemology*, edited by Heather Davis and Etienne Turpin. Open Humanities Press, Critical Climate Change series, 2015. http://openhumanitiespress.org/art-in-the-anthropocene.html. Accessed August 8, 2015.
Haraway, Donna, Catherine Lord, and Alexandra Juhasz. "Feminism, Technology, Transformation." Talks on the life and work of Beatriz da Costa. Laguna Art Museum, September 2013. FemTechNet. https://vimeo.com/80248724. 44:51 minutes. Accessed August 3, 2015.
Haraway, Donna, and Anna Tsing. "Tunneling in the Chthulucene." Joint keynote for the American Society for Literature and the Environment (ASLE), Moscow, Idaho, June 25, 2015. Posted by ASLE on October 1, 2015. 1:32:14 hours. https://www.youtube.com/watch?v=FkZSh8Wb-t8. Accessed March 20, 2016.
Harcourt, Wendy, and Ingrid Nelson, eds. *Practicing Feminist Political Ecologies*. London: Zed Books, 2015.
Harding, Susan. "Secular Trouble." Paper for the Conference on Religion and Politics in Anxious States, University of Kentucky, April 4, 2014.
Hartouni, Valerie. *Visualizing Atrocity: Arendt, Evil, and the Optics of Thoughtlessness*. New York: New York University Press, 2012.
Harvey, Graham, ed. *The Handbook of Contemporary Animism*. Durham, UK: Acumen, 2013.
Hawk Mountain. "American Kestrel." http://www.hawkmountain.org/raptorpedia/hawks-at-hawk-mountain/hawk-species-at-hawk-mountain/american-kestrel/page.aspx?id=498. Accessed September 1, 2015.
———. "Long-Term Study of American Kestrel Reproductive Ecology." http://www.hawkmountain.org/science/raptor-research-programs/american-kestrels/page.aspx?id=3469. Accessed September 1, 2015.
Hayward, Eva. "The Crochet Coral Reef Project Heightens Our Sense of Responsibility to the Oceans." *Independent Weekly*, August 1, 2012. http://www.indyweek.com/indyweek/the-crochet-coral-reef-project-heightens-our-sense-of-responsibility-to-the-oceans/Content?oid=3115925. Accessed August 8, 2015.
———. "FingeryEyes: Impressions of Cup Corals." *Cultural Anthropology* 24, no. 4 (2010): 577–99.
———. "Sensational Jellyfish: Aquarium Affects and the Matter of Immersion." *differences: A Journal of Feminist Cultural Studies* 23, no. 1 (2012): 161–96.
———. "SpiderCitySex." *Women and Performance: A Journal of Feminist Theory* 20, no. 3 (2010): 225–51.
Heil, Martin, Sabine Greiner, Harald Meimberg, Ralf Krüger, Jean-Louis Noyer, Günther Heubl, K. Eduard Linsenmair, and Wilhelm Boland. "Evolutionary Change from Induced to Constitutive Expression of an Indirect Plant Resistance." *Nature* 430 (July 8, 2004): 205–8.

Hesiod. *Theogony. Works and Days. Testimonia*. Edited and translated by Glenn W. Most. Loeb Classical Library no. 57. Cambridge, MA: Harvard University Press, 2007.

Hill, Lilian. "Hopi Tutskwa Permaculture." http://www.hopitutskwapermaculture.com/#!staff—teaching-team/c20ft. Accessed August 12, 2015.

Hill, M. A., N. Lopez, and O. Harriot. "Sponge-Specific Bacterial Symbionts in the Caribbean Sponge, *Chondrilla nucula* (Demospongiae, Chondrosida)." *Marine Biology* 148 (2006): 1221–30.

Hilty, Jodi, William Lidicker Jr., and Adina Merelender. *Corridor Ecology: The Science and Practice of Linking Landscapes for Biodiversity Conservation*. Washington, DC: Island, 2006.

Hird, Myra. *The Origins of Sociable Life: Evolution after Science Studies*. New York: Palgrave Macmillan, 2009.

Ho, Engseng. "Empire through Diasporic Eyes: A View from the Other Boat." *Society for Comparative Study of Society and History* (April 2004): 210–46.

———. *The Graves of Tarem: Genealogy and Mobility across the Indian Ocean*. Berkeley: University of California Press, 2006.

Hogan, Linda. *Power*. New York: W. W. Norton, 1998.

Hogness, Rusten. "California Bird Talk." www.hogradio.org/CalBirdTalk/. Accessed August 3, 2015.

Hölldobler, Bert, and E. O. Wilson. *The Ants*. Cambridge, MA: Harvard University Press, 1990.

———. *The Superorganism: The Beauty, Elegance, and Strangeness of Insect Societies*. New York: W. W. Norton, 2009.

"Holos." *Online Etymology Dictionary*. http://www.etymonline.com/index.php?term=holo-. Accessed August 9, 2015.

Horkheimer, Max, and Theodor Adorno. *Dialectic of Enlightenment*. Translated by Edmund Jephcott. Stanford, CA: Stanford University Press, 2002. First published (in German) 1944.

Hormiga, Gustavo. "A Revision and Cladistic Analysis of the Spider Family Pimoidae (Aranae: Araneae)." *Smithsonian Contributions to Zoology* 549 (1994): 1–104. doi:10.5479/si.00810282.549. Accessed August 6, 2015.

"Hormiga Laboratory." George Washington University. http://www.gwu.edu/~spiders/. Accessed August 6, 2015.

Hormone Health Network. "Emminen." http://www.hormone.org/Menopause/estrogen_timeline/timeline2.cfm. Accessed November 15, 2011. Not online August 13, 2015.

Horoshko, Sonia. "Rare Breed: Churro Sheep Are Critically Linked to Navajo Culture." *Four Corners Free Press*, November 4, 2013. http://fourcornersfreepress.com/?p=1694. Accessed August 12, 2015.

HorseAid. "What Are the Living Conditions of the Mares?" http://www.premarin.org/#. Accessed November 15, 2011. Not online August 13, 2015.

HorseAid Report. "PREgnant MARes' urINe, Curse or Cure?" *Equine Times News*, fall/winter 1988.

Horse Fund. "Fact Sheet." http://www.horsefund.org/pmu-fact-sheet.php. Accessed August 13, 2015.

Hubbell Trading Post. "History and Culture." http://www.nps.gov/hutr/learn/historyculture/upload/HUTR_adhi.pdf. Accessed August 11, 2015.

Hustak, Carla, and Natasha Myers. "Involutionary Momentum." *differences* 23, no. 3 (2012): 74–118.

Hutchinson, G. Evelyn. *The Kindly Fruits of the Earth*. New Haven, CT: Yale University Press, 1979.

IdleNoMore. "The Manifesto." http://www.idlenomore.ca/manifesto. Accessed March 19, 2016.

Indigenous Environmental Network. "Canadian Indigenous Tar Sands Campaign." http://www.ienearth.org/what-we-do/tar-sands/. Accessed August 8, 2015.

———. Postings 2015–16. http://www.ienearth.org/. Accessed March 20, 2016.

Ingold, Tim. *Lines, a Brief History*. New York: Routledge, 2007.

Intergovernmental Panel on Climate Change. *Climate Change 2014: Impacts, Adaptation, and Vulnerability: Summary for Policy Makers*. http://ipcc-wg2.gov/AR5/images/uploads/IPCC_WG2AR5_SPM_Approved.pdf. Accessed August 7, 2015.

———. *Climate Change 2014: Mitigation of Climate Change*. http://report.mitigation2014.org/spm/ipcc_wg3_ar5_summary-for-policymakers_approved.pdf. Accessed August 7, 2015.

ISUMA TV. *Inuit Knowledge and Climate Change*. World premier October 23, 2010, ImagineNative Media Arts and Film Festival, Toronto. 54:07 minutes. http://www.isuma.tv/inuit-knowledge-and-climate-change. Accessed March 20, 2016.

IUCN Red List of Threatened Species. "*Ambystoma barbouri*." http://www.iucnredlist.org/details/59053/0. Accessed September 1, 2015.

Iverson, Peter. *Diné: A History of the Navajos*. Photographs by Monty Roessel. Albuquerque: University of New Mexico Press, 2002.

Jacobs, Andrew. "China Fences in Its Nomads, and an Ancient Life Withers." *New York Times*, July 11, 2015. http://www.nytimes.com/2015/07/12/world/asia/china-fences-in-its-nomads-and-an-ancient-life-withers.html?_r=1. Accessed August 11, 2015.

Jacobsen, Thorkild. *The Treasures of Darkness: A History of Mesopotamian Religion*. New Haven, CT: Yale University Press, 1976.

Jayne, Caroline Furness. *String Figures and How to Make Them: A Study of Cat's Cradle in Many Lands*. New York: Charles Scribner & Sons, 1906.

Jepsen, Sarina, Scott Hoffman Black, Eric Mader, and Suzanne Granahan. "Western Monarchs at Risk." Xerces Society for Invertebrate Research. Copyright 2010. http://www.xerces.org/wp-content/uploads/2011/03/western-monarchs-factsheet.pdf. Accessed March 20, 2016.

Jerolmack, Colin. "Animal Practices, Ethnicity and Community: The Turkish Pigeon Handlers of Berlin." *American Sociological Review* 72, no. 6 (2007): 874–94.

———. *The Global Pigeon*. Chicago: University of Chicago Press, 2013.

———. "Primary Groups and Cosmopolitan Ties: The Rooftop Pigeon Flyers of New York City." *Ethnography* 10, no. 4 (2009): 435–57.

Johns, Wahleah. http://indigenousrising.org/our-delegates/wahleah-johns/. Accessed August 12, 2015.

Johns, Wahleah, and Enei Begay. Speech at Power Shift '09, Energy Action Coalition. Uploaded March 6, 2009. 6:05 minutes. https://www.youtube.com/watch?v=02f1nzY6_ro&index=3&list=PL043330BF525D3051. Accessed August 12, 2015. Not online March 20, 2016.

Johnson, Broderick H., ed. *Navajo Stories of the Long Walk Period*. Tsaile, AZ: Navajo Community College Press, 1973.

Johnson, Broderick H., and Ruth Roessel, eds. *Navajo Livestock Reduction: A National Disgrace*. Tsaile, AZ: Navajo Community College Press, 1974.

Jolly, Alison. *Lords and Lemurs: Mad Scientists, Kings with Spears, and the Survival of Diversity in Madagascar*. Boston: Houghton Mifflin, 2004.

———. *Thank You, Madagascar*. London: Zed Books, 2015.

Jolly, A., et al. "Territory as Bet-Hedging: *Lemur catta* in a Rich Forest and an Erratic Climate." In *Ring-tailed Lemur Biology*, edited by A. Jolly, R. W. Susman, N. Koyama, and H. Rasamimanana, 187–207. New York: Springer, 2006.

Jolly, Margaretta. "Alison Jolly and Hantanirina Rasamimanana: The Story of a Friendship." *Madagascar Conservation and Development* 5, no. 2 (2010): 44–45.

Jones, Dave. "Navajo Tapestries Capture the Soul of Her Land." UC Davis News and Information. January 4, 2013. http://dateline.ucdavis.edu/dl_detail.lasso?id=14307. Accessed August 11, 2015.

Jones, Elizabeth McDavid. *Night Flyers*. Middletown, WI: Pleasant Company, 1999.

Jones, Gwyneth. "True Life Science Fiction: Sexual Politics and the Lab Procedural." In *Tactical Biopolitics: Art, Activism, and Technoscience*, edited by Beatriz da Costa and Kavita Philips, 289–306. Cambridge, MA: MIT Press, 2008.

Justice, Daniel Heath. "Justice, Imagine Otherwise. The Kynship Chronicles." http://imagineotherwise.ca/creative.php?The-Kynship-Chronicles-2. Accessed October 8, 2015.

———. *The Way of Thorn and Thunder: The Kynship Chronicles*. Albuquerque: University of New Mexico Press, 2012.

Just Transition Alliance. http://www.jtalliance.org/docs/aboutjta.html. Accessed September 2, 2015.

Kaiser, Anna. "Who Is Marching for Pachamama? An Intersectional Analysis of Environmental Struggles in Bolivia under the Government of Evo Morales." PhD diss., Faculty of Social Sciences, Lund University, 2014.

Kaplan, Sarah. "Are Monarch Butterflies Really Being Massacred? A New Study Says It's a Lot More Complicated Than It Seems." *Washington Post*, August 5, 2015. http://www.washingtonpost.com/news/morning-mix/wp/2015/08/05/are-monarch-butterflies-really-being-massacred-a-new-study-says-its-a-lot-more-complicated-than-it-seems/. Accessed September 1, 2015.

Kazan, Elia, dir. *On the Waterfront*. Horizon Pictures. 1954.

Keediniihii (Katenay), NaBahe (Bahe). "The Big Mountain Dineh Resistance: Still a Cornerstone." February 26, 2015. http://sheepdognationrocks.blogspot.com/2015/02/the-big-mountain-dineh-resistance-still_38.html. Accessed August 12, 2015.

Keio University. "Pigeons Show Superior Self-recognition Abilities to Three Year Old Humans." *Science Daily*, June 14, 2008. www.sciencedaily.com/releases/2008/06/080613145535.htm. Accessed August 3, 2015.

Kenney, Martha. "Fables of Attention: Wonder in Feminist Theory and Scientific Practice." PhD diss., History of Consciousness Department, University of California at Santa Cruz, 2013.

Kiefel, Darcy. "Heifer Helps Navajos Bolster Sheep Herd." N.d. http://www.redshift.com/~bcbelknap/ashtlo/graphics/supplemental/supplfeb%2004/heifer_helps_navajos_bolst.htm. Accessed September 5, 2015.

King, Katie. "Attaching, for Climate Change: A Sympoiesis of Media." Book proposal, 2015.

———. "In Knots: Transdisciplinary Khipu." In *Object/Ecology*, special inaugural issue of *O-Zone: A Journal of Object Oriented Studies* 1, no. 1 (forthcoming). http://o-zone-journal.org/short-essay-cluster. Accessed March 20, 2016.

———. "A Naturalcultural Collection of Affections: Transdisciplinary Stories of Transmedia Ecologies." *S&F Online* 10, no. 3 (summer 2012). http://sfonline.barnard.edu/feminist-media-theory/a-naturalcultural-collection-of-affections-transdisciplinary-stories-of-transmedia-ecologies-learning/. Accessed August 6, 2015.

———. *Networked Reenactments: Stories Transdisciplinary Knowledges Tell*. Durham, NC: Duke University Press, 2011.

———. "Toward a Feminist Boundary Object-Oriented Ontology . . . or Should It Be a Boundary Object-Oriented Feminism? These Are Both Queer Methods." Paper for conference titled Queer Method, University of Pennsylvania, October 31, 2013. http://fembooo.blogspot.com. Accessed August 6, 2015.

King, Nicole. "King Lab: Choanoflagellates and the Origin of Animals." University of California, Berkeley. https://kinglab.berkeley.edu/. Accessed August 9, 2015.

Kingsolver, Barbara. *Flight Behavior*. New York: Harper, 2012.

Klain, Bennie. *Weaving Worlds*. Coproduced by Trickster Films, the Independent Television Service, and Native American Public Telecommunications. Navajo and English with English subtitles. 56:40 minutes. 2008. http://www.tricksterfilms.com/Weavi_Worlds.html. Accessed August 11, 2015.

Klare, Michael. *The Race for What's Left: The Global Scramble for the World's Last Resources*. New York: Picador, 2012.

———. "The Third Carbon Age." *Huffington Post*, August 8, 2013. http://www.huffingtonpost.com/michael-t-klare/renewable-energy_b_3725777.html. Accessed August 7, 2015.

———. "Welcome to a New Planet: Climate Change 'Tipping Points' and the Fate of the Earth." *TomDispatch*, October 8, 2015. http://www.tomdispatch.com/blog/176054/tomgram%3A_michael_klare%2C_tipping_points_and_the_question_of_civilizational_survival/. Accessed October 13, 2015.

———. "What's Big Energy Smoking?" *Common Dreams*, May 27, 2014. http://www.commondreams.org/views/2014/05/27/whats-big-energy-smoking. Accessed August 7, 2015.

Klein, Naomi. "How Science Is Telling Us All to Revolt." *New Statesman*, October 29, 2013. http://www.newstatesman.com/2013/10/science-says-revolt. Accessed August 7, 2015.

———. *The Shock Doctrine: The Rise of Disaster Capitalism*. New York: Macmillan/Picador, 2008.

Koelle, Sandra. "Rights of Way: Race, Place and Nation in the Northern Rockies." PhD diss., History of Consciousness, University of California at Santa Cruz, 2010.

Kohn, Eduardo. *How Forests Think: Toward an Anthropology beyond the Human*. Berkeley: University of California Press, 2013.

Kolbert, Elizabeth. *The Sixth Extinction: An Unnatural History*. New York: Henry Holt, 2014.

Kraker, Daniel. "The Real Sheep." *Living on Earth*, National Public Radio, October 28, 2005. 8:17 minutes. http://loe.org/shows/segments.html?programID=05-P13-00043&segmentID=5. Accessed August 12, 2015.

Kull, Christian. *Isle of Fire: The Political Ecology of Landscape Burning in Madagascar*. Chicago: University of Chicago Press, 2004.

LaBare, Joshua (Sha). "Farfetchings: On and in the SF Mode." PhD diss., History of Consciousness Department, University of California at Santa Cruz, 2010.

Lacerenza, Deborah. "An Historical Overview of the Navajo Relocation." *Cultural Survival* 12, no. 3 (1988). http://www.culturalsurvival.org/publications/cultural-survival-quarterly/united-states/historical-overview-navajo-relocation. Accessed August 10, 2015.

Laduke, Winona. *All Our Relations*. Boston: South End, 1999.

Lanno, Michael J., ed. *Amphibian Declines: The Conservation Status of United States Species*. Berkeley: University of California Press, 2005.

Latour, Bruno. "Facing Gaïa: Six Lectures on the Political Theology of Nature." Gifford Lectures, Edinburgh, February 18–28, 2013. Abstracts and videos. http://www.ed.ac.uk/schools-departments/humanities-soc-sci/news-events/lectures/gifford-lectures/archive/series-2012-2013/bruno-latour. Accessed August 7, 2015.

———. "War and Peace in an Age of Ecological Conflicts." Lecture for the Peter Wall Institute, Vancouver, BC, Canada, September 23, 2013. Video and abstract at http://www.bruno-latour.fr/node/527. Accessed August 7, 2015.

———. *We Have Never Been Modern*. Cambridge, MA: Harvard University Press, 1993.

———. "Why Has Critique Run Out of Steam? From Matters of Fact to Matters of Concern." *Critical Inquiry* 30, no. 2 (winter 2004): 225–48.

Lee, Erica. "Reconciling in the Apocalypse." *The Monitor*, March/April 2016. https://www.policyalternatives.ca/publications/monitor/reconciling-apocalypse. Posted March 1, 2016. Accessed March 19, 2016.

Le Guin, Ursula K. *Always Coming Home*. Berkeley: University of California Press, 1985.

———. "'The Author of Acacia Seeds' and Other Extracts from the *Journal of the Asso-*

ciation of Therolinguistics." In *Buffalo Gals and Other Animal Presences*, 167–78. New York: New American Library, 1988.
———. "The Carrier Bag Theory of Fiction." In *Dancing at the Edge of the World: Thoughts on Words, Women, Places*, 165–70. New York: Grove, 1989.
———. "A Non-Euclidean View of California as a Cold Place to Be." In *Dancing at the Edge of the World: Thoughts on Words, Women, Places*, 80–100. New York: Grove, 1989.
———. *A Wizard of Earthsea*. San Jose, CA: Parnassus, 1968.
———. *The Word for World Is Forest*. New York: Berkeley Medallion, 1976.
Lewis, Randolph. *Navajo Talking Picture: Cinema on Native Ground*. Lincoln: University of Nebraska Press, 2012.
"Library of Navajo String Games." © 2003 San Juan School District, Tucson, AZ. http://dine.sanjuan.k12.ut.us/string_games/games/index.html. Accessed August 3, 2015.
Lindeman, Raymond. "Trophic-Dynamic Aspect of Ecology." *Ecology* 32, no. 4 (1942): 399–417.
"List of Pigeon Breeds." https://en.wikipedia.org/wiki/List_of_pigeon_breeds. Modified January 22, 2016. Accessed March 20, 2016.
Lovecraft, H. P. *The Call of Cthulhu and Other Dark Tales*. New York: Barnes and Noble, 2009. First published in *Weird Tales* 11, no. 2 (February 1928): 159–78, 287.
Loveless, Natalie. "Acts of Pedagogy: Feminism, Psychoanalysis, Art, and Ethics." PhD diss., History of Consciousness Department, University of California at Santa Cruz, 2010.
Lovelock, James E. "Gaia as Seen through the Atmosphere." *Atmospheric Environment* 6, no. 8 (1967): 579–80.
Lovelock, James E., and Lynn Margulis. "Atmospheric Homeostasis by and for the Biosphere: The Gaia Hypothesis." *Tellus*, Series A (Stockholm: International Meteorological Institute) 26, nos. 1–2 (February 1, 1974): 2–10. http://tellusa.net/index.php/tellusa/article/view/9731. Accessed August 7, 2015.
Lustgarten, Abraham. "End of the Miracle Machines: Inside the Power Plant Fueling America's Drought." ProPublica, July 16, 2015. https://www.projects.propublica.org/killing-the-colorado/story/navajo-generating-station-colorado-river-drought. Accessed August 10, 2015.
———. "Killing the Colorado." Twelve parts. ProPublica, June 16, 2015. https://www.propublica.org/series/killing-the-colorado. Accessed August 10, 2015.
Lyons, Kristina. "Can There Be Peace with Poison?" *Cultural Anthropology Online*, April 30, 2015. http://www.culanth.org/fieldsights/679-can-there-be-peace-with-poison. Accessed September 1, 2015.
———. *Fresh Leaves*. Creative Ethnographic Non-fiction and Photographic Installation published by the Centre for Imaginative Ethnography's Galleria. York University, May 14, 2014. http://imaginativeethnography.apps01.yorku.ca/galleria/fresh-leaves-by-kristina-lyons/.
———. "Soil Science, Development, and the 'Elusive Nature' of Colombia's Amazonian Plains." *Journal of Latin American and Caribbean Anthropology* 19, no. 2 (July 2014): 212–36. doi: 10.1111/jlca.

———. "Soils and Peace: Imagining Dialogues between Soil Scientists and Farmers in Colombia." *Panoramas.* University of Pittsburgh, July 11, 2015. In English and Spanish. http://www.panoramas.pitt.edu/content/soils-and-peace-imagining-dialogues-between-soil-scientists-and-farmers-colombia. Accessed September 1, 2015.

Lyons, Oren R. "An Iroquois Perspective." In *American Indian Environments: Ecological Issues in Native American History*, edited by C. Vecsey and R. W. Venables. New York: Syracuse University Press, 1980.

Lyons, Oren, Donald Grinde, Robert Venables, John Mohawk, Howard Berman, Vine Deloria Jr., Laurence Hauptman, and Curtis Berkey. *Exiled in the Land of the Free: Democracy, Indian Nations and the U.S. Constitution.* Santa Fe, NM: Clear Light, 1998.

Main, Douglas. "Must See: Amazonian Butterflies Drink Turtle Tears." *Live Science*, September 11, 2013. http://www.livescience.com/39558-butterflies-drink-turtle-tears.html. Accessed September 2, 2015.

Mann, Adam. "Termites Help Build Savannah Societies." *Science Now*, May 25, 2010. http://news.sciencemag.org/sciencenow/2010/05/termites-help-build-savanna-soci.html. Accessed August 21, 2015.

Margulis, Lynn. "Archaeal-Eubacterial Mergers in the Origin of Eukarya: Phylogenetic Classification of Life." *Proceedings of the National Academy of Sciences* 93, no. 3 (1996): 1071–76.

———. "Biodiversity: Molecular Biological Domains, Symbiosis, and Kingdom Origins." *Biosystems* 27, no. 1 (1992): 39–51.

———. Faculty website, UMass Amherst. http://www.geo.umass.edu/faculty/margulis/. Accessed August 9, 2015.

———. "Gaia Hypothesis." Lecture for the National Aeronautic and Space Agency. Video recording. NASA, 1984. https://archive.org/details/gaia_hypothesis. Accessed August 7, 2015.

———. "Symbiogenesis and Symbionticism." In *Symbiosis as a Source of Evolutionary Innovation: Speciation and Morphogenesis*, edited by L. Margulis and R. Fester, 1–14. Cambridge, MA: MIT Press, 1991.

———. *Symbiotic Planet, a New Look at Evolution.* New York: Basic Books, 1999.

Margulis, Lynn, and Dorian Sagan. *Acquiring Genomes: A Theory of the Origin of Species.* New York: Basic Books, 2002.

———. "The Beast with Five Genomes." *Natural History*, June 2001. http://www.naturalhistorymag.com/htmlsite/master.html?http://www.naturalhistorymag.com/htmlsite/0601/0601_feature.html. Accessed August 9, 2015.

———. *Dazzle Gradually: Reflections on the Nature of Nature.* White River Junction, VT: Chelsea Green, 2007.

———. *Microcosmos: Four Billion Years of Microbial Evolution.* Berkeley: University of California Press, 1997.

"Mariposas que beben lágrimas de tortuga: Y no es el título de un poema, es la mágica realidad." *Diario ecologia.com.* N.d. http://diarioecologia.com/mariposas

-que-beben-lagrimas-de-tortuga-y-no-es-el-titulo-de-un-poema-es-la-magica-realidad/. Accessed September 3, 2015.

Mayr, Ernst. *Systematics and the Origin of Species from the Viewpoint of a Biologist*. Cambridge, MA: Harvard University Press, [1942] 1999.

"Mazahua People." *Wikipedia*. https://en.wikipedia.org/wiki/Mazahua_people. Accessed September 1, 2015.

Mazmanian, Sarkis. "Sarkis Lab." California Institute of Technology. http://sarkis.caltech.edu/Home.html. Accessed August 9, 2015.

Mazur, Susan. "Intimacy of Strangers and Natural Selection." *Scoop*, March 6, 2009. http://www.suzanmazur.com/?p=195. Accessed August 9, 2015.

McFall-Ngai, Margaret. "The Development of Cooperative Associations between Animals and Bacteria: Establishing Détente among Domains." *American Zoologist* 38, no. 4 (1998): 593–608.

———. "Divining the Essence of Symbiosis: Insights from the Squid-Vibrio Model." *PLOS Biology* 12, no. 2 (February 2014): e1001783. doi: 10.1371/journal.pbio.10017833. Accessed August 9, 2015.

———. "McFall-Ngai Lab." University of Wisconsin–Madison. http://labs.medmicro.wisc.edu/mcfall-ngai/research.html. Accessed August 9, 2015.

———. "Pacific Biosciences Research Center at the University of Hawai'i at Manoa." http://www.pbrc.hawaii.edu/index.php/margaret-mcfall-ngai. Accessed August 9, 2015.

———. "Unseen Forces: The Influence of Bacteria on Animal Development." *Developmental Biology* 242 (2002): 1–14.

McFall-Ngai, Margaret, et al. "Animals in a Bacterial World: A New Imperative for the Life Sciences." *Proceedings of the National Academy of Sciences* 110, no. 9 (February 26, 2013): 3229–36.

McGowan, Kat. "Where Animals Come From." *Quanta Magazine*, July 29, 2014. https://www.quantamagazine.org/20140729-where-animals-come-from/. Accessed August 9, 2015.

M'Closkey, Kathy. *Swept under the Rug: A Hidden History of Navajo Weaving*. Albuquerque: University of New Mexico Press, 2002.

M'Closkey, Kathy, and Carol Snyder Halberstadt. "The Fleecing of Navajo Weavers." *Cultural Survival Quarterly* 29, no. 3 (fall 2005). http://www.culturalsurvival.org/publications/cultural-survival-quarterly/united-states/fleecing-navajo-weavers. Accessed August 11, 2015.

McPherson, Robert. "Navajo Livestock Reduction in Southeastern Utah, 1933–46: History Repeats Itself." *American Indian Quarterly* 22, nos. 1–2 (winter–spring 1998): 1–18.

McSpadden, Russ. "Ecosexuals of the World Unite!" *Earth First! Newswire*, February 25, 2013. https://earthfirstnews.wordpress.com/2013/02/25/ecosexuals-of-the-world-unite-stop-mtr/. Accessed August 6, 2015.

"Medousa and Gorgones." *Theoi Greek Mythology*. http://www.theoi.com/Pontios/Gorgones.html. Accessed August 8, 2015.

Meloy, Ellen. *Eating Stone: Imagination and the Loss of the Wild*. New York: Random House, 2005.

Melville, Elinor G. K. *A Plague of Sheep: Environmental Consequences of the Conquest of Mexico*. Cambridge: Cambridge University Press, 1997.

Mereschkowsky, Konstantin. "Theorie der zwei Plasmaarten als Grundlage der Symbiogenesis, einer neuen Lehre von der Ent-stehung der Organismen." *Biologisches Zentralblatt*, Leipzig, 30 (1910): 353–67.

Merker, Daniel. "Breath Soul and Wind Owner: The Many and the One in Inuit Religion." *American Indian Quarterly* 7, no. 3 (1983): 23–39. doi: 10.2307/1184255. Accessed August 11, 2015.

Metcalf, Jacob. "Intimacy without Proximity: Encountering Grizzlies as a Companion Species." *Environmental Philosophy* 5, no. 2 (2008): 99–128.

Mindell, David. "Phylogenetic Consequences of Symbioses." *Biosystems* 27, no. 1 (1992): 53–62.

Minkler, Sam A. Photos for "Paatuaqatsi/Water Is Life." Website of Black Mesa Trust. http://www.blackmesatrust.org/?page_id=46. Accessed August 10, 2015.

Mirasol, Michael. "Commentary on *Nausicaä of the Valley of the Wind*." 11:42 minutes. Uploaded August 20, 2010. https://www.youtube.com/watch?v=tdAtYXzcZWE. Accessed September 1, 2015.

Miyazake, Hayao, writer and dir. Interview with Ryo Saitani. "The Finale of Nausicaä." *Comic Box,* special issue January 1995. http://www.comicbox.co.jp/e-nau/e-nau.html. Accessed September 1, 2015.

———. *Nausicaä of the Valley of the Wind*. Japanese anime film. Studio Ghibli. 1984.

Mock, Brentin. "Justice Matters." *Grist*. List for multiple posts, 2014–15. https://grist.org/author/brentin-mock/. Accessed March 17, 2016.

Molina, Marta. "Zapatistas' First School Opens for Session." *Waging Nonviolence*. August 12, 2013. http://wagingnonviolence.org/feature/the-zapatistas-first-escuelita-for-freedom-begins-today/. Accessed September 2, 2015.

"Monarch Butterfly." *Wikipedia*. https://en.wikipedia.org/wiki/Monarch_butterfly. Accessed September 1, 2015.

"Monarch Butterfly Biosphere Reserve." *Wikipedia*. http://en.wikipedia.org/wiki/Monarch_Butterfly_Biosphere_Reserve. Accessed September 1, 2015.

"Monarch Butterfly Conservation in California." *Wikipedia*. https://en.wikipedia.org/wiki/Monarch_butterfly_conservation_in_California. Accessed September 1, 2015.

Monterey Bay Aquarium. "Tentacles: The Astounding Lives of Octopuses, Squids, and Cuttlefish." Exhibit, 2014–15. http://www.montereybayaquaum.org/animals-and-experiences/exhibits/tentacles. Accessed August 10, 2015.

Monument Valley High School. "Ndahoo'aah Relearning/New Learning Navajo Crafts/Computer Design." Monument Valley, 1996. http://www.math.utah.edu/~macarthu/Ndahooah/overview.html. Accessed August 11, 2015.

Moore, Jason W. "Anthropocene, Capitalocene, and the Myth of Industrialization." June 16, 2013. https://jasonwmoore.wordpress.com/2013/06/16/anthropocene-capitalocene-the-myth-of-industrialization/. Accessed August 7, 2015.

———. "Anthropocene or Capitalocene, Part III." May 19, 2013. http://jasonwmoore.wordpress.com/2013/05/19/anthropocene-or-capitalocene-part-iii/. Accessed August 8, 2015.

———. *Capitalism and the Web of Life: Ecology and the Accumulation of Capital*. London: Verso, 2015.

———, ed. *Anthropocene or Capitalocene?* Oakland, CA: PM Press, 2016.

Moran, Nancy. "Nancy Moran's Lab." University of Texas at Austin. http://web.biosci.utexas.edu/moran/. Accessed August 9, 2015.

Morgan, Eleanor. "Sticky Tales: Spiders, Silk, and Human Attachments." *Dandelion* 2, no. 2 (2011). http://dandelionjournal.org/index.php/dandelion/article/view/78/98. Accessed August 10, 2015.

———. Website. http://www.eleanormorgan.com/filter/Spider/About. Accessed August 10, 2015.

Morley, David, and Kuan-Hsing Chen, eds. *Stuart Hall: Critical Dialogues in Cultural Studies*. London: Routledge, 1996.

Morrison, Toni. *Paradise*. New York: Knopf, 1997.

"Mountain Justice Summer Convergence, 2015." https://www.mountainjustice.org/. Accessed September 1, 2015.

"Mountaintop Removal Mining." *Wikipedia*. http://en.wikipedia.org/wiki/Mountaintop_removal_mining. Accessed September 1, 2015.

Muir, Jim. "The Pigeon Fanciers of Baghdad." BBC, March 20, 2009. 2:20 minutes. http://news.bbc.co.uk/2/hi/middle_east/7954499.stm. Accessed March 20, 2016.

Murphy, Michelle. "Thinking against Population and with Distributed Futures." Paper for "Make Kin Not Babies" panel at the meetings of the Society for Social Studies of Science, Denver, November 14, 2015.

National Oceanic and Atmospheric Administration. Fisheries. "Green Turtles." http://www.nmfs.noaa.gov/pr/species/turtles/green.htm. Updated August 26, 2015. Accessed March 20, 2016.

"Nausicaä: Character." *Wikipedia*. https://en.wikipedia.org/wiki/Nausica%C3%A4_%28character%29. Accessed September 1, 2015.

"Nausicaä of the Valley of the Wind." *Wikipedia*. http://en.wikipedia.org/wiki/Nausica%C3%A4_of_the_Valley_of_the_Wind_%28film%29. Accessed September 1, 2015.

Navajo Sheep Project. http://navajosheepproject.com/intro.html. Accessed August 10, 2015.

———. "History." http://navajosheepproject.com/nsphistory.html. Accessed August 12, 2015.

"Navajo String Games by Grandma Margaret." Posted by Daybreakwarrior, November 27, 2008. 5:35 minutes. http://www.youtube.com/watch?v=5qdcG7Ztn3c. Accessed August 3, 2015.

Needham, Joseph. *The Grand Titration: Science and Society in East and West*. London: Routledge, [1969] 2013.

Nelson, Diane M. *Who Counts? The Mathematics of Death and Life after Genocide*. Durham, NC: Duke University Press, 2015.

Never Alone (Kisima Ingitchuna). http://neveralonegame.com/game/. Accessed August 9, 2015.

Never Alone. Announcement Trailer. Posted May 8, 2014 by IGN. 2:24 minutes. https://www.youtube.com/watch?v=G2C3aIVeL-A. Accessed March 20, 2016.

"*Never Alone* Cultural Insights—Sila Has a Soul." Posted May 23, 2015 by Ahnnoty. 1:30 minutes. https://www.youtube.com/watch?v=sd5etFc_Py4, May 23, 2015. Accessed March 20, 2016.

Nies, Judith. "The Black Mesa Syndrome: Indian Lands, Black Gold." *Orion*, summer 1998. https://orionmagazine.org/article/the-black-mesa-syndrome/. Accessed August 10, 2015.

———. *Unreal City: Las Vegas, Black Mesa, and the Fate of the West*. New York: Nation Books, 2014.

North American Equine Ranching Information Council. "About the Equine Ranching Industry." http://www.naeric.org/about.asp?strNav=11&strBtn. Accessed August 13, 2015.

———. "Equine Veterinarians' Consensus Report on the Care of Horses on PMU Ranches." http://www.naeric.org/about.asp?strNav=0&strBtn=5. Accessed August 13, 2015.

Oberhauser, Karen S., and Michelle J. Solensky, eds. *The Monarch Butterfly: Biology and Conservation*. Ithaca, NY: Cornell University Press, 2004.

Olsson, L., G. S. Levit, and U. Hossfeld. "Evolutionary Developmental Biology: Its Concepts and History with a Focus on Russian and German Contributions." *Naturwissenschaften* 97, no. 11 (2010): 951–69.

Oodshourn, Nelly. *Beyond the Natural Body: An Archaeology of Sex Hormones*. London: Routledge, 1994.

The Original People. "Speakers for the Dead: Documentary about the Original Black Settlers of Princeville, Ontario, Canada." 50 minutes. Posted May 6, 2012. https://www.youtube.com/watch?v=rofbINBjb6I. Accessed March 20, 2016.

"Pacific Islands Ecosystems at Risk." http://www.hear.org/pier/species/acacia_mearnsii.htm. Accessed August 21, 2015.

Paget-Clarke, Nic. "An Interview with Wahleah Johns and Lilian Hill." *Motion Magazine*, June 13, 2004. Kykotsmovi, Hopi Nation, Arizona. http://www.inmotionmagazine.com/global/wj_lh_int.html. Accessed August 12, 2015.

Palese, Blair. "It's Not Just Indigenous Australians v. Adani over a Coal Mine. We Should All Join this Fight." *Guardian*, April 3, 2015. http://www.theguardian.com/commentisfree/2015/apr/03/its-not-just-indigenous-australians-v-adani-over-a-coal-mine-we-should-all-join-this-fight. Accessed August 31, 2015.

Pan-American Society for Evolutionary Developmental Biology. Inaugural Meetings. University of California at Berkeley, August 5–9, 2015. http://www.evodevopanam.org/meetings—events.html. Accessed August 2, 2015.

"Patricia Wright." http://www.patriciawright.org/. Accessed August 11, 2015.

Peabody Energy. "Factsheet: Kayenta." https://mscusppegrs01.blob.core.windows.net/mmfiles/files/factsheets/kayenta.pdf. Accessed August 10, 2015.

———. “Peabody in China.” http://www.peabodyenergy.com/content/145/peabody-in-china. Accessed August 10, 2015.

———. “Powder River Basin and Southwest.” http://www.peabodyenergy.com/content/247/us-mining/powder-river-basin-and-southwest. Accessed August 10, 2015.

Peace Fleece. “Irene Benalley.” http://www.peacefleece.com/irene_bennalley.html. Accessed August 12, 2015.

———. “The Story.” http://www.peacefleece.com/thestory.htm. Accessed August 12, 2015.

Pembina Institute. “Alberta’s Oil Sands.” http://www.pembina.org/oil-sands/os101/alberta. Accessed August 7, 2015.

———. “Oil Sands Solutions.” http://www.pembina.org/oil-sands/solutions. Accessed August 7, 2015.

Perley, Bernard. “Zombie Linguistics: Experts, Endangered Languages and the Curse of Undead Voices.” *Anthropological Forum* 22, no. 2 (2012): 133–49.

Petras, Kathryn. “Making Sense of HRT. Natural? Synthetic? What’s What?” http://www.earlymenopause.com/makingsenseofhrt.htm. Accessed August 13, 2015.

Pfennig, David. “Pfennig Lab.” University of North Carolina at Chapel Hill. http://labs.bio.unc.edu/pfennig/LabSite/Research.html. Accessed August 9, 2015.

Piercy, Marge. *Woman on the Edge of Time.* New York: Knopf, 1976.

PigeonBlog. http://www.pigeonblog.mapyourcity.net/. ISEA ZeroOne San Jose. http://2006.01sj.org/content/view/810/52/. Accessed February 17, 2012. Not online March 20, 2016.

PigeonBlog 2006–2008. http://nideffer.net/shaniweb/pigeonblog.php. Accessed March 20, 2016.

Pignarre, Philippe, and Isabelle Stengers. *La sorcellerie capitaliste: Pratiques de désenvoûtement.* Paris: Découverte, 2005.

“*Pimoa cthulhu.*” *Wikipedia.* https://en.wikipedia.org/wiki/Pimoa_cthulhu. Accessed August 6, 2015.

“Planet of the Ood.” Episode of *Dr. Who*, series 4, April 19, 2008. https://en.wikipedia.org/wiki/Planet_of_the_Ood. Accessed August 8, 2015.

Porcher, Jocelyne. *Vivre avec les animaux: Une utopie pour le XXIe Siècle.* Paris: Découverte, 2011.

Potnia Theron, Kameiros, Rhodes, circa 600 BCE. http://commons.wikimedia.org/wiki/File:Gorgon_Kameiros_BM_GR1860.4-4.2_n2.jpg. Accessed August 8, 2015.

Potts, Annie, in conversation with Donna Haraway. “Kiwi Chicken Advocate Talks with Californian Dog Companion.” In “Feminism, Psychology and Nonhuman Animals,” edited by Annie Potts, special issue, *Feminism and Psychology* 20, no. 3 (August 2010): 318–36.

Poulsen, Michael, et al. “Complementary Symbiont Contributions to Plant Decomposition in a Fungus Farming Termite.” *Proceedings of the National Academy of Sciences* 111, no. 40 (2013): 14500–14505. http://www.pnas.org/content/111/40/14500. Accessed August 9, 2015.

"Premarin Controversy." *Wikipedia*. http://en.wikipedia.org/wiki/Premarin#Controversy. Accessed August 13, 2015.

Prigogine, Ilya, and Isabelle Stengers. *Order Out of Chaos*. New York: Bantam, 1984.

Prior, Helmut, Ariane Schwarz, and Onur Güntürkün. "Mirror-Induced Behavior in the Magpie (*Pica pica*): Evidence of Self-Recognition." *PLoS Biology* 6, no. 8 (2008): e202. doi: 10.1371/journal.pbio.0060202. Accessed August 3, 2015.

Prosek, James. *Eels: An Exploration from New Zealand to the Sargasso, of the World's Most Mysterious Fish*. New York: Harper, 2011.

"Protein Packing: Inner Life of a Cell." Harvard University and XVIVO with BioVisions. Posted by XVIVO Scientific Animation. 2:51 minutes. https://www.youtube.com/user/XVIVOAnimation. Accessed March 20, 2016.

Puig de la Bellacasa, María. "Encountering Bioinfrastructure: Ecological Movements and the Sciences of Soil. *Social Epistemology* 28, no. 1 (2014): 26–40.

———. "Ethical Doings in Naturecultures." *Ethics, Place and Environment* 13, no. 2 (2010): 151–69.

———. "Matters of Care in Technoscience: Assembling Neglected Things." *Social Studies of Science* 41, no. 1 (2011): 85–106.

———. *Matters of Care: Speculative Ethics in More Than Human Worlds*. Minneapolis: University of Minnesota Press, forthcoming 2016.

———. *Penser nous devons: Politiques féminists et construction des saviors*. Paris: Harmattan, 2013.

———. "Touching Technologies, Touching Visions: The Reclaiming of Sensorial Experience and the Politics of Speculative Thinking." *Subjectivity* 28, no. 1 (2009): 297–315.

Pullman, Philip. *His Dark Materials Omnibus: The Golden Compass, The Subtle Knife, The Amber Spyglass*. New York: Knopf, 2007.

Pyle, Robert Michael. *Chasing Monarchs: Migrating with the Butterflies of Passage*. New Haven, CT: Yale University Press, [1999] 2014.

"Racing Pigeon-Post." http://www.articles.racing-pigeon-post.org/directory/articles_index.php. Accessed February 17, 2012.

Raffles, Hugh. *Insectopedia*. New York: Random House, 2010.

Ramberg, Lucinda. *Given to the Goddess: South Indian Devadasis and the Sexuality of Religion*. Durham, NC: Duke University Press, 2014.

———. "Troubling Kinship: Sacred Marriage and Gender Configuration in South India." *American Ethnologist* 40, no. 4 (2013): 661–75.

Raun, A. P., and R. L. Preston. "History of Diethylstilbestrol Use in Cattle." *American Society of Animal Science*, 2002. https://www.asas.org/docs/publications/raunhist.pdf?sfvrsn=0. Accessed August 13, 2015.

Rea, Ba, Karen Oberhauser, and Michael Quinn. *Milkweed, Monarchs and More*. 2nd ed. Union, WV: Bas Relief, 2010.

Reed, Donna, and Starhawk. *Signs out of Time: The Story of Archaeologist Marija Gimbutas*. Documentary film, Belili Productions, 2004. 59 minutes. https://www.youtube.com/watch?v=whfGbPFAy4w. Accessed August 8, 2015.

Ren, C., et al. "Modulation of Peanut-Induced Allergic Immune Responses by Oral Lactic Acid Bacteria-Based Vaccines in Mice." *Applied Microbiological Biotechnology* 98, no. 14 (2014): 6353–64. doi: 10.1007/s00253-014-5678-7.

Rendón-Salinas, E., and G. Tavera-Alonso. "Forest Surface Occupied by Monarch Butterfly Hibernation Colonies in December 2013." Report for the World Wildlife Fund-Mexico.

Robinson, Kim Stanley. *2312*. New York: Orbit/Hatchette, 2012.

Rocheleau, Dianne. "Networked, Rooted and Territorial: Green Grabbing and Resistance in Chiapas." *Journal of Peasant Studies* 42, nos. 3–4 (2015): 695–723.

Rocheleau, Dianne, and David Edmunds. "Women, Men and Trees: Gender, Power and Property in Forest and Agrarian Landscapes." *World Development* 25, no. 8 (1997): 1351–71.

Rohwer, Forest, Victor Seguritan, Farooq Azam, and Nancy Knowlton. "Diversity and Distribution of Coral-Associated Bacteria." *Marine Ecology Progress Series* 243 (2002): 1–10.

Roosth, Sophia. "Evolutionary Yarns in Seahorse Valley: Living Tissues, Wooly Textiles, Theoretical Biologies." *differences* 25, no. 5 (2012): 9–41.

Rose, Deborah Bird. *Reports from a Wild Country: Ethics for Decolonisation*. Sydney: University of New South Wales Press, 2004.

———. "What If the Angel of History Were a Dog?" *Cultural Studies Review* 12, no. 1 (2006): 67–78.

Rosen, Ruth. "Pat Cody: Berkeley's Famous Bookstore Owner and Feminist Health Activist (1923–2010)." *Journal of Women's History* website (online only). Posted 2011 by JWH, Binghamton University, State University of New York. http://bingdev.binghamton.edu/jwh/?page_id=363. Accessed March 20, 2016.

Ross, Alison. "Devilish Ants Control the Garden." BBC News. http://news.bbc.co.uk/2/hi/science/nature/4269544.stm. Accessed August 21, 2015.

Ross, Deborah. "Deborah Ross Arts." http://www.deborahrossarts.com/. Accessed August 11, 2015.

Rowe, Claudia. "Coal Mining on Navajo Nation in Arizona Takes Heavy Toll." *Huffington Post*, June 6, 2013. http://www.huffingtonpost.com/2013/06/06/coal-mining-navajo-nation_n_3397118.html. Accessed August 10, 2015.

Russ, Joanna. *The Adventures of Alyx*. New York: Gregg, 1976.

———. *The Female Man*. New York: Bantam Books, 1975.

Sagan, Lynn. "On the Origin of Mitosing Cells." *Journal of Theoretical Biology* 14, no. 3 (1967): 225–74.

Salomon, F. *The Cord Keepers: Khipus and Cultural Life in a Peruvian Village*. Durham, NC: Duke University Press, 2004.

San Jose Museum of Quilts and Textiles. "Black Mesa Blanket: Enduring Vision, Sustaining Community." http://www.sjquiltmuseum.org/learnmore_BlackMesa.html. Accessed August 11, 2015.

Schmitt, Carl. *The Nomos of the Earth in the International Law of the Jus Publicum Europaeum*. Translated by G. L. Ulmen. Candor, NY: Telos, [1950] 2003.

Scottoline, Lisa. *The Vendetta Defense*. New York: Harper, 2001.

Seaman, Barbara. "Health Activism, American Feminist." *Jewish Women: A Compre-*

hensive Historical Encyclopedia. March 20, 2009. Jewish Women's Archive. http://jwa.org/encyclopedia/article/health-activism-american-feminist. Accessed August 13, 2015.

"Short History of Big Mountain-Black Mesa." Posted by American Indian Cultural Support (AICS) and Mike Wicks, 1998–2006. http://www.aics.org/BM/bm.html. Accessed August 10, 2015.

"Sierra Club Sponsors 'Water Is Life' Forum with Tribal Partners." January 5, 2012. http://blogs.sierraclub.org/scrapbook/2012/01/sierra-club-co-sponsors-water-is-life-forum-with-tribal-partners.html. Accessed August 10, 2015.

Simpson, George Gaylord. *Tempo and Mode in Evolution*. Columbia Classics in Evolution. New York: Columbia University Press, [1944] 1984.

Skurnick, Lizzie. *That Should Be a Word*. New York: Workman, 2015.

The Soufan Group. "TSG IntelBrief: Geostrategic Competition in the Arctic: Routes and Resources." March 6, 2014. http://soufangroup.com/tsg-intelbrief-geostrategic-competition-in-the-arctic-routes-and-resources/. Accessed August 8, 2015.

Soulé, Michael, and John Terborgh, eds. *Continental Conservation: Scientific Foundations of Regional Reserve Networks*. Washington, DC: Island, 1999.

Starhawk. *Truth or Dare: Encounters with Power, Authority, and Mystery*. San Francisco: Harper, 1990.

Starkey, Daniel. "*Never Alone* Review: It's Cold Outside." Eurogamer.net. November 20, 2014. http://www.eurogamer.net/articles/2014-11-20-never-alone. Accessed August 11, 2015.

Steffen, Will, Wendy Broadgate, Lisa Deutsch, Owen Gaffney, and Cornelia Ludwig. "The Trajectory of the Anthropocene: The Great Acceleration." *The Anthropocene Review*, January 16, 2015. doi: 10.1177/2053019614564785. Accessed March 16, 2016.

Stengers, Isabelle. *Au temps des catastrophes: Résister à la barbarie qui vient*. Paris: Découverte, 2009.

———. "The Cosmopolitical Proposal." In *Making Things Public*, edited by Bruno Latour and Peter Weibel, 994–1003. Cambridge, MA: MIT Press, 2005.

———. *Cosmopolitics I* and *Cosmopolitics II*. Translated by Robert Bononno. Minneapolis: University of Minnesota Press, 2010 and 2011.

———. *Hypnose entre magie et science*. Paris: Les Empêcheurs de penser en rond, 2002.

———. "Relaying a War Machine?" In *The Guattari Effect*, edited by Éric Alliez and Andrew Goffey, 134–55. London: Continuum, 2011.

Stengers, Isabelle, and Vinciane Despret. *Les faiseuses d'histoires: Que font les femmes à la pensée?* Paris: Découverte, 2011.

———. *Women Who Make a Fuss: The Unfaithful Daughters of Virginia Woolf*. Translated by April Knutson. Minneapolis: Univocal, 2014.

Stengers, Isabelle, in conversation with Heather Davis and Etienne Turpin. "Matters of Cosmopolitics: On the Provocations of Gaïa." In *Architecture in the Anthropocene: Encounters among Design, Deep Time, Science and Philosophy*, edited by Etienne Turpin, 171–82. London: Open Humanities, 2013.

Stephens, Beth, with Annie Sprinkle. *Goodbye Gauley Mountain: An Ecosexual Love Story*. http://goodbyegauleymountain.org/. Accessed August 6, 2015.

Stephens, Elizabeth. "Goodbye Gauley Mountain." http://elizabethstephens.org/good-bye-gauley-mountain/. Accessed September 1, 2015.

Stories for Change. Site sponsored by massIMPACT. http://storiesforchange.net/. Accessed March 19, 2016.

Strathern, Marilyn. *The Gender of the Gift: Problems with Women and Problems with Society in Melanesia*. Berkeley: University of California Press, 1990.

———. *Kinship, Law and the Unexpected: Relatives Are Always a Surprise*. Cambridge: Cambridge University Press, 2005.

———. *Partial Connections*. Lanham, MD: Rowman and Littlefield, 1991.

———. *The Relation: Issues in Complexity and Scale*. Cambridge, UK: Prickly Pear, 1995.

———. *Reproducing the Future*. Manchester, UK: Manchester University Press, 1992.

———. "Shifting Relations." Paper for the Emerging Worlds Workshop, University of California at Santa Cruz, February 8, 2013.

Strawn, Susan, and Mary Littrel. "Returning Navajo-Churro Sheep for Weaving." *Textile* 5 (2007): 300–319.

Street Art SF Team. "The Bird Man of the Mission." October 7, 2014. http://www.streetartsf.com/blog/the-bird-man-of-the-mission/. Accessed September 28, 2015.

Styger, Erica, Harivelo M. Rakotondramasy, Max J. Pfeffer, Erick C. M. Fernandes, and David M. Bates. "Influence of Slash-and-Burn Farming Practices on Fallow Succession and Land Degradation in the Rainforest Region of Madagascar." *Agriculture, Ecosystems, and Environment* 119 (2007): 257–69.

"Survival and Revival of the String Figures of Yirrkala." http://australianmuseum.net.au/Survival-and-Revival-of-the-String-Figures-of-Yirrkala. Updated March 19, 2015. Accessed August 3, 2015.

Survival International. "Shifting Cultivation." http://www.survivalinternational.org/about/swidden. Accessed August 11, 2015.

Svenson-Arveland, et al. "The Human Fetal Placenta Promotes Tolerance against the Semiallogenic Fetus by Producing Regulatory T Cells and Homeostatic M_2 Macrophages." *Journal of Immunology* 194, no. 4 (February 15, 2015): 1534–44. http://www.jimmunol.org/content/194/4/1534. Accessed September 1, 2015.

Tagaq, Tanya. "Animism." http://tanyatagaq.com/. Accessed September 3, 2015.

———. "*Animism*—Album Trailer." May 5, 2014. 1:25 minutes. https://www.youtube.com/watch?v=ItYoFr3LpDw&feature=youtu.be. Accessed September 3, 2015.

———. "Tagaq Brings Animism to Studio Q." Interview with Jian Gomeshi. Posted by Q on CBC on May 27, 2014. 16:50 minutes. https://www.youtube.com/watch?v=ZuTIySphv2w. Accessed March 20, 2016.

———. "Tanya Tagaq's Polaris Prize Performance and Introduction." Polaris Music Prize Gala, September 27, 2014. Full show 3:52:36. http://tanyatagaq.com/2014/09/tanya-tagaqs-polaris-prize-performance-introduction/. Accessed March 20, 2016.

Takahashi, Dean. "After *Never Alone*, E-Line Media and Alaska Native Group See Big Opportunity in 'World Games.'" *GamesBeat*. February 5, 2015. http://venturebeat.com/2015/02/05/after-never-alone-e-line-media-and-alaska-native-group-see-big-opportunity-in-world-games/. Accessed August 11, 2015.

Talen, Reverend Billie. "Beware of the Robobee, Monsanto and DARPA." June 4, 2014. http://www.revbilly.com/beware_of_the_robobee_monsanto_and_darpa. Accessed August 8, 2015.

TallBear, Kim. http://www.kimtallbear.com/. Accessed September 24, 2015.

———. "Failed Settler Kinship, Truth and Reconciliation, and Science." Posted March 3, 2016. http://www.kimtallbear.com/homeblog/failed-settler-kinship-truth-and-reconciliation-and-science. Accessed March 17, 2016.

———. "Making Love and Relations Beyond Settler Sexualities." Lecture given for the Social Justice Institute, University of British Columbia. Published February 24, 2016. 55:39 minutes. https://www.youtube.com/watch?v=zfdo2ujRUv8. Accessed March 19, 2016.

Tao, Leiling, Camden D. Gowler, Aamina Ahmad, Mark D. Hunter, and Jacobus C. de Roode. "Disease Ecology across Soil Boundaries: Effects of Below-Ground Fungi on Above-Ground Host–Parasite Interactions." *Proceedings of the Royal Society B* 282, no. 1817 (October 22, 2015). doi: 10.1098/rspb.2015.1993. Accessed March 20, 2016.

Tar Sands Solutions Network. http://tarsandssolutions.org/about/. Accessed August 7, 2015.

Tate, Andrew, Hanno Fischer, Andrea Leigh, and Keith Kendrick. "Behavioural and Neurophysiological Evidence for Face Identity and Face Emotion Processing in Animals." *Philosophical Transactions of the Royal Society B* 361, no. 1476 (2006): 2155–72. doi: 10.1098/rstb.2006.1937. Accessed August 12, 2015.

Tauber, Alfred. "Reframing Developmental Biology and Building Evolutionary Theory's New Synthesis." *Perspectives in Biology and Medicine* 53, no. 2 (2010): 257–70. doi: 10.1353/pbm.0.0149. Accessed August 2, 2015.

Teller, Terry. "So Naal Kaah, Navajo Astronomy." http://www.angelfire.com/rock3/countryboy79/navajo_astronomy.html. Accessed August 3, 2015.

Terranova, Fabrizio. 2016. *Donna Haraway: Story Telling for Earthly Survival*. 86 minutes. l'Atelier Graphoui and Spectres Production. Premier May 2016 at le KunstenFestivaldesArts, Brussels.

"The Thousand Names of Gaia/Os Mil Nomes de Gaia: From the Anthropocene to the Age of the Earth." Conference in Rio de Janeiro, September 15–19, 2014. https://thethousandnamesofgaia.wordpress.com/. Accessed August 8, 2015.

———. Videos. https://www.youtube.com/c/osmilnomesdegaia. Accessed August 8, 2015.

Toda, Koji, and Shigeru Watanabe. "Discrimination of Moving Video Images of Self by Pigeons (*Columba livia*)." *Animal Cognition* 11, no. 4 (2008): 699–705. doi: 10.1007/s10071-008-0161-4. Accessed August 12, 2015.

"A Tribute to Barbara Seaman: Triggering a Revolution in Women's Health Care." *On the Issues Magazine* (fall 2012). http://www.ontheissuesmagazine.com/11spring/2011spring_tribute.php. Accessed August 13, 2015.

Trujillo, Juan. "The World Water Forum: A Dispute over Life." *The Narcosphere*. Posted in Spanish, March 17, 2006; in English, March 23, 2006. http://narcosphere.narconews.com/notebook/juan-trujillo/2006/03/the-world-water-forum-a-dispute-over-life. Accessed September 3, 2015.

Tsing, Anna. "Feral Biologies." Paper for "Anthropological Visions of Sustainable Futures" conference, University College London, February 12–14, 2015.

———. *Friction: An Ethnography of Global Connection*. Princeton, NJ: Princeton University Press, 2005.

———. *The Mushroom at the End of the World: On the Possibility of Life in Capitalist Ruins*. Princeton, NJ: Princeton University Press, 2015.

———. "A Threat to Holocene Resurgence Is a Threat to Livability." Unpublished manuscript, 2015.

———. "Unruly Edges: Mushrooms as Companion Species." *Environmental Humanities* 1 (2012): 141–54.

Tsing, Anna, Nils Bubandt, Elaine Gan, and Heather Anne Swanson, eds. *Arts of Living on a Damaged Planet: Stories from the Anthropocene*. Minneapolis: University of Minnesota, forthcoming.

Tsing, Anna, Nils Bubandt, Noboru Ishikawa, Donna Haraway, Scott F. Gilbert, and Kenneth Olwig. "Anthropologists Are Talking about the Anthropocene." *Ethnos* 81, no. 4 (2016): 1–30. doi: 10.1080/00141844.2015.1105838. Accessed March 20, 2016.

Tsutsumi Chunagon Monogatari. https://en.wikipedia.org/wiki/Tsutsumi_Ch%C5%ABnagon_Monogatari. Modified February 23, 2016. Accessed March 20, 2016.

Tucker, Catherine M. "Community Institutions and Forest Management in Mexico's Monarch Butterfly Reserve." *Society and Natural Resources* 17 (2004): 569–87.

"Turkish Tumblers.com." http://turkishtumblers.com/. Accessed August 3, 2015.

United Nations. "World Population Prospects: Key Findings and Advance Tables, 2015 Revision." Population Division of the Department of Economic and Social Affairs. http://esa.un.org/unpd/wpp/Publications/Files/Key_Findings_WPP_2015.pdf. Accessed September 29, 2015.

University of Alaska Fairbanks. "Alaska Native Language Center." "Inupiaq." https://www.uaf.edu/anlc/languages/i/. Revised January 1, 2007. Accessed September 25, 2015.

U.S. Coast Guard. "Pigeon Search and Rescue Project, Project Sea Hunt." http://www.uscg.mil/history/articles/PigeonSARProject.asp. Modified January 12, 2016. Accessed March 20, 2016.

U.S. Fish and Wildlife Service. "The American Eel." April 29, 2014. http://www.fws.gov/northeast/newsroom/eels.html. Accessed September 1, 2015.

Utah Indian Curriculum Guide. "We Shall Remain: Utah Indian Elementary Curriculum Guide—The Goshutes: The Use of Storytelling in the Transmission of Goshute Culture." Digitized 2009. http://content.lib.utah.edu/cdm/ref/collection/uaida/id/17874. Accessed March 20, 2016.

ValBio. "ICTE-Centre ValBio Publications." www.stonybrook.edu/commcms/centre-valbio/research/publications.html. Accessed August 24, 2015.

Vance, Dwight A. "Premarin: The Intriguing History of a Controversial Drug." *International Journal of Pharmaceutical Compounding* (July/August 2007): 282–86. http://www.ijpc.com/abstracts/abstract.cfm?ABS=2619. Accessed August 13, 2015.

van Dooren, Thom. *Flight Ways: Life at the Edge of Extinction*. New York: Columbia University Press, 2014.

———. "Keeping Faith with Death: Mourning and De-extinction." November 10, 2013. http://extinctionstudies.org/2013/11/10/keeping-faith-with-death-mourning-and-de-extinction/. Accessed August 6, 2015.

van Dooren, Thom, and Vinciane Despret. "Evolution: Lessons from Some Cooperative Ravens." In *The Edinburgh Companion to Animal Studies*, edited by Lynn Turner, Ron Broglio, and Undine Sellbach. In progress.

van Dooren, Thom, and Deborah Bird Rose. "Storied-Places in a Multispecies City." *Humanimalia: A Journal of Human/Animal Interface Studies* 3, no. 2 (2012): 1–27.

———. "Unloved Others: Death of the Disregarded in the Time of Extinctions." Special issue of *Australian Humanities Review* 50 (May 2011).

Varley, John. Gaea trilogy: *Titan* (1979), *Wizard* (1980), and *Demon* (1984). New York: Berkeley Books.

Vidal, Omar, José López-Garcia, and Eduardo Rendón-Salinas. "Trends in Deforestation and Forest Degradation in the Monarch Butterfly Biosphere Reserve in Mexico." *Conservation Biology* 28, no. 1 (2013): 177–86.

Voices for Biodiversity. "The Sixth Great Extinction." http://newswatch.nationalgeographic.com/2012/03/28/the-sixth-great-extinction-a-silent-extermination/. Accessed August 7, 2015.

Walcott, Charles. "Pigeon Homing: Observations, Experiments and Confusions." *Journal of Experimental Biology* 199 (1996): 21–27. http://jeb.biologists.org/content/199/1/21.full.pdf. Accessed August 3, 2015.

Walters, Sarah. "Holobionts and the Hologenome Theory." *Investigate: A Research and Science Blog*. September 4, 2013. http://www.intellectualventureslab.com/investigate/holobionts-and-the-hologenome-theory. Accessed August 9, 2015.

Watanabe, Shigeru, Junko Sakamoto, and Masumi Wakita. "Pigeons' Discrimination of Paintings by Monet and Picasso." *Journal of the Experimental Analysis of Behavior* 63, no. 2 (March 1995): 165–74. doi: 10.1901/eab.1995.63-165. Accessed August 3, 2015.

"Water Management in Greater Mexico City." *Wikipedia*. http://en.wikipedia.org/wiki/Water_management_in_Greater_Mexico_City. Modified February 27, 2016. Accessed March 20, 2016.

Weaver, Harlan. "'Becoming in Kind': Race, Class, Gender, and Nation in Cultures of Dog Rescue and Dogfighting." *American Quarterly* 65, no. 3 (2013): 689–709.

———. "Trans Species." *Transgender Studies Quarterly* 1, nos. 1–2 (2014): 253–54. doi: 10.1215/23289252-2400100.

"Weaving in Beauty." Posted April 2, 2009 by Mary Walker. http://weavinginbeauty.com/its-all-about-the-rugs/2009-heard-museum-guild-indian-market-dy-begay-and-berdina-charley. Accessed August 11, 2015.

Weber, Bob. "Rebuilding Land Destroyed by Oil Sands May Not Restore It, Researchers Say." *Globe and Mail*, March 11, 2012. http://www.theglobeandmail.com/news/national/rebuilding-land-destroyed-by-oil-sands-may-not-restore-it-researchers-say/article552879/. Accessed August 7, 2015.

Weisiger, Marsha. *Dreaming of Sheep in Navajo Country*. Seattle: University of Washington Press, 2009.

———. "Gendered Injustice: Navajo Livestock Reduction in the New Deal Era." *Western Historical Quarterly* 38, no. 4 (winter 2007): 437–55.

Weller, Frank. *Equine Angels: Stories of Rescue, Love, and Hope*. Guilford, CT: Lyons, 2008.

Wertheim, Christine. "CalArts Faculty Staff Directory." https://directory.calarts.edu/directory/christine-wertheim. Accessed August 11, 2015.

Wertheim, Margaret. "The Beautiful Math of Coral." TED video. 15:33 minutes. Posted February 2009. http://www.ted.com/talks/margaret_wertheim_crochets_the_coral_reef?language=en. Accessed August 11, 2015.

———. *A Field Guide to Hyperbolic Space*. Los Angeles: Institute for Figuring, 2007.

Wertheim, Margaret, and Christine Wertheim. *Crochet Coral Reef: A Project by the Institute for Figuring*. Los Angeles: IFF, 2015.

West Virginia Department of Natural Resources. "Rare, Threatened, and Endangered Animals." http://www.wvdnr.gov/Wildlife/PDFFiles/RTE_Animals_2012.pdf. Accessed September 1, 2015.

"West Virginia State Butterfly." http://www.netstate.com/states/symb/butterflies/wv_monarch_butterfly.htm. Accessed September 1, 2015.

White, Richard. *The Roots of Dependency: Subsistence, Environment, and Social Change among the Choctaws, Pawnees, and Navajos*. Lincoln: University of Nebraska Press, 1983.

Whitehead, Alfred North. *The Adventures of Ideas*. New York: Macmillan, 1933.

Wickstrom, Stephanie. "Cultural Politics and the Essence of Life: Who Controls the Water?" In *Environmental Justice in Latin America: Problems, Promise, and Practice*, edited by David V. Carruthers. Cambridge, MA: MIT Press, 2008.

Wilks, John. "The Comparative Potencies of Birth Control and Menopausal Hormone Drug Use." *Life Issues.net*. http://www.lifeissues.net/writers/wilks/wilks_06hormonaldruguse.html. Accessed August 13, 2015.

Willink, Roseann S., and Paul G. Zolbrod. *Weaving a World: Textiles and the Navajo Way of Seeing*. Santa Fe: Museum of New Mexico Press, 1996.

Wilson, Kalpana. "The 'New' Global Population Control Policies: Fueling India's Sterilization Atrocities." *Different Takes* (winter 2015). http://popdev.hampshire.edu/projects/dt/87. Accessed August 12, 2015.

Witherspoon, Gary, and Glen Peterson. *Dynamic Symmetry and Holistic Asymmetry*. New York: Peter Lang, 1995.

Women's Health Initiative. "Risks and Benefits of Estrogen Plus Progestin in Healthy Postmenopausal Women." *Journal of the American Medical Association* 288 (2002): 321–33.

World-Ecology Research Network. https://www.facebook.com/pages/World-Ecology-Research-Network/174713375900335. Accessed August 7, 2015.

"World Market in Pigeons." http://www.euro.rml-international.org/World_Market.html. Accessed August 3, 2015.

World Wildlife Fund. "Living Blue Planet: Crisis in Global Oceans as Marine Species Halve in Size since 1970." September 15, 2015. http://assets.wwf.org.uk/custom/stories/living_blue_planet/. Accessed October 13, 2015.

Wright, P. C., and B. A. Andriamihaja. "Making a Rain Forest National Park Work in Madagascar: Ranomafana National Park and Its Long-Term Commitment." In *Making Parks Work: Strategies for Preserving Tropical Nature*, edited by J. Terborgh et al., 112–36. Washington, DC: Island, 2002.

"Wurundjeri." *Wikipedia*. http://en.wikipedia.org/wiki/Wurundjeri. Modified July 9, 2015. Accessed August 3, 2015.

Xena Warrior Princess. "Dreamworker." Series 1, September 18, 1995. http://www.imdb.com/title/tt0751475/. Accessed August 10, 2015.

xkcd. "Bee Orchid." https://xkcd.com/1259/. Accessed August 10, 2015.

Yellowstone to Yukon Conservation Initiative. http://y2y.net/work/what-hot-projects. Accessed August 31, 2015.

Yong, Ed. "Bacteria Transform the Closest Living Relatives of Animals from Single Cells into Colonies." *Discover*, August 6, 2012. http://blogs.discovermagazine.com/notrocketscience/2012/08/06/bacteria-transform-the-closest-living-relatives-of-animals-form-single-cells-into-colonies/#.VXYon6YVpFU. Accessed August 9, 2015.

———. "Consider the Sponge." *New Yorker*, April 24, 2015. http://www.newyorker.com/tech/elements/consider-the-sponge. Accessed August 9, 2015.

———. "The Guts That Scrape the Skies." *Phenomena: Not Exactly Rocket Science*. September 23, 2014. http://phenomena.nationalgeographic.com/2014/09/23/the-guts-that-scrape-the-skies/. Accessed August 9, 2015.

Youth, Howard. "Pigeons: Masters of Pomp and Circumstance." Smithsonian National Zoological Park. *Zoogoer* 27 (1998). http://nationalzoo.si.edu/Publications/ZooGoer/1998/6/pigeons.cfm. Accessed February 17, 2012.

Zalasiewicz, Jan, et al. "Are We Now Living in the Anthropocene?" *GSA (Geophysical Society of America) Today* 18, no. 2 (2008): 4–8.

"Zapatista Army of Mazahua Women in Defence of Water in the Cutzamala Region: Testimonies." *Development* 54, no. 4 (2011): 499–504.

Zebich-Knos, Michele. "A Good Neighbor Policy? Ecotourism, Park Systems and Environmental Justice in Latin America." Working paper presented at the 2006 Meeting of the Latin American Studies Association, San Juan, Puerto Rico, March 15–18, 2006.

Zimmer, Carl. "Watch Proteins Do the Jitterbug." *New York Times*, April 10, 2014. http://www.nytimes.com/2014/04/10/science/watch-proteins-do-the-jitterbug.html?_r=1. Accessed August 6, 2015.

Zolbrod, Paul G. *Diné Bahane': The Navajo Creation Story*. Albuquerque: University of New Mexico Press, 1984.

Zoutini, Benedikte, Lucienne Strivay, and Fabrizio Terranova. "Les enfants du compost, les enfants des monarques: Retour sur l'atelier 'Narrations spéculatives.'" In *Gestes spéculatifs*, edited by Isabelle Stengers. Paris: Hermann, 2015.

Index

Page numbers followed by *f* refer to illustrations.

Aboriginal peoples, 27, 80, 165, 184n47, 214n17, 218n11; languages of, 26, 173n33. *See also under names of specific peoples*

abundance: Children of Compost on, 136; new kinds of, 147; of oddkin, 145; of futures, 221n17

acacias: *Acacia verticulata*, 7, 123, 124, 214n13; ants and, 7, 122, 124, 125; Le Guin's story of, 121–24; types of, 118, 123

Adam/Adama, 12, 170n3

Advanced Energy for Life, 194n37

Aegis, 54

aerial surveillance vehicles, 22

Africa, 15, 72, 118, 123, 124; South, 177n24, 214n13. *See also names of specific countries*

African Americans, 24, 205n93, 207n12, 219n11. *See also* blacks

agential realism, 34

agriculture and agribusiness, 93, 99, 114, 181n42, 213n9, 222n22; Big, 115; in Camille stories, 152; in Communities of Compost, 141, 160; DES and Premarin and, 109, 115; shifting cultivation, 82, 196–97n58

air pollution, 20, 21, 74; PigeonBlog and, 20–24

Ako Project, 71, 81–85, 195n56

Alaska: native languages of, 192n29; *Never Alone* and, 71, 87

Alberta, Canada, 44; tar sands oil extraction in, 183–84n47, 184n48

alignment: as metaphor, 41–42

altruism, by babblers, 127–28

American Geophysical Union Meetings, 47, 187n63

American Racing Pigeon Union, 24

ancestors, 74, 103; in Communities of Compost, 154, 156, 157

Anidjar, Gil, 179n35

animal, animals, 65–66, 108, 161*f*; abuse of, 23; in art, 23; bacteria and, 67;

animal, (*continued*)
becoming-with, 28; bonding with Children of Compost, 218n8; breeders of, 113, 129; communication, 122, 129, 151; as coshapers with humans and cyborgs, 20; as daemons, 162; domestication of, 181n42; flourishing of, 111; genomes of, 65; industrial production of, 113; keystone, 223n22; lovers of, 111; modern, 65; multicellularity of, 64–65; murder of, 92, 93; partners and companions with humans, 16, 23, 24, 89, 109–10, 114, 118, 129, 140–41, 146, 149, 151, 166, 221n21; rights of, 7, 24, 113, 172n30; self-recognition of, 18; slaughter of, 109; as symbionts, 8, 67, 139–41, 144, 146–47, 149; as weapons, 22. *See also* animism; companion species; creatures; critters; *and under names of specific animals*
"Animals in a Bacterial World," 67, 191n21, 192n23
animism, 89, 97, 162, 199n68; Tagaq and experimental practice of, 164, 165; sensible materialism and, 88, 165
Anthropocene, 5, 44–47, 52, 72, 100, 174n4, 184n50, 198n65; Arctic warming in, 73, 88; as boundary event, 100, 102, 206n4; burning and fire in, 44, 45*f*, 46, 90, 184n48, 186n61; conceptualization of, 62; etymology of, 44; human-induced climate change in, 44, 45, 56; colonial and imperial, 202n80; compared to Capitalocene, 47, 55, 56; chthonic ones and, 53; Chthulucene and, 55, 76; dying coral reefs in, 56, 72; damaged worlds of, 33, 137; dating of, 181n42; destruction of refuge in, 192n28; discomfort in, 180n36; discourse of, 185n53; Earthbound in, 50; effects of man on earth during, 44, 47, 74; evolution in, 62; extinctions and exterminations of, 86; fossil burning in, 55, 193n36; Gaia and, 51, 175n11, 189n6; game over in, 56; globalizing extravaganza of, 45, 182n45; Holocene vs., 100, 178n31; horrors of, 3, 36, 57; icon for, 45*f*; imperial, 71; inflection point of, 100, 206n5; Latour on, 40–43; naming of, 31, 36, 99, 179–80n35; Native Climate Wisdom for, 87; objections to, 49–50; panics in, 55; recuperation in, 47; scandals of, 2; science art activism in, 71, 79; Stengers on, 43; stories of, 49; symbiogenesis in, 57; sympoiesis in, 57, 88; transformative time of, 31; trash of, 57; urgencies of, 7, 35, 67, 69, 223n26
Anthropocene Working Group, 182n44
Anthropogenic changes, 45, 52, 76, 99, 181n42, 208n18
Anthropos, 2, 39, 174n4; Children of Compost and, 149; epoch and, 47, 48, 55, 57; as fossil-making man, 46; fracking and, 47; god-like, 50; meeting chthonic ones, 54; myth system and, 49; not humus, 149; sky-gazing, 53; as term, 30, 184n45
Anthropo-zoo-genesis, 127–29, 132
ants: acacias and, 7, 122, 124, 125; Le Guin's story of, 121–24; Myrmex and, 213–14n9; SF stories and, 120; species of, 124
apocalypse, 3, 4, 55, 231n8; myths and narrative of, 35, 37, 150
Appalachian region, 141, 221–23n22, 225n31. *See also* West Virginia
Arabian babblers, 7, 127–28, 130
archaea and bacteria, 61, 190n16; Margulis on, 60–61, 64
Arctic, 198n65; resources and security in, 184n48; sympoiesis in, 88; warming of, 73, 76
arctic fox, in *Never Alone*, 87, 88, 198n64
Arendt, Hannah, 5, 36–37, 177n18; on visiting, 127, 130, 177n19
Arizona, 71, 74–75, 95, 170n11, 213n9, 226n39. *See also Black Mesa entries*
art, 89; in action, 21; animals in, 23; needed by biologies and politics, 98; collaborative, 78; in Compost Communities, 140, 150; da Costa's research and, 24; design-activist practices in, 133; fan fiction as, 136; of fiber, 76, 78, 79; markets for, 92, 201n77; Navajo weaving

as, 89–91; Batman Park's pigeon loft as, 25; research and, 24, 27; science worldings of, 64, 67, 69, 71–72, 86, 97; in SF, 81; of feminist speculative fabulation, 12; terraforming as, 11; thinking and, 200n75; vegetative, 122
Artemis, 52
art/science activism, 5, 175n10; art activism and, 176–77n17; Crochet Coral Reef as, 79; da Costa as, 26; PigeonBlog, 20–22, 132; pigeon loft as, 133; pigeons as collaborators in, 5
"arts of living on a damaged planet," 85, 87, 152; Tsing on, 37, 87, 136
Ashevak, Kenojuak: *Animals of Land and Sea*, 161*f*
assemblages, 42, 43, 78, 103, 192n28, 218n8; ecological, 58; fungi and, 37; multispecies, 101, 193n34; natural-cultural, 38; of species, 99–100, 178n31; symbiotic, 60
Athena, 54, 213–14n9
attachment sites, 10, 14, 116, 144, 190n16
attachment theory (Bowlby), 19, 20
attunement, 7, 10, 39, 98, 126, 128, 227n41, 228n49
Australia, 177n24, 217n6; Great Barrier Reef, 71, 79, 80, 90; string figures in, 172–73n33
autopoiesis, autopoietic systems, 33, 43, 47, 49, 189n6; defined, 58; Margulis and, 61, 189n4; sympoiesis and, 176n13, 180n36, 180n38
Ayerest, 111

babies, 102, 208n16, 208–9n18, 217n7, 227n41; in Camille stories, 134, 136, 139, 140, 146, 147, 154. *See also* "Make Kin Not Babies"
bacteria, 102, 185n56, 190n16; acacia gum and, 123; animals and, 67; in choanoflagellate-bacteria model, 64–65; as communicators, 122; fungus and, 123; as greatest terraformers, 99; Hawaiian bobtail squid and symbiont, 66; Margulis and, 60–61; string figures and, 66; symbiotic, 224n30; symbiogenesis, 60–62. *See also* archaea and bacteria
"banality of evil" (Arendt), 36
Barad, Karen, 34; *Meeting the Universe Halfway*, 175n12, 205n1
Barash, David, 175–76n12
Batman, John, 27
Batman Park: effort to deal with pigeon-human conflict, 28; pigeon loft in, 26, 27, 28*f*
Bechtel Corporation, 73, 74
"becoming involved in one another's lives," 69, 71, 76
becoming-with, 11, 12, 25, 60, 78, 107, 119, 125, 212n1; animals and bacteria, 65; Camille stories and, 137, 148, 150, 152, 168; in Chthulucene, 55, 188n69; companion species and, 13, 110; Despret and, 12, 25; SF and, 3, 71; multispecies and, 63, 71, 78; necessity of, 4; pigeons and humans, 15–20, 22, 24, 28; stories of, 40, 119; symanimagenesis and, 154; technoscience and, 104
bee, bees: acacias and, 123; communication and, 68; Deborah and, 187n63; extinction of, 5, 98; orchids and, 69, 70*f*, 192n27; Potnia Melissa and, 52; Potnia Theron and, 186n61; robo-, 187n67
"Bee Orchid," 69, 70*f*
Begay, 192n2
Begay, D. Y., 201–2n78
Begay, Glenna, 95
Begay, Jay, 204n88
belief, 36, 42, 51, 88, 178n34, 199–200n72
Berenty Primate Reserve (Madagascar), ring-tailed lemurs in, 81
Berokoff, Tanya, 19, 20
Betsimisaraka people, 197n58
Big Capital, 160, 186n57
Big Energy, 160, 183n46
big-enough stories (Clifford), 50, 52, 54, 101, 185n54
Big Mountain, 73, 92, 203n81. *See also* Black Mesa; Dzil ni Staa
Big Pharma, 7, 108, 110, 115
biocapital, 106

biodiversity, 44, 49, 82, 100, 218n9; acacias and, 123; in Appalachians, 222–23n22; of coral reefs, 72, 79; destruction of, 47, 180n37; education for, 197n61; loss of, 180n37, 197n58; in Madagascar, 82, 83, 85, 197n58; marine, 193n32; refuges for, 206n5; terra and, 55, 56
biologists, 5, 7, 60, 68, 72, 123, 185n53, 190n10
biology, 33, 62–63, 98, 154, 189n7, 190n10; cell, 60, 65; contemporary, 174n7; developmental, 64, 66; domains of field, 66, 79; historians of, 191n21; marine, 66, 78, 79; models and, 63. *See also* EcoDev; EcoEvoDevo; ecological evolutionary developmental biology; EvoDevo
biopolitics, 6, 99, 106, 170n3, 211n6
Bird, Elizabeth, 216n4
birds, 39, 54, 127–28, 146, 222–23n22; extinction of species of, 27, 180n37, 221n22. *See also* pigeons; *and under specific names of other bird species*
birth rates, 102, 145, 209n18
blackbird singing, 126, 127, 130
#Black Lives Matter, 207n12
Black Mesa, 73–76, 225n31; art/science activisms in, 5, 95; forced removal from, 203n81; in Navajo cosmology, 96–97; Navajo weaving and, 89; Peabody Energy and, 73–76, 193–94n37; sedentarization and, 196n58; as world worth fighting for, 97. *See also* Big Mountain; Dzil ni Staa
Black Mesa Blankets, 200n74
Black Mesa coalitional work, 96, 97, 217n6; mine closed by, 74, 193n35; non-Native allies and, 203n81. *See also under names of specific organizations*
Black Mesa Indigenous Support, 71, 192n30, 203n81
Black Mesa Trust (Hopi), 71, 192n30
Black Mesa Water Coalition (BMWC), 96–97, 192n30, 196n58, 200n74, 204n91
Black Mesa Weavers for Life and Land, 71, 95, 192n30, 200n74
blacks, 24, 207n12, 228n45. *See also* African Americans
Blue Ridge Mountains, 219n11
bodily modifications, in Communities of Compost, 140–41
boundary, boundaries, 33, 43, 61, 176n13, 195n58
boundary event, 100, 102, 206n4
bounded individualism, 5, 30, 33, 49
bounded units, 60, 62, 67
Bowlby, John, 19, 20
Boyden, John, 75, 194n39
Bubandt, Nils Ole, 185n53, 206n5
Buddhism, 175–76n12
bullying: in Camille stories, 150, 160, pigeons and, 20
Burning Man Festival, 182–83n45
Butler, Octavia, 7, 118–19; Parable novels of, 119–20, 213n3, 213n8
butterflies drinking turtle tears, 156, 157, 227n41

California, 31, 44, 52, 94, 119, 213n8; monarch butterflies in, 142, 153*f*, 153–54, 219n12
California, Southern, 20, 21, 124, 203n81, 219n12, 226n39; pigeon racing, 10, 16–20, 132–33; PigeonBlog, 20–24, 23*f*
camels, 131–32
Camille 1, 142–54; American kestrels and, 222n22; as charmed, 224n29; compostist and, 148; gender and, 221n18; heritage of, 225n31; Kess as friend, 149; nature and, 153, 154; *Nausicaä of the Valley of the Wind*, 151, 152; response-ability by parent, 143; response-ability of, 152; as symbiogenetic with Monarch butterfly, 142, 148, 154
Camille 2, 152–58, 225n32; chin implants of, 152; cosmopolitics and, 157; education by Mazahua activist women, 155, 156, 224n30; introduced to Monarcas, 154; “Soy Mazahua,” 156
Camille 3, 159–62; cosmopolitics and, 161; teaches Camille 4, 162; travels of, 160

Camille 4, 162–65, 227n44; mentoring Camille 5, 162–64
Camille 5, 143, 166–67; initiation of, 163; mentored by Camille 4, 162–64; as Speaker for the Dead, 164, 166
Camille stories, 134–68, 216–17n4, 218n8; agribusiness and, 152; animisms in, 162; becoming-with and, 137, 148, 150, 152, 168; as a Child of Compost, 134, 136; rapid climate change in, 159; collaborations in, 154; flourishing in, 168; holobionts in, 148; humanity as humus in, 160; kin making in, 159, 160, 162, 216–17n4; knots and, 143; Mazahua, 156, 157, 167; naturalsocial becoming-with in, 148; non-symbionts felt threatened in, 149; origin of, 136, 138; people move to West Virginia, 141; relays of, 140; reproductive choice in, 142; resistance in, 137, 155; speculative fabulation and, 8, 134, 136; stories of, 144–68; string figures and, 144; sympoiesis in, 155, 160; visiting in, 152; world of, 137–43
Canada, 195n56, 222n22; Arctic and, 184n48; forest fires in, 45*f*; Inuit language of, 192n29; monarch butterflies in, 8, 142, 152; Premarin and, 111–12; tar sands extraction in, 183n47, 184n48, 217n6, 225n31
cancer: adenocarcinoma, 106; breast, 107, 108, 109, 112, 210n2; ovarian, 211n4; prostate, 107. *See also* carcinogens
capital, capitalism, 2, 7, 50, 73, 93, 176n12, 182n43, 185n52, 208n18; anti-, 6; Big Capital and, 160, 186n57; consumer, 223n26; global, 47, 123, 206n5; heterogeneous, 36; nature and, 185n52; undifferentiated, 37. *See also* Capitalocene
Capitalocene, 5, 47–51, 55, 99, 100, 102, 176n12, 180n36; Arctic in, 73; art activism in, 71, 79; chthonic ones not friends with, 53; Chthulucene and, 76, 185–86n56; colonial and imperial, 202n80; displacement in, 206n5; extinctions in, 57, 86, 159, 162, 164; game over in, 56, 57; globalization of, 48; Great Acceleration of, 144; horrors of, 3; Icon for the Capitalocene: Sea Ice Clearing from the Northwest Passage, 48*f*; response-ability in, 176n12; scandals of, 2; sustainability and, 183n46; as time of burning, 90; urgencies of, 7, 69, 137; unworldings during, 56, 57; as word, 184–85n50
Capsule (Crasset), 25, 26, 133
carbon: as burden, 183n45; from fossil fuels, 46, 73; Third Age of, 47, 48
carcinogens, 108, 210n2, 211n4. *See also* cancer
Card, Orson Scott: *Ender's Game*, 227–28n45; *Speaker for the Dead*, 69, 101, 227–28n45
care, caring: for animals, 111, 113, 129; for country, 27, 49, 80, 96; Hawaiian babblers and, 128; for kin, 103; lack of, 36; material play and, 79; matters of, 36, 41; patterns of, 89; practices of, 56; and nuturing of symbionts, 140, 146; relationships and, 19; response-ability and, 22, 105; for sheep, 91, 92, 95; storytelling and, 37, 116; sympoietic, 178n34; for other worldings, 50, 55
carrier bag narratives, 119–23, 125, 213n8; Le Guin and theory of, 7, 39, 42, 118, 120, 125, 177n27, 213n4
Carson, Kit, 91–92
categories: of authority, 42; of art, 89; of belief, 88; of city dwellers, 25; of financialization, 208n17; of human-animal work, 23; of kin, 216n4; of modern/traditional, 165; of nature, culture, and biology of the Camille Stories, 154, 161; of people, 21; of pigeons, 15; of religious/secular, 165; of science, 89; of thought, 43
cat's cradle, 205n94; games, 14, 26, 176n17, 184n50; multispecies, 9*f*, paintings of, 35, 35*f*, 176–77n17; parable of camels and, 131; thinking-with, 34
cattle and cows, 92, 109, 115, 129
Cayenne (Hot Pepper), 7, 105–8, 110, 114, 115, 116

cells: choanoflagellates as, 65; models of evolution of, 64, 65, 190n16; health of, 107; in holobionts, 60, 62, 64; Margulis on, 60, 61; nucleated, 189n4; sympoietic arrangements of, 58
-cene, as suffix, 55, 206n7
Cenestin, 211n6
Cerisy colloquiae, 132, 134
Cervantes, Julio Garduño, 156
change, 176n13; Anthropogenic, 45; biological, 141; collaborations and, 154; going on and, 40; mutation as, 60; rates and distributions of, 73, 99; stories and, 40, 43, 103, 144; in worlding, 160. *See also* climate change
chaos, 174n4; Gaia and, 51, 180n36, 181n38; reformulating order out of, 189n6
Charnas, Suzy McKee, Holdfast Chronicles, 188n67
chicken, chickens (*Gallus gallus*), 109, 113, 118; as biological model, 63
Children of Compost, 134–68; animals bonding with, 218n8; biogenetic reproduction of, 139; Camille as, 134, 136; community creation of, 138; EcoEvoDevo biological changes in, 141; gatherings of, 217–18n7; litter of, 215n1 (ch. 8); as pilot project, 136, 215–16n3; rendering-capable and, 8, 136; reproduction of, 143; stories and, 136; stories of, 144–68; symbionts and, 139–41; sympoiesis in, 141, 153, 154
China, 184n48, 194n37, 196n58, 217n6; Chinese language and, 195n56
Chisolm, Kami, 203n81
choanoflagellates, 64, 65, 190n16
Christianity, Christians: belief and, 88; climate change and, 6, 200n72, 208n18; kinship systems in, 203n82; Mary in, 186–87n62
chthonic ones, chthonic powers, 2, 53, 71, 173–74n4, 175n11; as allies, 186n57; as dreadful, 50, 54; etymology of, 173n4; Gaia and, 181n38; Gorgons as, 53, 54; ongoing, 180n36; Potnia Theron as reminder of, 52; of Terra, 31, 179n35, 186n62. *See also* symchthonic ones
Chthulu, 31, 33. *See also* Chthulucene; Cthul(h)u
Chthulucene: 2, 5, 51–57, 71, 88, 98, 101, 192n28; becoming-with in, 55, 188n69; Anthropocene/Capitalocene and, 55, 76, 159, 185–86n56; denizens of, 81; Diné and, 94; Earthbound and, 43; etymology of, 31, 53; icon of, 186n61; Lovecraft's Cthulhu vs., 101, 169n2 (intro.), 174n4; making compost in, 57; making kin in, 4, 89; as netbag, 54, 55, 121; nurturing of, 193n36; ongoingness in, 3, 71, 180n36; as real and possible timespace, 101; resurgence in, 94, 97; slogans of, 102; stories of, 31, 55, 56, 88; string figures in, 79; as sympoietic, 33, 55, 98, 168; tentacular tasks of, 42, 90, 97, 159; terrors of, 174n4; urgencies of, 7
Church of Stop Shopping, 187n63
circumpolar North, 5, 71, 73. *See also* Arctic
citizen science, 21, 24, 27
Clarke, Adele, 216n4
Cleveland, Robert, 198n64
Clifford, James (Jim), 101, 173n3, 185n54
climate change, 100, 159; in Anthropocene, 44; Christians and, 6, 208n18; Climate Change summit, 80; deniers of, 208n18; Inuit and, 198n65, 198n67; irreversible, 140; modeling of, 45, 181–82n43; putative benefits of, 194n37; summit on, 80; violent, 145
climate justice, 97, 193n32, 204n91
Climate Justice Alliance, 97
cnidarians, 32; corals and, 45, 54, 56, 72
coal industry and coal mining: in Appalachia and West Virginia, 218–19n11; in Australia, 217n6; Black Mesa and, 73–76, 94, 203n81, 217n6; in China, 217n6; Coconino aquifer and, 74; in Communities of Compost, 141, 153; emissions from, 194n37; Mohave Generating Station and, 73–74, 96, 193n35; mountaintop removal and, 144, 175n10; Navajo aquifer and, 73–74; Navajo Generating

Station and, 74; sedentarization and, 196n58; transition from, 96
Cody, Pat, 210n2
collaboration, collaborations, 5, 102; in art activism, 5; between breeders and their animals, 22, 129; with Camille stories, 136, 140, 147, 150, 154; in Chthulucene, 56, 101; computer world games and, 87; cosmopolitics and, 23; between da Costa and engineer, 22; between humans and babblers, 128–29, with Inupiat, 87; oddkin and, 4; between racing pigeons and humans, 16–20, 25; between peoples, 16; research, 63, 66, 111; response-ability and, 21, 22, 25, 28; story-making and, 136, 143; survival and, 37; sympoietic, 66, 87, 102; webs of, 206n6
Colombia, Putumayo region of, 158*f*, 217n6, 225n32
colonialism, colonialists, 5, 94; anthropogenic devastation and, 76; appropriation and, 199n68; in Madagascar, 82, 197n58; neo-, 165; post-, 15, 82, 121, 125, 138, 150, 154, 221n17; settler, 96; state, 196n58. *See also* colonization, colonizers; decolonialism and decolonization
colonization, colonizers, 13, 26, 73, 82, 87, 88, 89, 199n68, 200n72; Columbian Exchange and, 223n26; creatures of empire and, 203–4n83; non-, 57; U.S., 92. *See also* decolonialism and decolonization
Colorado Plateau, 73–76, 91–94, 204n83
Colorado River, 193n36; transbasin water transfers and, 74, 226n39
Colorado Water Users Association, 226n39
Columbidae family, 170n13. *See also* pigeons
communication, 104, 187n65, 228n45; animal, 122; bacteria and, 66; of butterflies, 224n30; multispecies, 65, 68, 123–24, 214n14, 228n45; with pigeons, 18–22; plants and, 122. *See also* language, languages
Communities of Compost, 8, 138, 147, 217n4; agriculture in, 141, 160; in Amazon, 227n41; ancestors in, 154, 156, 157; art in, 140–50; binding sym and nonsym, 160–61; biodiversity in, 159; birthing practices in, 138–39; birth rates in, 145, 146, 147, 159; bodily modifications in, 146, 149; bullying in, 150, 160; *caracoles* and, 155; in Chthulucene, 159, 168; conservation practices of, 155; corridor thinking of, 218n9; cosmopolitics in, 157, 160; critters in, 160; culture in, 149; early days of, 149; ecojustice in, 154, 155, 160; education in, 150; extinctions in, 143, 159, 162; gender choice in, 140; Great Dithering and, 144, 145; as humus, 140, 160; hymn of, 228n49; immigration in, 147; industrial agriculture in, 141; making kin in, 138, 147, 160; in Mexico, 154; mining and, 217n6; modernity and, 157; monarch butterflies in, 141, 160, 162; population control of, 220–21n17; population of, 147, 159; ravaged habitations of, 141, 147, 160; rehabilitating land in, 154, 155; scientists of, 146, 157; SF and, 227–28n45; sociality in, 145; stories in, 161; symanimagenesis in, 154, 160; symbionts in, 139–40, 144–49, 159, 218n8; symbiogenetic children in, 154, 159, 160; sympoiesis in, 146, 147, 157, 160; Tagaq's effect on, 165; training minds in, 140, 160; worlding in, 160; Zapatistas and, 155–57
companion species, 11, 110, 115; accountable, 125; actions of, 65; ant-acacia associations, 124, 125; becoming-with and, 13, 110; *cum panis*, 29, 55, 115; infection by, 29, 115; instruction by, 117; plants and, 122; playing string figures with, 9–29, 132, practices of, 10–16; sowing worlds and, 118; stories of, 40, 119; symbiosis and, 124
compost, 134–68, 187n62; collaborations as, 4; made by Chthulucene, 57; making oddkin as, 4; not posthuman, 11, 55, 97, 101; piles of, 97, 180n38; Strathern on, 34. *See also* Children of Compost; Communities of Compost

compostist, 150; archives of, 162; Camille 1 and, 148; citizenship rights of, 147, 159; environmental justice and, 159; plant sympoiesis and, 147; practices of, 159, 160; not posthuman(ist), 97, 101–2; Pullman's stories and, 161, 162; self-described, 145; symbiont development and, 149, 162; work to restore, 156
compounding pharmacies, 107, 108, 109, 211n4
conglobulation, 110, 114
conjugating, 110; kin, 110–14
Connery, Chris, 188n69, 217n6
conservation: in Communities of Compost, 146, 155; efforts for Hawaiian crows, 38; Jolly and, 81, 85; Madagascar and, 85; through fire suppression, 196n58
containers, 14, 39, 91; importance of, 40, 118; time of, 11
contraception and birth control, 6, 27, 209–10n18, 211n6. *See also* reproductive freedom and rights
Cook Inlet Tribal Council, 71, 86*f*, 87, 198n64, 199n68. See also *Never Alone*
copresence, 4, 132, 136
coproduction, coproducers, 18, 21, 22, 49, 179n35
coral and coral reefs, 55, 56, 188n68, 205n94; art/science activisms and, 5, 71, 72, 78, 194n43; dying coral symbioses and, 45, 56, 183n45; as ecosystems, 64, 73, 78, 193n32, 194n43; Great Barrier Reef, 71, 79, 80, 90. *See also* Crochet Coral Reef
Cornell University Laboratory of Ornithology, 24, 27
corridors, 87, 138, 143, 152, 154, 160, 168; as central task of communities, 140; conservation of, 146, 222n22; ecologists of, 153; thinking in, 218n9
cosmology, cosmological performance, 14, 89–91, 97; Navajo weaving as, 96, 201n77, 202n79
cosmopolitics, 176n12, 207n12; Camille 2 and, 157; Camille 3 and, 160, 161; composing by storytelling, 15; Da Costa and, 23; globalizing, 11; Stengers and, 12, 98; Strathern on, 12
country, 26; care of, 27, 49, 80, 96
coupling, linear, 180n36; nonlinear, 43; structural, 176n13
coyote, coyotes, 13, 14*f*, 74, 91, 202–3n80
Crasset, Matali, 25, 26; *Capsule*, 26*f*, 26, 133
crayfish, 146, 222–23n22
creatures, 32, 61, 189n2; critters vs., 169n1 (intro.); of empire, 5, 15, 26, 203n83. *See also* critters
Crist, Eileen, 49–50, 185n53
critical theory, theorists, 3, 20, 51, 178n32, 211n18
critters, 61, 98, 169–70n3, 183n46, 196n58, 205n94; in Ako Project, 83; alliances with, 50; ants as, 124; art-design-activist practices joining, 133; associations with ants and acacias, 125; autopoietic systems and, 33; on Black Mesa, 73, 74, 96; in Chthulucene, 101, 186n61, 192n28; in coral reefs, 56, 72; cyborgs as, 104–5; on edge of disappearing, 8; endangered, 221–22n22; entanglements and, 49, 60; exterminations of, 55; health of, 20, 21; humans as, 99; humus and, 2; immiseration of, 4; interpenetration of, 58; invertebrates, 54; IT, 32; kin and, 216n4; of Madagascar, 72; marine, 78, 183n45; mathematics, 68; memory and, 69; metabolic transformations of, 56; moral, 101; mortal, 1; novel, 218n8; people and, 3; and pigeon racing, 10; precarity and, 37; primitive, 190n16; recuperating people and, 133; rendering-capable and, 7; self-recognition and, 18, 19; shared flesh of, 103; slave gardens and, 206n5; stories and, 15; sympoiesis among those in Communities of Compost, 140; as term, 169n1 (intro.); of Terra, 97
Crochet Coral Reef, 76–81, 89, 194n43, 202n79; as art/science activism, 71, 79; beaded jellyfish for, 77*f*; becoming-with and, 78; IFF and, 71, 77*f*, 78; string figures and, 79

crocheting, models of hyperbolic planes and, 76, 78
Crownpoint Navajo Rug Auction, 96, 200n76
Crutzen, Paul, 44, 45, 183n45
Cthulhu (Lovecraft), 101, 169n2 (intro.), 174n4
culture, 13, 52, 79, 95, 104, 149, 190n7
curiosity, curious practice, 58, 63, 68, 83, 133, 147, 150, 168; Despret and, 7–8, 127, 130–31; Tsing and, 37
cuttlefish, 55, 188n69
cyborg, cyborgs: as coshapers, 20; cocktail of, 211n7; defined, 104; feminist, 118; as holoents, 210n1; as littermates, 7, 31, 104–5, 109–10; as kin, 104; speculative fabulation and, 105; tools of, 63; worlding, 115
Cyborgs for Earthly Survival, 102, 114, 117, 216n4
cynicism, 3, 38, 51, 56, 97
Czerneda, Julie, 161

da Costa, Beatriz, 25, 172n24; art and research of, 21, 24, 26; cosmopolitics, 23; Crasset and, 25; death of, 171n23; PETA vs., 23, 24; PigeonBlog and, 20, 22, 132, 172n28
Danainae, 227n43; extinction of, 162–63
Danowski, Déborah, 52
DARPA (Defense Advanced Research Projects Agency), 22
Darwin, Charles, 16, 50, 68, 192n24; neo-Darwinians and, 68, 176n12, 190n16, 191n21; *On the Origin of Species*, 62
databases, 24, 63, 214n13
Davis, Angela, 223n25
dead, as active presence, 8, 69, 132. *See also* speakers for the dead
de Castro, Eduardo Viveiros, 52, 88, 165
decolonialism and decolonization, 6, 94, 138, 154, 218n9, 221n17, 225n32; alliances and coalitions and, 71, 193n35, 203n81, 209n18; indigenous and, 71, 80; literatures of, 154; multispecies studies and worlding and, 150, 225n32. *See also* colonialism, colonialists; colonization, colonizers
Defense Advanced Research Project Agency, 22
deforestation, in Madagascar, 196–97n58
deities, 54, 188n69; kin making and, 216n4. *See also* God; goddesses; gods; sky gods; *and under names of individual deities*
de la Cadena, Marisol, 227n41
Deleuze, Gilles, 12
demography, 197n58, 209n18; demographic abstractions and, 210n18. *See also* human numbers; population
Dempster, M. Beth: on sympoiesis, 33, 61, 176n13, 189–90n7
denunciation, 50, 56, 127, 185n55
de Roode, Jaap, 163*f*, 167*f*; Laboratory, 223n24
DES (diethylstilbesterol): in agriculture, 109–10; anxiety over, 106, 108; bans on, 109; feminist women's health activism against, 7, 210n2; history of, 107–10, 210–11n4; manufacture and marketing of, 211n6, 211n7; response-ability and, 104–16
DES daughters, 105, 210n2; multispecies, 109
DeSmet (Trumbull), Raissa, 186n61
despair, 4, 51, 56, 130; resurgence vs., 71, 97. *See also* hope
Despret, Vinciane, 128, 170n7, 177n25; becoming-with and, 12, 25; Camille stories and, 134; da Costa and, 25; on parable of camels, 131, 132; on pigeons, 25, 132, 133; on philosophical ethology, 132; on Crasset's pigeon loft, 25; teachings of, 101, 130; thinking-with and, 7–8, 126; on virtue of politeness, 127–29; with Porcher, 129; with Stengers, 130, 131; work of, 129, 131, 132
developmental biology, 63–64, 66, 205n94. *See also* EcoDevo; EcoEvoDevo; ecological evolutionary developmental biology; EvoDevo
deVries, Karen: "Prodigal Knowledge," 178–79n34

Dewey, Ryan, 181n42
Día de los Muertos, 154, 156
Di Chiro, Giovanna, 205n94
diethylstilbesterol. *See* DES
digital cultures: indigenous, 199n66; digital world and, 215–16n3
Diné (Navajo), 225n31; activists, 71, 75–76; Black Mesa and, 73–76; disorder and greed, 91; forced removal of, 75; matrifocalism of, 93; resistance of, 203n81; rug weaving and, 96, 202n79. *See also Navajo entries*
Diné be'iiná (The Navajo Lifeway), 71, 95
Diné College (Tsaile, AZ), 95, 202n78
Dinetah, 92
discernment, sympoietic, 178n34
discomfort, 179n35, 180n36
displacement, 125, 133, 157, 206n5, 209n18
Dithering, The, 102. *See also* Great Dithering, The
DNA (Deoxyribonucleic acid), 37, 61, 66, 78
Doctor Who, Ood in, 187n65
dogs: at Burning Man, 182n45; estrogen and, 6, 7, 105–8, 110, 114–16, 210n3, 211n4
double death, 44, 47, 49, 164, 177n24, 180n37, 193n32; explained, 163*f*, 214n17. *See also* extinction
Dovey, Ceridwen: *Only the Animals*, 227n44
Dubiner, Shoshanah: *Endosymbiosis*, 59*f*, 188–89n2
Dutch, 44; language, 174n4; Old, 174n7
Dzil ni Staa, 92, 203n81. *See also* Big Mountain; Black Mesa

Earth, earth: abiotic and biotic factors of, 55; chthonic ones of, 2, 71; damaged, 2, 4, 10, 37, 52; earthlings and, 4, 58, 103; etymology of, 174n4; flourishing on, 10, 40; habitation threatened, 43, 132; history of, 118; as bumptious holobiome, 98; other names of, 33, 98; reserves of, 100; symbiosis on, 124; synchroncity on, 73; as Venus, 188n68. *See also* Gaia; Terra, terra
Earthbound (Latour), 41–43, 175n11, 178n31, 179n35; in Anthropocene, 41, 42, 175n11; Athena as traitor to, 54; chthonic ones and, 53, 56; Humans in History vs., 42, 50
Earthseed community, 119–20, 125
eating, 65, 73, 165, 170n4, 223n22: sex and, 190n16
EcoDevo (ecological developmental biology), 190n7, 191n18
EcoEvoDevo (ecological evolutionary developmental biology), 5, 7, 97, 98, 176n12, 190n11, 205n94, 218n10; biological changes in, 141; stories of, 122; symbiosis model systems and, 64, 191n18; transformative processes of, 64
EcoEvoDevoTechnoHistoPsycho (ecological evolutionary developmental historical ethnographic technological psychological studies), 150
ecojustice, 102, 154, 157, 208n18
ecology, 97; of practices, 34, 178–79n34, 228n49
ecosexual practices, ecosexuals, 32, 102, 175n10, 218n11
ecosystems, 27, 67, 76, 96, 159; as bounded unit, 62; collapse of, 46, 47, 100, 109, 159; of coral reefs, 45, 64, 72, 78; forest, 81; marine, 72; as sympoietic, 183n47, 190n7
education, 6, 7, 28, 209n18; biodiversity, 197n61 in Communities of Compost, 140, 150, 153
eels, American (*Anguilla rostrata*), 146, 222n22
Eglash, Ron, 202n78
Egypt, ancient, 26, 173n4
Eichmann, Adolf, 36, 37, 47
Ejército de Mujeres Zapatistas en Defensa del Agua (Zapatista Army of Mazahua Women in Defense of Water), 155–57
Eli Lily, as DES manufacturer, 107
embryology, embryos, 62, 176n12, 205n94
emergence of complex cells, 64–65; fusions of genomes and, 60–61

Emmenin®, 111, 211n7
"empty land," 138
encountering, encounters, 43, 56, 68, 154; actual, 7, 126, 127; beings-in-, 13
Endangered Species Act (ESA), 222–23n22
Ender Wiggin, 227–28n45
Endosymbiosis, 58, 59*f*, 189n2
energy: renewable, 46, 96, 183n46. *See also* coal industry and coal mining; fossil fuel extraction and mining; Peabody Energy
English (language), 87, 90, 103, 169n2 (intro.), 195n56; Old, 174n4
entanglement, entanglements, 4, 13, 34, 36, 39, 68, 91, 125; Chthulucene and, 101, 159; collaborative, 63, 129; critters and, 60; daemons and, 162; dis-, 68; of holobionts and holobiomes, 63; Indo-European tangles and, 174n4; of multispecies lives, 115; Navajo weaving and, 91, 96; string figures and, 34; with tentacular ones, 43, 71, 90, 159; of times, 11; transcontextual transdisciplinary tangles and, 202n79; Van Dooren on, 173n2. *See also* tangling, tangles
environmentalism, nonimperialist, 182n43
environmental justice, 96, 159, 205n94, 207n12; activists for, 61; compostist and, 159; indigenous and, 160, 226n39; Le Guin and, 213n3; multispecies, 8, 96, 155, 175n1, 193n32; pigeons and, 20
Equine Angels Rescue Sanctuary, 212n15
Erinyes, 54
estrogens and steroids, 6, 211n7; artificial, 211n6; dogs and, 105, 210n3; equine, 211n6; heart disease and, 112; human, 211n6; molecules of, 109; natural, 107; women's deficiencies of, 105–6
ethnologists, 13
Euclidean geometry, 52, 192n24
Eurocentrism, 52, 185n51
European Society for Evolutionary Developmental Biology, 191n18
EvoDevo (evolutionary developmental biology), 191n18
evolutionary theory, 49, 97; biological, 62; carrier bag theory and, 177–78n27, 213n4; Christians and, 200n72; involution and, 68; Margulis and, 60; neo-Darwinian, 191n21; systems and, 176n13. *See also* Modern Synthesis; New Synthesis
exceptionalism, human, 11, 13, 30, 38, 39, 49–50, 57, 137; dating of Anthropocene and, 181n42; pretensions of, 216n4; Tsing and, 212n1
experimental life forms, 68, 78, 195n45
extended evolutionary synthesis, 64–67, 176n12, 190n11
extinction, extinctions: accelerating, 46, 132, 137; in Anthropocene and Capitalocene, 35, 46, 55, 86; in Communities of Compost, 140, 143, 145, 148, 159, 162, 163, 167; edge of, 102; Madagascar and, 72; mass, 130, 148; as multispecies genocides, 37, 55, 130; sixth great extinction event, 4, 43, 180n37; Van Dooren on, 38, 173n2
Extinction Studies Working Group, 177n24
extractionism, extractions, 4, 82, 100, 120, 132, 137, 155, 196n58; accelerating, 46; in Anthropocene and Capitalocene, 90; fossil fuels and, 5, 75, 141, 183nn46,47, 217n6, 225n31

fables and wild facts, 13, 88, 131, 152, 187n65
fabulation, 120, 136. *See also* speculative fabulation
family, 2, 103, 208n16; as word, 216n4
fan fiction, Camille stories and, 136
"Farfetchings," 213n8
female: estrogens for aging, 105; research for, 114. *See also* woman, women
feminism, feminists, 6, 102, 205n94; biologists as, 108; collective thinking-with of, 173 epigraph 2n; communitarian anarchism and, 12; DES and, 7, 210n2;

feminism, (*continued*)
gens and, 208n17; geo-, 225n32; horses and, 111, 188n67; indigenous public work and, 207n12; political ecology and, 225n32; reproduction and kin and, 216n4; science studies and, 111; speculative, 2, 10, 31, 53, 101, 105, 150, 174n7, 177–78n27, 213n4; speculative fabulation and, 34, 212n1, 213n4
fiber arts, 76, 78, 79, 89, 202n79
Fifth International Prosimian Congress (Madagascar), 85
figures, 51–52, 55. *See also* string figures, string figuring
fingery eyes, 31–32, 174n6, 183n45. *See also* tentacularity, tentacular ones
Fire God, 13
First Nation peoples, 153, 184nn47,48. *See also* Aboriginal peoples; Native Americans
Fisher, Elizabeth, *Women's Creation*, 177n27, 213n4
Floria, Maria, *Broken Rainbow*, 194n40
flourishing: in Communities of Compost, 140; generative, 101; kin making increases, 138; multispecies, 116; of racing pigeons and humans, 16–20; reseeding, 117; response-ability and, 56; speculative fabulation for, 81
forest habitat, 5, 193n34; in Madagascar, 5, 72–73, 81–85, 193n34, 195–97n58
fossil burning, 46, 55, 193n36
fossil fuel extraction and mining, 5, 52, 56, 73, 141, 194n38; in Arctic, 184n48; on Black Mesa/Big Mountain, 75, 203n81; fracking, 46, 47, 56, 194n37, 219n11; investment in, 183n46; pollution and, 225n31; from tar sands, 183–84nn47,48, 217n6. *See also* coal industry and coal mining
Foucault, Michel, 108
Fourth World Water Forum, 227n40
Fredeen, Amy, 198n64, 199n68
French (language), 1, 169n1 (ch. 1), 172n32, 175n11
friendship, friends, 36, 38, 149, 168, 205n94; dogs and, 105; horses and, 54; kin making and, 145; pigeons as, 17*f*; practice of, 150, 221n17; scientists and, 71, 82, 83, 107, 192n23
frog (*Xenopus laevis*), as biological model, 63; coyote and, 202–3n80
fruit flies (*Drosophila melanogaster*), as biological model, 63
Fukushima disaster, 183n46
fungus, fungi: as communicators, 122; fungal associates and, 212n1, 218n8; fungal mycorrhizal symbionts and, 123, 227n43; metaphors and, 102; termites and, 62
future, 42, 46, 57, 71, 142, 148, 149, 168, 184n47; fragile, 140; futurism and, 2, 4, 6; imagined, 1

Gaia (Gaea), 98, 101; Anthropocene and, 51, 189n6; belief and, 178n34; chaos and, 180n36, 181n38, 186n58; Chthulucene, 51; geostories and stories of, 49, 175n11; gut of, 179n35; Hesiod on, 180–81n38, 186n58; intrusion of, 180n36; Latour on, 40–43, 175n11; Margulis and, 61; Stengers on, 43–44, 52, 54, 175n11, 180–81n38; Thousand Names of (Os Mil Nomes de), 33, 52, 98; Varley on, 175n11
gambling, on pigeon racing and, 171n13
game, games: computer and video, 58, 198nn64,65, 199nn66,67; of life, 176n17; making of, 198n64; of string figures, 10, 12, 13, 14, 25, 55, 88, 132, 170n11, 201n77; theory of, 191n21; world, 202n79. See also *Never Alone*
"game over," 57; attitude of, 2–4; cynicism and, 2; "too late" and, 56
Garza, Alicia, 207n12
Gearon, Jihan, 204n91, 204–5n93
gender, genders, 174n4; bonds of sex and gender, 102, 139; in Communities of Compost, 138–40; conventional, 137; current four, 144, 221n18; linguistics and, 170n3; structure of, 171n21
generations, 220n16; blasting of, 214n17; in "The Camille Stories," 8, 134, 140–44;

making futures for, 1, 134; reconnecting, 95; "series of interlaced trails," 32
generative justice, 202n78
genocide, 10, 36–37, 42, 48, 55, 100, 130, 132, 137, 147, 209n18; in United States, 76, 86, 91–92, 203n81
gens, 103, 208n17
geology, geologists, 44, 46, 60, 178n31; epochs of, 44, 46, 60, 181n42, 182n44
German, 111, 115; language, 169n3, 174n4
Gestes spéculatifs, 134
"getting on together," 10, 28, 29
Gift from Talking God, A, 204n88
Gilbert, Scott F., 191n18, 191n21, 206n5; on Anthropocene, 206n4; Camille 2's butterfly antennae beard and, 224n30; education of, 192n23; on humans as symbionts, 173 epigraph 1n; holobiont theory of, 191n21; on model organisms, 63–64; "We Have Never Been Individuals," 67
Gimbutas, Marija, 186n58
Global Invasive Species Database, 214n13
globalization, global capitalism, 11, 44, 45, 48, 185n51
global knowledge, 181–82n43
global warming, 73, 78, 222n22. *See also* climate change
God, 3, 4, 41; as change, 120; Virgin Mary and, 186n62. *See also* goddesses; gods; sky gods
goddesses, 213n9; riverine, 186n61; virgin, 54; winged, 52, 53*f*. *See also under names of individual deities*
godkin, 2
gods, 170n3, 181n38, 183n45, 195n44, 216n4; chief, 173n4; sky gods, 186n57; subgods, 31; tricks of, 40, 42; vanished, 47. *See also* deities; God; goddesses; *and under names of individual deities*
going-on-with, 8, 187n65
"going too far," 126, 130
Goldenthal, Baila, 177n17; paintings of, 35, 35*f*, 176–77n17
Goldman, Emma, 125
good-enough, 33, 60, 73, 179n35
good questions, 127, 128
Gordon, Deborah, 7, 213–14n9
Gorgons: description of, 53–54; face of, 53*f*; Hesiod on, 54; Medusa as, 5, 52, 54, 186–87n62; as word, 52–53. *See also* Medusa
Gorgonians, 54
Goshute people, 31, 203n80
Gosiute people, 202n80
Gosiute/Shoshoni Project, University of Utah, 202–3n80
Goslinga, Gillian, 216n4
Gray, Minnie, 198n64
Great Acceleration, great accelerations, 4, 6, 144, 181n42, 217–18n7, 223n26. *See also* human numbers; population
Great Dithering, The, 144, 145, 221n19
Greek language and mythology, 2, 101, 169n2 (intro.), 175n11, 186n58; ancient, 173n4; "Anthropos" in, 183n45; Myrmex, 213n9
greenhouse effect, 73, 188n68
Greening Earth Society, 194n37
Green turtles (*Chelonia mydas*), in Great Barrier Reef, 80*f*, 80
Guattari, Félix, 34
guman: humus and, 169–70n3; Terrapolis and, 11, 12, 169–70n3

Hadfield, Michael, 191n21, 192n23, 205n94
Halberstadt, Carol, 95
Hall, Stuart, 186n56
Hamel, Gildas, 173–74n4
Hainish species (Le Guin), 120–21
Haraway, Donna: "Cyborg Manifesto," 107; as feminist science studies scholar, 111; photograph by, 90*f*; *Primate Visions*, 83; *When Species Meet*, 107
Harding, Susan, 164
Harpies, 54
Hartouni, Valerie, 36
Harvard University, Micro Robotics Laboratories, 187n63
Haudenosaunee (Iroquois), 144; Great Binding Law of, 220n16
Hawaii, 57*f*, 66, 123, 191n21, 192n23, 205n94

Hawaiian crows, ʻalalā, 38–39
Hawk Mountain raptor sanctuary, 222n22
Hayward, Eva, 188n68; on tentacularity, 174nn5,6
Hebrew language, 170n3, 187n63
Hesiod, 54, 180–81n38, 186n58
Hinduism, 186n61
Hogness, Rusten, 13, 32, 90*f*, 106, 209n18, 216n4
holding open space for another, 38, 85, 133, 143, 160, 163, 195n54
"holo-" prefix, 60, 189n3
holobiome, holobiomes, 8, 52, 62, 71; of coral reefs, 5, 56, 64, 71, 80; crumbling of, 69; earth as, 98; holobionts and, 63; string figural, 227n43
holobiont, holobionts, 62, 67, 190n16; breaking apart, 69; etymology of, 60; Gilbert and, 191n21; holobiomes and, 63; knots and, 60; Margulis and, 189n3; squid-bacteria, 66
Holocene, 41; superseded by Anthropocene, 44–45, 100; Tsing on, 178n31, 192n28
holoents, 60, 62, 71, 207n12, 210n1
hologenomes, 189n3
homeostasis, 33, 61, 67, 189n6
Homer, 151, 181n38
Homo, 2, 11, 32, 55, 149
Homo sapiens, 30, 99, 181n42; not humus, 149
hope, 3, 150; despair vs., 4; erosion of, 46; new sources of, 182n43; as virus, 114
Hopi on Black Mesa, 5, 71, 73–76, 196n58, 225n31; as activists, 74–75, 193n35; coal digging and, 74–75; land partition laws and, 203n81
Hormiga, Gustavo, 32*f*, 174n4
hormone replacement therapy (HRT), 110–11, 113, 211n7
hormones, 107, 111, 112; sex, 108, 115; synthesized, 114; nomenclature of, 211n6
horse, horses: in Charnas Holdfast Chronicles, 188n67; estrogen and, 7; part of the pattern of Navajo pastoralism, 92; Premarin and, 111–14, 116; slaughter of, 112, 113, 114; treatment of, 111, 113–14; urine of, 211n6. *See also* pregnant mare urine
horse advocacy groups, 113–14, 115
HorseAid, 113, 114
host, hosts, 13, 31, 42, 67, 154, 179n35, 223n24, 224n30; symbiont and, 60, 67, 189n3
hostage, *hostis*, 42–43, 179n35, 182n43
hózhó, 76, 96; carrying capacity vs., 93; decapitalization of Navajo as a people and, 93–94; Navajo weaving and, 90, 201n77; restoration of, 14, 91, 93, 94; string games and, 14
Hubbell Trading Post, 201n77
human, humans, 20, 21; becoming-with pigeons, 15–20, 24, 28; of earth, 55; as environment-making species, 185n52; etymology of, 169–70n3; Humans in History, 40, 42; Latour on, 178n31, 179n35; matters of, 55; as partners with animals, 221n21; rendering-capable, 16–23; Species Man, 47–49; as symbionts, 173; as tentacular critter, 32, 55
human-butterfly worldings, 152, 155
human exceptionalism, 216n4; anthropocene discourse and, 50; Camille stories and, 137; companion species refusing, 13; dating of Anthropocene and, 181n42; Heideggerian worlding and, 11; making history and, 49, 50; mourning and, 38; scarcity and, 50; stories and, 39; Terrapolis against, 11; Tsing and, 212n1; becomes unthinkable, 30, 57
humanimal worlds, 118
humanism: Arendt's, 177n18; post-, 11, 13, 32, staying with the trouble and, 6
humanities, 32, 45; as humusities, 32, 57, 97, 216n4
humanity, 13, 50, 187n65, 207n12; humus as, 149, 160
human numbers, 102, 103, 208–9n18, 223n26; Great Acceleration of, 6, 217–18n7; increase of, 4, 6–7, 210n18; multispecies environmental justice and, 8. *See also* demography; population

human rights, 75, 203n81
humus, 169n3; becoming-with, 119; Children of Compost and, 140; guman and, 169–70n3; as humanity, 149, 160; multicritter, 2; SF and 150; sympoiesis and, 141; technological innovation and, 160; Terrapolis and, 11
humusities, not humanities, 32, 57, 97, 216n4
Hustak, Carla, "Involutionary Momentum," 67–68
Hwéeldi: as Long Walk of the People, 92; killing the animals (1863), 91–92, 93–94; originary trauma of, 92; second great, 93–94, 203n81
hyperbolic space, 68, 76, 192n24

ice melt, 5, 46, 73, 76
Idso, Craig, 194n37
IFF. *See* Institute for Figuring
immigration, 147, 209n18
imperialism, 6, 13, 29, 30, 71, 73, 120, 206n5, 208n18, 223n26
India, 52, 177n17
indigenous cosmopolitics, 157, 218n9, 227n41
Indigenous Environmental Movement, 194n37
indigenous media, Shoshoni Claymation video, 202–3n80. See also *Never Alone*
indigenous peoples, 5; belief among, 88; Black Mesa Indigenous Support of, 71; in Communities of Compost, 137, 142, 152, 155, 157; decolonial allies of, 71, 193n35; digital cultures of, 199n66, 199n68; destruction of, 120; environmentalists, 74, 96, 97; on exterminations, 86; genocides of, 48; herding people, 71; language learning and, 203n80; movements of, 221n17; production of, 89; stories, 87; storytellers, 86, 87; world games and, 86
individualism, 60; bounded, 5, 30, 33; neoliberal as bounded, 33; possessive, 60; self-recognition and, 18, 19; utilitarian as bounded, 49, 57
Indo-European language, 174n4, 208n17
information, 21, 23, 36, 61, 88; technologies, 33, 104
inheriting, inheritance, 2, 27, 50, 89, 125, 130, 131–32, 138, 140, 150
"In love and rage" (Goldman), 175n10
Institute for Figuring (IFF), 71, 77*f*, 78
Intergovernmental Panel on Climate Change (IPCC): reports of, 41, 46, 182n43
International Energy Agency (IEA), 183n46
International Fund for Horses, 113–14, 212n14
International Tribunal for the Rights of Nature, 80
International Union of Geological Sciences, 182n44
Internet, 21, 87, 113, 203n80
Inter-Society for Electronic Arts, 20
"intimacy of strangers" (Margulis), 60
"intimacy without proximity" (Metcalf), 79
intra-action, 24, 34, 60, 205n1; interactions and, 63, 64, 99, 208n17
Inuit, 89, 164, 165; climate change knowledge of, 198n65; language of, 192n29, 199n67
Inuktitut language film, 198n65
Inupiaq, Inupiat, 71, 87, 198n64; defined, 192n29
Inupiat, Alaska, 58, 71, 87, 192n29, 198n64
invasive species, 193n34, 214n13
invertebrate marine biology, 66, 192n23
"Involutionary Momentum," 67; of crochet coral reef, 78; orchid-bee entanglements, 67–68, 98; sensible materialisms of, 88
Iowa State College, 109
Iran, 171n13, 172n28
Iroquois Nation, *Constitution of the Iroquois Nations*, 144
Israelis, 127, 128, 187n63, 204n90

Japan, 223nn27–28
jellyfish, 5, 32, 48, 77*f*
Johns, Wahleah, 204n91

Jolly, Alison: Ako Project and, 81*f*, 82–85, 195n56, 197n61; daughter of, 195n57; death of, 81, 188n; interviews of, 83; lemurs and, 81–83; *Thank You, Madagascar*, 195nn57–58; Wright and, 85, 195n58
Jolly, Margaretta, 81*f*, 84*f*, 195n57, 197n59
Just Transition, 96, 97, 226n39

Kady, Roy, 204n88
kainos (now), 2, 55, 206n7
Kanawha River, 141, 144, 218n11
Kenney, Martha, 187n65
kestrels, American (*Falco sparverius*), 146, 149, 222n22
khipu, 202n79
khthôn (earth), 2, 173n4
kin, 102, 174n4, 208n16; ancestors as, 103; arboreal, 120; broadening term of, 102–3; conjugating, 110–14; cyborgs as, 104; kind and, 103, 109; innovative enduring, 209n18; get, 120; non-natal, 130; polytemporal and polyspatial, 60; reproduction and, 216n4; selection of, 128; stories of, 114; as wild category, 2; as word, 216n4. *See also* kinship; making kin; oddkin
King, Katie, 33, 174n6, 176n13, 202n79
King, Nicole, 64–65, 190n16
King Coal, 219n11, 225n31
kinnovation, kinnovator, 208n16, 209n18
kinship, 2, 52, 89, 91, 116, 216n4; cross-species, 106, 107; invitation to, 176n17; false universal, 207n12; systems of, 203n82; urgent questions concerning, 2. *See also* kin
Klare, Michael, 46, 183n46, 193n32
knots and unknotting, 180n36; attachment sites and, 14; Camille stories and, 143; Gaia and, 180n38; holobiont and, 60; living and dying and, 175n12; models and, 63; multispecies response-ability and, 16; Peace Fleece and, 204n90; pigeons and, 15, 16; sympoietic, 78; tentacular ones and, 31; weaving and, 91
knowledge, knowledges, 199n67; global, 181–82n43; practices of, 178n34, 218n10; privilege and, 111; situated, 97; systems of, 196n58; trans-, 174n6; worlds of, 202n78
K-Pg boundary, 100, 206n4
Kull, Christian, *Isle of Fire*, 196n58
Kunuk, Zacharias, 198n65

LaBare, Sha, 213n8
La Base de Loisirs de Caudry, 26
Laboratory of Comparative Cognitive Neuroscience, 19
language, languages, 11, 71, 122, 150, 151, 174n5; Aboriginal, 26, 173n33, 192n29; generativity of, 178n31; of Goshutes, 31; indigenous, 203n80; Latour on, 41; Navajo, 13, 95; pigeons in, 15; poverty of, 226n38; Shoshone and Shoshoni, 202–3n80; across taxa, 66. *See also* communication; linguistics; *and names of specific languages*
Latin, Latinate (Roman), 11, 31, 33, 174n4, 175n11
Latin American Water Tribunal, 227n40
Latour, Bruno, 5, 180n36, 181n43; on critique, 178n32; discourses of denunciation rejected by, 185n55; on Earthbound, 175n11, 178n31, 179n35; Gaia geostories and, 40–43, 175n11; geopolitics and, 40–43; Parliament of Things and, 181n41; thinking as speculative fabulation, 42; "War and Peace in an Age of Ecological Conflicts," 178n31; on war, 42–43
Leave It in the Ground, 194n37
Lee, Erica, 207n12
Le Guin, Ursula, 5, 43, 50, 118, 119, 121, 178n31; *Always Coming Home*, 87, 101, 213n8; "The Author of Acacia Seeds," 213n3; carrier bag theory, 7, 39–40, 118, 123, 125, 177n27, 213n4; *Earthsea*, 125; on Hainish, 121; LaBare and, 213n8; *Earthsea*, 151; *The Word for World Is Forest*, 120, 213n3
Lemur Conservation Foundation, 81*f*, 84*f*, 195n56

lemurs, 5, 56, 72, 195n56: lemur-human histories; ring-tailed, 81*f*, 81, 85; names of species of, 82–83
lichens, 30, 56, 117, 125, 191n21; in Capitalocene, 56; lyrics, 7, 57; symbiosis of, 72; "We are all," 56, 72, 179–80n35
life, living: game of, 29, 100, 116, 124, 176n17; along lines, 32; symbiotic view of, 67, 124
life and death, living and dying, 10, 16, 25, 33, 41, 42, 58, 116, 129, 141, 173n2; conditions and configurations of, 15, 137; cyborg, 104–5; with endangered critters, 8; in *hózhó*, 91, 94; involution and, 68; messiness of, 42, 182n43; multispecies, 98, 124, 137, 143, 204n83; ongoing, 56; patterns of, 105, 138, 140, 162, 175n12; practices of, 73, 159; response-able, 2, 28, 36, 38, 69; in the ruins, 37, 47, 138, 143; shared, 39, 40, 119, 144; string figures and, 49, 55; well, 29, 43, 51, 56, 86, 98, 105, 116, 132
life story, 40, 43, 119
linguistics: gender and, 170n3; phyto-, 122; precincts of, 178n31; therolinguists and, 122–23, 125. *See also* language, languages
looping: in Chthulucene, 188n69; of love and rage, 79
love, and pigeons: agape love and, 20; of breeders, 22; of pigeon fanciers, 25, 27
love and rage: germs of healing in, 137; Goldman and, 125; looping of, 79
Lovecraft, H. P., Cthulhu and, 101, 169n2 (intro.), 174n4
Loveless, Natalie, 200n75
Lovelock, James E., Lynn Margulis and, 43, 51, 61, 180n36, 180n38
Lyons, Kristina, 158*f*, 225n32

Madagascar, 71–73; Ako Project of, 81–86; beings of, 72, 81; schools of, 83
Madagascar forests, 72–73; conservation of, 195–97n58; destruction of, 196–97n58; diversity in, 83, 85, lemurs in, 5, 81; reforestation of, 193n34; tavy burning in, 82, 196n58; UNICEF and, 195n58
Ma'ii Ats'áá' Yílwoí (Coyotes Running Opposite Ways), 13, 14*f*
"Make Kin Not Babies," 5–6, 102, 103, 137, 160, 208–9n18, 216n4; images of, 139*f*, 164*f*; as project, 215–16n3
making kin, 1, 2, 12, 99–103; in Chthulucene, 3, 89, 102; Communities of Compost and, 154, 160; ethnographers and, 216n4; by Native Americans, 92, 203n82; popularity of, 208n16; purposes of, 4–5; respect and, 207n12; by Stephens and Sprinkle, 218n11; stories of, 4–5; sympoietically, 221n17. *See also* kin; kinship
making-with, 5, 58, 71, 137. *See also* sympoiesis, sympoieses
Malagasy: art workshops for, 197n59; conservation and, 195–97n58; -English book projects, 71, 82–83; language of, 85, 195n56; as shifting cultivators, 82, 196n58; as scientists, 85; vulnerability of, 82
Malm, Andreas, 101, 184n50, 206n6
Manitoba, 112, 115, 212n15
Man the Hunter, 39, 40, 48, 118
Margulis, Lynn, 5, 59*f*, 63, 188n, 188–89n2; autopoiesis and, 180–81n38; background of, 60; on becoming-with, 60; *Dazzle Gradually*, 188n2; death of, 189n2; Gaia theory and, 43, 61; Lovelock and, 43, 51, 61, 180n36, 180n38; on *Mixotricha paradoxa*, 61–62, 64; rejection of papers of, 189n4; Sagan and, 188n2, 189n4; symbiogenesis and, 97
Mariposa mask, 135*f*
marriage, 93, 171n21, 218n11
Marx, Karl, 50, 111, 185n56; Marxism, Marxists and, 50, 176n12, 185–86n56, 208n17
Mary, Virgin, 186–87n62
materialism, materialities, 2, 8, 11, 63, 69, 97, 132, 143; animism and, 88, 165; of global warming, 78; hyphae and, 2; im-, 36; Latour and, 42; libraries and, 150; loopy, 79; radical, 43; sensible, 88, 165; in science, 12; of visual culture, 195n54

material semiotics, 4, 23, 44, 88; chemistry and, 66; generativity and, 206n5; practices of, 200n74
mathematics, mathematicians, 12, 62, 63, 68, 120, 192n24, 201–2n78; fiber arts and, 76–79, 89, 101, 202n79
matsutake mushrooms, 37
Matsuyama, Bob, 22, 24, 172n24
matter, 120, 121, 125; "of care," 36, 41
Mazahua people, 142, 219–20n14; dead of, 154; history of, 154–55, 219–20n14; as hosts, 154; land and water struggles of, 226n40; language of, 220n14, 226n37; monarchs and, 155–56, 167; movement of, 227n40; women of, 155, 157, 226n40
McCain, John, 75
McFall-Ngai, Margaret, 66, 191n18, 191n21, 192n23
McIntyre, Vonda N., 77*f*
M'Closkey, Kathy, 200n74, 201n77
McNeal, Lyle, 95, 204n88
Medusa, 53, 101; as Mistress of Animals, 5; as mortal Gorgon, 5, 52, 54, 186–87n62; Pegasus and, 54, 188n68; Perseus vs., 54. *See also* Melissa; Potnia Theron
Medusa (Pacific day octopus), 5
Melville, Elinor G. K., 203–4n83
menopause, 111; cyborgs and, 115; hormone treatments for, 110, 112, 211n7
Mereschkowsky, Konstantin, 63
Métis, 153, 184nn47–48
Mexico: Chiapas, 225n32, 226n35; Mexico City, 157; monarch butterflies and, 8, 225n31; sheep in, 203n83; state of, 219n14; transbasin water projects, 225n31; volcanoes in, 141, 152, 154, 160, 167, 225n31
Michoacán, 141, 152, 154, 167, 219–20n14
microbes, 5, 43, 49, 73, 120, 180n35; coral and, 72; as critters, 169n1; sponges and, 190n16; study of, 60, 66; tangles of, 32; worlding of, 63
microbiomes, 115, 148, 170n3
migration, 148; Camille and, 152, 217n6; immigration, 147; in-, 147; Mazahua and, 155; Monarch butterfly and, 8, 141–43, 152–55, 160, 162, 163; out-, 147; of refugees, 145
milkweeds (*Asclepias meadii*) (*Asclepias syriaca*), 142, 145*f*, 148, 167*f*, 219n13
mining, 100; of Black Mesa, 193n35; Communities of Compost and, 141, 217n6. *See also* coal industry and coal mining; fossil fuel extraction and mining
misogyny, 101, 208n18, 210n18, 220n17
Mistress of the Animals (Potnia Theron), 5, 52
Mistress of the Bees (Potnia Melissa), 52
Mixotricha paradoxa, 61–62, 64
Miyazake, Hayao, 151, 223–24n29
model, models: of climate change, 181–82n43, 194n37; for developmental symbiosis, 64, 66; knots and, 63; new models for EcoEvoDevo, 191n18; scale, 182n43; systems of, for sympoietic, multispecies thinking and worlding, 71–72, 82, 87, 89; for transformation of organizational patterning, 64; as work object, 63–64
modernism, modernizing, 6, 208n18; categories of, 178n34; critique of, 185n56; progress and, 62; secular, 88; modernity and, 41, 50, 157
Modern Synthesis, 49, 62–63, 190n11
monarch butterfly (*Danus plexippus*), 8, 143, 145*f*, 153*f*, 167*f*; antennae of, 224n30; children bonding with, 8, 220n15; extinction of, 162, 163; migrations of, 8, 141, 142, 153, 154, 160, 162, 219n12, 225n31; Mazahua reserve and, 155, 167; milkweeds and, 142; overwinter in Michoacán and Mexico, 141; parasites of, 163*f*, 226n43; Camille's symbiont as, 142, 143, 148, 149, 154, 160, 167; sympoiesis and, 154, 155, 157; in Camille's West Virginia, 141
monotheism, 2, 162
Monsanto, 142
monsters, 2, 101, 169n2
Monterey Bay Aquarium *Tentacles* exhibition, 188n69

Moore, Jason, 100, 185n50; Capitalocene and, 101, 184–85n50, 185n52, 206n6
Moran, Nancy, 66–67, 191n18
Morgan, Eleanor: spider silk art of, 174n5
Morrison, Toni: *Paradise*, 92
mortal worlds, 33, 131
Mother Earth, Gaia as, 186n58
motley, 110, 115, 119, 123, 137, 217n5
mountaintop removal, 175n10; habitat destruction by, 222–23n22
mourning, grieving loss, 38–39, 51, 86, 101; practices of, 38–39, 51, 132, 150, 164; as representing, 166
mouse, mice (*Mus musculis*), 107, 191n18; as biological model, 63
movements for environmental justice, 61; indigenous, 160, 226n39; multispecies, 8, 96, 155, 175n1, 193n320
mud, muddle: belief and, 178–79n34; etymology of, 174n7; for SF, 179–80n35
Mudd, Victoria: *Broken Rainbow*, 194n40
multigender flourishing, 221n18
multinaturalism, 88, 154, 165, 218n9
multiplacetime recuperation, 213n8
multispecies (adj.): art in action, 21, 26; becoming-with, 63, 71, 78; communication, 214n14; environmental justice, 8, 96, 155, 175n10, 193n320; flourishing, 3, 221n18; imagining, 134; immiseration, 4, 37, 46, 137, 208n18, 221n19; kinship and, 2, 29, 145; Le Guin's *The Word for World Is Forest*, 120; recuperation, 8, 26, 27, 114, 213n8; responsible for, 29, 56; 116, 130; response-ability, 16, 56, 132; resurgence, 5, 8; SF, 20, 25; socialities, 218n8; storytelling, 10–16; struggles in nature, 40; trust, 22; urban, 27; urgency, 35; worlding, 10, 105, 225n32
multispecies living and dying, 58, 98, 124, 137, 143; on Colorado Plateau, 204n83; stories of, 10
multitemporal geo-political zones, 202n79
mural of Putumayo landscape, 158*f*
Murphy, Michelle, 210n18, 216n4
Mushroom at the End of the World, The (Tsing), 37–38, 182n43, 208n7
mushrooms, 37–38, 62, 212n1; as guide for living in ruins, 182n43. *See also* fungus, fungi
mustard (*Arabidopsis thaliana*), as biological model, 63
mutation, 44, 60, 69, 114, 139
Myers, Natasha: "Involutionary Momentum," 67–68
Myrmex stories, 122, 213–14n9

na'atl'o' (Navajo string figure games), 13–14, 26, 169n1 (ch. 1), 201n77
Naegle, Buster, 95
Narration Spéculative workshop, 134
narrative, narratives, 7, 42, 101, 128, 143; of apocalypse, 150; carrier bag, 119, 121, 123, 177–78n27, 213n4; in evolutionary theory, 213n4; SF, 102, 207n15. *See also* stories; storytelling, storytellers
narrative netbag, 119, 121, 123, 177–78n27, 213n4
National Evolutionary Synthesis Center (NESC), 191n21
nationalism, 6, 56, 209n18
National Public Radio (NPR), 198n64, 199n68
National Women's Health Network, 210n2
Native American Rights Fund, 75
Native Americans: in Blue Ridge Mountains, 218–19n11; making kin by, 203n82; video games and, 89. *See also under names of specific peoples*
naturalcultural: assemblages as, 38; degradation of woodlands, 142; differences, 122; ecology as, 211n6; histories, 28, 39, 118, 125; multispecies trouble, 121; places, 225n32; response-able, 125; to think-with, 40
natural gas, 122, 184n48; fracking and, 46, 194n37, 219n11
natural selection, 60, 69
nature, 7; Ako Project and, 83; Camille 1 and, 153, 154; capital and power and, 185n52; cheap, 100; divisions of, 93, 118; Latour on, 40, 41; Le Guin on, 40; multispecies struggles and, 40; niche as, 18;

nature, (*continued*)
Children of Compost reproduction and, 160; reserves of, 133; selva vs., 225n32; in worldings, 13, 50
Nausicaä of the Valley of the Wind (anime), 223n27, 223–24n29; Camille I's favorite, 151; as fable of companionship, 152
Navajo (language), 169n1 (ch. 1)
Navajo, Diné, 5, 196n58; Black Mesa and, 73–76; cosmology of, 90; culture of, 201–2n78; forced removal of, 92; land partition laws and, 203n81; sheep herding of, 91–93, 203n81; string figures and games, 14, 170n11, 201n77
Navajo-Churro sheep, 73–74, 204n88; Black Mesa Blankets and, 89, 200n74; carrying capacity of, 93, 94; as companions, 91; decapitalization and, 93–94; exterminations of in New Deal, 92–93, 203–4n83; herders of, 71, 91–92; history of, 91–92; hunting of, 94–95; ongoing Diné Bahane' and, 94; organizations restoring, 95–96; as partial healing, resurgence of, 95–96; second Hwéeldi and, 93–94; in Southwest, 202n80; at Southwest Range and Sheep Breeding Laboratory, 94; survival of, 94
Navajo creation story (*Diné bahane'*), 94, 201n76, 201n77
Navajo-Hopi Land Settlement Act (1974), 75, 203n81
Navajoland (Diné Bikéyah), 170n11
Navajo Nation (Naabeehó Bináhásdzo), 74, 89, 170n11, 200–201n76
Navajo on Black Mesa, 71; as activists, 95, 193n35, 217n6; relocation of, 194n40. *See also under Black Mesa entries*
Navajo Sheep Project, 95, 192n30, 203n83
Navajo weaving, 89–97, 90*f*, 204n88; as art, 89–91; as commodity, 200n74, 200–201n76, 201n77; cosmology of, 91, 96, 201n77; exploitation of, 201n77; in museums, 201–2n78; response-ability and, 89
Ndahoo'aah summer program, 201–2n78
n-dimensional niche space, 10–11, 16
Needham, Joseph: *The Grand Titration*, 176n12
Negev desert, 7, 127
nematode (*Caenorhabditis elegans*), as biological model, 63
neo-imperialism, 208n18
neoliberalism, 33, 181n43, 208n18
netbags for Capitalocene, 176n12, 184n50. *See also* narrative netbags
networked reenactments, 174n6
networks, multispecies, 21, 32, 48, 137, 214n14
Nevada, 73, 182–83n45, 202n80. *See also* Mohave Generating Station
Never Alone (*Kisima Ingitchuna*) computer game (E-line Media and Cook Inlet Tribal Council), 58, 71, 86–89, 199n66, 199n70, 200n72, 202n79; cover image for, 86*f*; creators of, 87–88; English subtitles of, 87; history of game making and, 87–88; Inupiaq narration in, 87; Nuna in, 87, 88, 198n64, 198n65; *Sila* in, 199n67
"New Age" cultures, 87, 165, 175n11
New Gauley, 144–55, 222n22, 223n26, 225n31; scientists in, 224–25n30
"New New Synthesis," string figures and, 63
New Synthesis, 62
Nez, Lena, 95
Nideffer, Robert, 172n24
nomads, 196n58
nongovernmental organizations (NGOs), 197n61
nonhumans, 23, 71, 149, 153; estrogens of, 110; human numbers and, 4, 107; humans and, 14, 16, 44, 48–50, 55, 63, 72, 91, 96, 137, 140, 143–45, 148, 163, 164; in Madagascar, 72
North America, 15, 73; circumpolar lands and seas of, 5; eastern, 142; petrotoxic lakes of, 184n48; West of, 44. *See also under names of countries in*
North American Equine Ranching Information Council (NAERIC), 112–13, 212n14

Northwest Passage, 184n48
nothing-but-critique, 178n32
nuclear energy, 183n46, 202n80; Nuclear Age, 47; pollution from, 100; testing for, as dating of Anthropocene, 181n42; uranium mining and, 75, 94

Oaxaca, 201n77
oceans, 80*f*, 80; acidification of, 56, 72; warming of, 45, 72; refugees from, 188n68. *See also* coral and coral reefs
Octopi Wall Street!, 51*f*
octopuses: *Octopus cyanea*, 55, 57*f*; Pacific Day octopus, 5, 57*f*, 188n69; Standard Oil as, 186n57; tentacularity of, 55
oddkin, 2, 209n18; generative, 3; making, 4, 216–17n4; as term, 221n20. *See also* kin
Ohmu, 151, 224n29
oil, age of, 46
oil palm trees, 193n34, 206n5
Olympiad, Olympians, 53, 54, 181n38, 186n58; Olympus and, 181n38
one-child policy, 220n17
ongoingness: ants and acacias and, 125; of Chthulucene, 2, 3, 101, 202n80; definition of, 132; healing and, 138, 167; killing of, 1, 44, 163*f*, 177n24, 214n17; living and dying and, 38, 144; SF and, 3, 179n35; at stake, 101; stories in Anthropocene not about, 49; symbiosis and, 140; sympoesis and, 125; threatened, 89, 137
Onondaga Nation, *Constitution of the Iroquois Nations*, 144
Ood (*Doctor Who*), 187n65
orchid (*Ophrys apifera*), 5, 68–69, 70*f*, 98, 192n27
organicism, 119, 176n12
organisms, 30; cyborgs and, 104; as part of holobiont, 60; involution and, 68; no longer supported, 30, 33; as models, 64; part of Modern Synthesis, 62; resurgence and, 193n34; in state of symbiosis, 218n8; sympoietic arrangements of, 58
organization, 61, 176n13, 218n8
orientation, 18, 176n13
Original People, 228n45
originary trauma, 92, 93, 203n81
"our bodies ourselves," 110
Ouroboros, 118, 173n4

Palmer, Fred, as Peabody Energy lobbyist, 194n37
Palomar Racing Pigeon Club, 19
Pan-American Society for Evolutionary Developmental Biology, 191n18
Papua New Guinea, 12, 34, 193n32
Parable of the Sower (Butler), 119, 213n3, 213n8
paradox, necessity of, 176n12
parasites, 162, 163*f*, 179n35, 227n43
partial connections, 13, 65, 104
pastoralism, pastoralists, 91–94, 147, 196n58
patriarchy, 6; language and, 208n17; patriarchal mode and, 174n4
patterns: in Camille stories, 159; Crochet Coral Reef and, 78; cyborg and, 105; developmental, 66; knottings of companion species in, 13; multispecies and, 29; of Navajo blankets, 201n77; Navajo pastoralism, 92; Navajo weaving and, 91; response-ability and, 16, 34, 35; with SF, 20; string figures and, 3, 10, 12; storied cosmos and, 91; in writings, 12
Peabody Energy: bankruptcy, 194n37, Black Mesa and, 73–74, 193–94n37, 194n39, 225n31; BMWC vs., 96; Hopi and, 75; Kayenta mine, 74
Peace Fleece, 96, 204n90
Pegasus, 54, 188n68
Pendleton Woolen Mills, 200n74
People for the Ethical Treatment of Animals (PETA), 23, 24
People's Climate Movement, 194n37
Peoples Tribunal, 204n91
per, as gender-neutral pronoun, 225n32, 221n21
permaculture: pigeon lofts and, 28; Terrapolis and, 11
Perseus, 54
Persian (language), 171n13, 174n4

Peru, 217n6, 219n11; Amazonian region of, 227n41
Pfizer, 112
pharmacy, pharmacies: Big Pharma, 7, 108, 110, 115; compounding, 107, 108, 109, 211n4; homeopathy, 107, 108
pharmakon, 105
philopatry (love of home), 39, 223n23
philosophical ethology (Despret), 132
philosophy: bounded individualism and, 5, 30; as cat's cradle, 132; compost and, 97; Despret's, 131–32; holist ecological, 173n2; human exceptionalism and, 30; modernity and, 157; independent organisms and, 33; self-recognition and, 18–19; stories and, 161; Whitehead's, 12; at stake to each other, 132
phylogenetic relationships, 189n3, 190n16
phytolinguistics, 122
Piercy, Marge, *Woman on the Edge of Time*, 221n21
PigeonBlog, 20–24, 23*f*, 132, 172n24, 172nn27–28; as art/science activism, 20–22, 132; as speculative fabulation, 22
pigeon fanciers (colombopiles), 27, 132, 133; in Baghdad, 171n13; Crasset's loft and, 25, 133; in La Défense, 26; PigeonBlog and, 21, 22, 24; for racing, 18
pigeon lofts: as art, 25, 133; Batman Park, Melbourne, 26–28; Crasset's *Capsule*, 25, 26, 133; municipal, 28, 29
pigeon racing: in California, 10, 14, 16–20, 133; difficult conditions in, 133; in England, 192n24; in Iran, 133, 171n13; in *On the Waterfront*, 171n15; in popular culture, 171n15; wives and, 171n21; as working-class men's sport, 21, 133
pigeons: agape love and, 20; backpacks for, 22; becoming-with humans, 15–20, 24, 28; *Bird Man of the Mission*, 17*f*; bullying, 20; as companions, 15, 16, 17*f*, 20; collaborations and response-ability, 21, 22, 25, 28; communications and, 21; diversity of, 5; domestic (*Columba livia domestica*), 15*f*, 170–71n13; feral, 15, 16, 18, 25, 27; hatching control, 27, 28; history, 15–18; homing, 18, 25, 171n15; knots and, 15, 16; make history, 29; mirror test and, 19, 171n17; as pests, 15; Project Sea Hunt, 18; racing pigeons, 10, 16–22, 24, 25, 132; "rats of the sky," 27; "rats with wings," 24; rendering-capable, 16–23; and humans being response-able, 20; rock doves and, 15, 16, 170n13; self-recognition by, 18, 19; in social experiments, 20; species of, 170n13; as spies, 21, 172n28; worlding of, 16
pig industry, 215n5
Pignarre, Philippe, 50, 51
Pimoa cthulhu to *Pimoa chthulhu*, 31, 32*f*, 52, 54, 55
Pitchfork, 28
Plantationocene, 99, 100, 137, 162; as boundary event, 206n4; as Great Catastrophe, 137, 166; living and dying in, 164; naming, 99, 206n5
plantation system: reforestation and, 193n34; slaves and, 206n5
plants: art and, 122; companion species and, 122; phytolinguistics and, 122; symbionts in Communities of Compost, 147, 149; spread of, 99; sympoiesis of, 147
play: Ako Project and, 85; art as, 23; Camilles and butterflies in, 143; Camille Stories as object of, 136; Children of Compost and, 139, 145, 150, 159; in Chthulucene, 55, 56, 101; of collective making, 89; companion species and, 13, 15; Crasset's pigeon loft and, 26; Crochet Coral Reef and, 78, 79; dogs and, 107; IFF and, 78; kin stories and, 114; Latour and, 42; material, 78, 79; *Never Alone* and, 87, 88; realm of, 23; string figures and, 4, 13; "working together" and, 129
PMU (pregnant mare urine), 110, 112–14, 211n6, 212nn14–15; history of, 111–12
poiesis, 31, 33
Polaris Music Prize, 165
political economy, 30, 208n17
politics: in Anthropocene, 57; biologies and art needed by, 98; collaboration for an-

imals in, 23; cyborgs and, 104; DES and, 108; of inclusion, 12; bounded individualism and, 5; Latour on, 41, 42; livable, 90; and oddkin, 145; pigeons as, 15; SF and, 150; stories and, 161; as a string figure, 10; Tsing on, 38
politeness, 127
polymorphy, polymorphism, 186n58, 188n69
Ponto, Kevin, 20, 172n24
poor people, 4, 6, 72, 74, 83, 159, 194n37
population: category trouble with, 208n18; of Communities of Compost, 147, 159; explosion of, 208n18; of Madagascar, 197n58; policies to control, 6, 210n18, 220–21n17; world, 220n17; World Population Prospects, 196–97n58. *See also* demography; human numbers
Porcher, Jocelyn, 129, 215n5
post-, as prefix, 212n2
posthuman, posthumanism, 13, 87, 134; in Anthropocene, 50; in Chthulucene, 55; compostist not, 11, 32, 55, 97, 101–2
Potnia Theron (Potnia Melissa), 52–54, 53*f*, 186nn61–62
poultry, DES and, 109
precarity, 37
preciousness, 138, 145, 208n16, 228n49
predictability, 19, 33, 61, 112, 176n13
pregnancies, 111; abnormal outcomes of, 109; DES and, 106
pregnant mare urine (PMU), 110, 112–14, 211n6, 212nn14–15; history of, 111–12
Premarin, 211n6, 212n14; conjugating kin with, 110–14; history of, 111–12; agribusiness and, 115; response-ability and, 104–16
prestige, 128, 130
"The Princess Who Loved Insects," 151
privilege, 38, 83, 151; knowledge and, 111
prodigal daughter, 178–79n34
progress: critique of, 185n56; Latour on, 40, 41; Marxism and, 51; modernism and, 15, 37, 50, 62; political progressives and, 3
Project PigeonWatch, 24
pronatalism, 209–10n18
pronouns, gender-neutral, 221n21
proportionality (Latour), 178n31
propositions, 67, 68, 89, 128, 185n53
Protestant vs. Catholic semiotics, 179n35
protists, 61, 218n8
protozoan, protozoa, 162, 163*f*, 188–89n2, 227n43
Puig de la Bellacasa, María, 170n3, 173n epigraph 2, 174n7, 175n9, 178n32
Pullman, Philip, 161, 162

queer, 6, 54, 105, 106, 138, 175n11, 209n18

racial purity fantasies, 209n18
Racing Pigeon Post, 19
racism, 6, 24, 208n18, 219n11
Raffles, Hugh: *Insectopedia*, 223n28
Ramberg, Lucinda, 216–17n4
Ranomafana National Park (Madagascar), 82, 85, 195–97n58
Rasamimanana, Hantanirina, Ako Project and, 81*f*, 83, 84*f*, 85, 195n56, 197n61
Ratsiraka, Didier, 83
Ratu Kidul, 181n61
reciprocal induction: becoming-with and, 40, 119, 212n1
reconciliation, 10, 207n12, 212n2
recuperation, 25, 101, 212n2, 214n17; acacias and, 123; across differences, 21; Batman Park and, 27; biological-cultural-political-technological, 101; in Communities of Compost, 140, 164; ecological, 43; histories of, 15; multispecies and, 8, 26, 27, 50, 114, 117; multispecies art for, 21; partial, 38; partial connection to *hózhó*, 14; possible, 117, 125; realm of play and, 23, 24; SF premise of, 24; SF games and, 16, 20; space for, 25; staying with the trouble and, 20; storytelling and, 10; terraforming and, 213n8; terran, 47
Reed, Donna: *Signs out of Time*, 186n58
reforestation, 193n34, 222n22, 226n40
refugees, 102, 120; without refuge, 100, 145; from oceans, 188n68

refugia, refuges: of Communities of Compost, 147; coral reefs as, 72, 193n33; of Holocene, 178n31, 192n28; reconstitute, 101; slave gardens as, 206n5; wiping out of, 100
Regional Chamber of the Rights of Nature Tribunal (Far North Queensland, Australia), 80
rehabilitation, 33, 71, 155
relationality, 68, 91, 165, 175n12
relatives, logical and familial, 65, 103, 138, 202n80, 216n4, 221n20
relay, 3, 130, 134, 140; return and, 10, 12, 34; string figures and, 12, 13, 25, 33, 105
religion: belief and, 178n34, 200n72; colonialism and, 157; desecration of, 203n81; Earthseed, 119–20; monotheisms, 2; secular vs., 165; technofixes and, 3; revivalism and, 138. *See also* God
remembering, 2, 3, 7, 25, 28–31, 54, 69, 92, 106, 116, 164; memory and, 25, 26, 69, 86, 136, 166
re-membering, 25, 27: of chthonic ones, 175n11; of Potnia Theron, 186n61
rendering capable: birds and observers and, 128; Communities of Compost and, 8, 136; critters and, 7; Despret and, 126; partners and, 12, 141; of pigeons and humans, 16–23; worlding for flourishing by, 96
renewable energy, 96
reproductive freedom and rights, 217–18n7; feminists and, 6; as choosing symbiont, 8, 139; reconfiguring, 209n18, 217–18n7
resilience, 86, 212n2, 225n31
resistance: of Big Mountain Diné, 202n81; Camille stories and, 137, 155; histories of, 15; new practices of, 51
response-ability, 2, 110, 111, 189n6, 199n67; absence and, 132; Barash on, 175–76n12; becoming more, 98; in Camille stories, 143, 144, 152; as capacity to respond, 78; in Capitalocene, 176n12; and collaboration between pigeons and people, 20, 21, 22, 25, 28; cultivating, 34–36, 71, 78, 130, 132, 178n32; DES and Premarin in, 104–16; differences in, 29; ecology inspired by, 68; feminist ethic of, 68; flourishing and, 56; of life and death, 69; naturecultures and, 125; Navajo weaving and, 89; patterns of, 16; as praxis of care and response, 105; SF game of, 11; SF and, 12; Speaker of the Dead and, 69; staying with the trouble and, 28; stories and, 29, 115–16; symbiogenesis and, 125; Terrapolis and, 11; Tsing on, 38; Van Dooren on, 38; viral, 114–16
restoration, 212n2; biologists, 123; of Navajo-Churro, 95; of wetlands, 226n40
resurgence, 76, 86, 192n28, 212n2, 228n49; BMWC and, 97; coalitional work for, 97, 193n32; on Colorado Plateau, 94, 95; earth's power of, 73; environmental, 213n3; in Malagasy worlds, 82; memory and, 6; multispecies, 5, 8, 178n31; of Navajo Churro sheep; *Never Alone* and, 88; possible, 71; regrowing forest as, 193n34; yearning for, 89
revolt: action of, 49; Great Dithering and, 145; movements of, 47, 51; Octopi Wall Street, 51*f*
Right, 208–10n18
Rights of Nature Tribunal, Regional Chamber, Far North Queensland (2015), 80
Robinson, Kim Stanley: *2312*, 102, 207n15, 221n19
robobees, 187n63
Rocheleau, Dianne, 225n32
Rockefeller, John D., 186n57
Ross, Deborah, Ako Poject and, 81*f*, 83, 84*f*, 195n56, 197n59
Rose, Deborah Bird, 177n24, 212n2, 214n17
Rowell, Thelma, 127
Ruby, Edward (Ned), 66
Russ, Joanna: *The Adventures of Alyx*, 161; *The Female Man*, 187n63
Russia, 63, 184n48, 191n18, 204n90, 209n18. *See also* Soviet Union

Sagan, Lynn (Lynn Margulis), 188n2, 189n4

salamanders, Appalachian, 146, 222n22
Sanctuarío de la Biosfera Mariposa Monarca (Monarch Butterfly Biosphere Reserve), 141
San Francisco, Burning Man and, 182–83n45
San Jose Museum of Quilts and Textiles, 200n74
Sapp, Jan, 67, 191n21
Saudi Arabia: oil reserves of, 184nn47–48
Schmitt, Carl, 42–43, 178n31, 179n35
science, sciences, 89; agriculture as, 181n42; Big, 115; knowledge practices and, 218n10; as not modern, 226n38; religion and, 200n72; as sensible practices, 199–200n72; Western, 176n12
science art/activist worldings, 64, 67, 69, 97, 98; examples of, 86; staying with the trouble and, 71–72
science fact, 105; science fiction and, 7; as SF, 2; speculative fabulation and, 3
science fiction, 105; feminist, 187n63; science fact and, 7; as SF, 2, 7
Science News, 19
scientists: becoming-with birds, 127–29; cynicism of, 3; of Communities of Compost, 146; of New Gauley, 224n30
Seaman, Barbara, 210n2
Seaweed sisterhood, 205n94
secularism, 138, 157, 185n56
secular modernism, 88
sedentarization, 196n58
seedbag, 119, 150, 214n10
self-recognition, 18, 19, 96
selva, 225n32
Seminar in Experimental Critical Theory, 20
semiotics, 4, 122, 177n26; Protestant v. Catholic, 179n35. *See also* material semiotics
settlers, 219n11; alliances with Native Americans by, 193n35; categories by, 154; colonialism by, 50, 96, 138, 155, 193n35; as environmentalists, 74; culture split and, 93; heritage of, 154; sexualities of, 203n82, 207n12; women fiber workers as, 89
Seven Days of Art and Interconnectivity, 20
sex, 114, 148, 188n67; eating and, 190n16; gender and, 102, 139, 216n4; orchids and, 68; settler sexualities and, 203n82, 207n12; sex hormones and, 108, 115; sex workers and, 175n10; sexual freedom and, 6. *See also* ecosexual practices, ecosexuals; reproductive freedom and rights
SF, 2–3, 7, 10, 105, 188n67, 224n30; art and, 81; autopoietic systems and, 33; becoming with and, 3, 71; Camille stories as writing practice, 134; compost pile of, 177n18; da Costa and Despret committed to, 25; facts, 187n65; Gaia in, 175n11; genres of, 212n1; humus and, 150; LaBare and, 213n8; Lovecraft and, 101; as mode, 120, 213n8; muddles and, 174n7, 179–80n35; narrative of, 102; novel, 221n19; patterns of, 20, 76, 174n5; pigeons and, 18; practice of, 20, 177n27, 213n4; response-ability and, 11, 12; terra in, 175n11; speculative fabulation and, 2, 10, 12, 14, 105, 150, 212n1; storytelling as, 39; string figures as, 10, 12, 31, 41, 71, 105, 150; sympoietic, 79; worlding of, 14, 170n3, 178n32. *See also* science fiction; speculative fabulation; speculative feminism; science fact; so far; string figures
sheep, 5: as companions, 91; DES and, 109; face recognition skills of, 96, 204n89; Peace Fleece and, 96, 204n90; Soay, 127; in U.S. Southwest, 203–4n83. *See also* Navajo-Churro sheep
Shenhua Coal Group, 194n37
shifting cultivation, 82, 196–97n58
Shoshone people, 202n80
Shoshoni Claymation video, 202–3n80
Sierra Club, 193n35
sign and signifier, 179n35; resignification and, 186n57, 216n4
Sila (Inuit term), 87, 199n67
situated knowledges, 97
Skinner, B. F., 19
sky gods, 53, 57; arrogance of, 56; enemies of, 54; image of, 39, 118; offspring of, 31; technoid, 186n57

slaves: gardens of, 206n5; plantation system and, 206n5
slogans: Composting is so hot!, 32, 102; Cyborgs for Earthly Survival!, 117; Make Kin Not Babies, 102, 103, 137, 139*f*, 164*f*; Run Fast, Bite Hard!, 117; Shut Up and Train!, 117; Stay with the Trouble!, 117; Think We Must!, 30, 34, 36, 40, 47, 130, 174n6, 175n10
Soay sheep, 127
sociality, 11, 73, 149; in Communities of Compost, 145; multispecies, 218n8
so far, as SF, 2
soil biota, 227n43
Southwest United States: in art, 177n17; Navajo in, 170n11, 200n76; peoples of, 202n80; sheep in, 203–4n83
Soviet Union, 83, 85, 204n90. *See also* Russia
"Soy mazahua," 156, 226n37
Speaker for the Dead (Orson Scott Card), 101, 227–28n45; story arc of, 69
speakers for the dead, 8, 101, 164, 166, 167, 168, 227–28n45; task of, 69
species: anthropogenic effects on, 99; assemblages of, 100, 103; bioactivity across, 211n6; citizen science and, 24; coevolving, 134; endangered, 143; extinction of, 38, 102, 167; Margulis and, 60; migratory, 140; play and, 150; preference for migrants chosen as symbionts, 8, 140, 146; social, 15, 147; string figure games and, 13; survival plans for, 38; threatened, 43; Tsing and, 37. *See also* multispecies flourishing
Species Man, 47–49
speculative fabulation, 8, 81, 105, 213n4; art of feminist, 12; "The Camille Stories," as collection of, 8, 134, 136; *chthonios* as muddle for, 53; cyborgs and, 105; for flourishing, 81; Latour's thinking as, 42; PigeonBlog as, 22; real stories and, 10; science fact and, 3; SF, 2, 10, 12, 14, 150; staying with the trouble in, 133; Strathern as example of, 12, 34; tentacular ones and, 31; Terrapolis and, 10, 11; the thousand names and, 101; Tsing and, 212n1
speculative feminism, 105, 213n4; courage of, 174n7; as SF, 2, 177–78n27; speculative fabulation and science fact and, 3; Strathern's work in, 12, 34
spiders: *Pimoa cthulhu*, 31, 52, 54, 55; silk webs of, 174n5; sympoiesis as, 33; as tentacular critter, 2, 32
Spider Woman, 91, 101, 170n11; Holy Twins and, 91, 201n77
spirits, kin making and, 216n4
sponges, 174n5, 190n16
Sprinkle, Annie, 32, 102, 139*f*, 175n10, 218n11
squid (*Euprymna scolopes*), 185n56, 188n69; -bacteria (*Vibrio fischeri*) model, 64, 65, 66; string figures and, 66; as tentacular critter, 32, 55
Stanford torus, 175n11
Starhawk: as neopagan, 186n58; *Signs out of Time*, 186n58; song of, 166, 228n49
Star Trek, 187n65
state, biopolitical: population control and, 6
staying with the trouble, 1, 2, 3, 31, 76, 89, 96, 116, 133, 150, 167; acacias and, 214n13; Chthulucene and, 54, 55; commitment to, 114; Despret and, 7; how to address, 6; imaginative worlding and, 185n53; in Madagascar, 196n58; partial recuperation and, 10; requires making oddkin, 4; recuperation and, 20, 27; requirements of, 4; response-ability and, 28, 114; SF and, 12, 13, 20; string figures and, 27; sympoiesis and, 58–98, 125; on terra, 63; Terrapolis and, 12; worlding and, 29, 116
"Stay with the Trouble!," 117; speculative fabulation, 133
Stengers, Isabelle, 5, 12, 51, 181n43; cat's cradle and, 34; on capitalism, 50; cosmopolitics of, 12, 98, 175n11, 176n12; Despret and, 130, 131; discourses of

denunciation rejected by, 185n55; on ecology of practices, 228n49; *Gestes Spéculatifs*, 134; Gaia of, 43–44, 54, 175n11, 180n36; as materialist, 42; on science and beliefs, 199–200n72
Stephens, Beth, 32, 102, 139*f*; *Goodbye Gauley Mountain*, 175n10, 218–19n11
sterilization, 209n18
Stoermer, Eugene, 44, 183n45
stories, 67, 71, 199n67; of Anthropocene, 49; becoming-with, 40, 119; big, 101; of Camille, 8; carrier bag, 39, 179n35; for Chthulucene, 88; in computer games, 86, 198n65; of critters and collaborators, 15, 172n27; of companion species, 40, 119; Diné creation, 170n11; fables and, 187n65; feminist theory and, 213n4; force of, 35, 37; of Gaia, 40–43, 175n11; indigenous, 86, 199n68; of lemurs, 83; of making kin, 4–5, 114; old, 179n35; Navajo string games and, 14; Ouroboros's, 118; ownership of, 199n68; prick stories and tales, 39, 40, 46, 118, 181n38; response-ability and, 29, 115–16; risk of listening to, 132; scientists and, 69; SF adventure, 7; from social justice movements, 213n3; speculative fabulations and, 10; strengthening response-abilities, 29; string figures and, 10, 144; symbionts in, 218n8; symchthonic, 76; sym fiction, 216n3; systemic, 49; Terrapolis and, 10; urgencies and, 37; Ursula Le Guin and, 39–40; from visiting, 127; weaving and, 91; worlds and, 12. *See also* narrative, narratives
storytelling, storytellers: apparatus of, 39, 199n68; carrier bag theory of, 39; chthonic critters and, 54; cosmopolitics composed by, 15; as compostist practice, 150; digital, 216n3; Inupiat, 71; Jolly and Rasamimanana as, 85; master, 198n64; multispecies, 10–18; using string figures, 10; urban penguins and, 39
Strathern, Marilyn, 5, 65, 102, 103, 225n32; on cosmopolitics, 12; as ethnographer of thinking practices, 12, 34; *The Gender of the Gift*, 12, 216–17n4
string figure games: caring for worldings in, 55; with companion species, 132; collected by ethnologists, 13; Despret, da Costa, and Crasset playing, 25; Navajo, 14, 170n11; *Never Alone* and, 88; patterns and, 10, 12; as played, 13; restoring *hózhó* with, 14
string figures, string figuring, 5, 105, 175n11, 182n43; approach of, 217n4; in Australia, 172–73n33; Camille Stories and, 144; cat's cradle, 9*f*, 14, 26, 34, 35*f*, 131, 132, 169n1; *Cat's Cradle/String Theory*, 35*f*; Chthulucene and, 79; concerns and, 41; Crochet Coral Reef and, 79; cyborg litters and, 104, 105; da Costa and Despret committed to, 25; examples of complex, 202n79; French *jeux de ficelles*, 14, 26; like stories, 10; naturalcultural history and, 28; *na'atl'o'* in Navajo, 13, 14; Navajo, 14, 170n11; Navajo figure *Ma'ii Ats'áá' Yílwoí* (Coyotes Running Opposite Ways), 13, 14*f*; *matjka-wuma* in Yirrkala, 26; "New New Synthesis" and, 63; PigeonBlog and, 22; patterns of, 22, 87; playing with companion species, 9–29; as SF, 2, 3, 10, 12, 41, 150; sympoietic multispecies and, 49, 66; tentacular ones and, 31, 32, 35; as term, 169n1 (ch. 1); think-with, 31; women and, 205n94
Styger, Erica, 196–97n58
subversion, Latour on, 178n31
Sumeria SF worlding, 173n4
surface mining, strip mining, 73, 74, 219n11
sustainability: in Anthropocene, 33; in Capitalocene, 183n46
Sustainable Resource Alberta, 44
Swept under the Rug (M'Closkey), 200n74, 201n77
sym, as pronoun, 221n21
symanimagenesis, 98; becoming-with and, 154
symanimagenic, 8, 72, 88, 160, 167

symanimants, 168
symbiogenesis, 58–67, 88, 118; 179n35, 218nn7–8; in Anthropocene, 57; becoming-with and, 125; as powerful framework, 66; Nicole King and, 64–65; Margulis and, 97; response-ability and, 125; science/art worldings and, 67; Stengers and, 180n38; stories of, 122; as troublemaker, 61
symbiogenetic, 63–65, 71–72, 88, 118, 154, 160, 162, 167; joins, 8, 148; kin making, 159
symbiont, symbionts: bacterial, 66, 123, 191n18, 205n94; broken, 69; Camille and, 142, 143, 152; cnidarian, 56; in Communities of Compost, 8, 146, 139, 140, 141, 146, 147, 149, 159, 166, 168; development of, 149; diversity of, 66; fungal, 123, 162; holobionts and, 60; homeostasis and, 67; humans as, 173n epigraph 1; lichen and, 56; microbial, 67; plants and, 147; in stories, 218n8; symbiont children, 140, 141, 146, 149
symbiosis, symbioses, 60, 61, 98, 221n21, 227n41; Anthropocene and, 49; Camille and, 143, 148 152, 167; Communities of Compost and, 140, 144, 146, 147, 148, 149; companion species and, 124; coral holiobiome and, 54, 72; entanglements of holobionts and holobiomes and, 63, 64; labs working on, 191n18; Margulis and, 60, 61, 189n3; *Mixotricha paradoxa*, 61; monarch-human, 220n15, 224n30; mycorrhizal, 174n4; New Synthesis and, 62–63; pea aphid with *Buchnera*, 66; scales of, 67; squid-bacterial, 64, 66; stories of, 122; termites and, 62. *See also* making-with
symbiotes, 218n8
symchthonic ones, 71, 102, 192n28
symchthonia, 136
sym fiction (sympoiesis, symchthonia), 136, 216n3
sympoiesis, 97, 98, 179n35, 190n7; Ako Project and, 81–82; in Anthropocene, 57; aspects of, 65; autopoiesis and, 58, 61, 125, 176n13; becoming-with and, 125; Black Mesa Blankets and, 94, 96, 200n74; Camille born for, 137; as carrier bag for ongoingness, 176n12; Children of Compost and, 136, 138, 140, 146; computer world games and, 86; defined, 58; Dempster on, 33, 61; humus and, 141; as making with, 5; Margulis and, 61; models for developmental, 65–66; of *Never Alone*, 71, 88; Potnia Melissa and, 52; of practical coalition, 89–90; radical, 221n21; renewing terra as, 55; SF and, 40; stories of, 49; string figures and, 34, 66; symbiogenetic, 88; Tsing and, 37; as word, 184n50
sympoietic collaborators, 102
sympoietic knotting, 180n36, 180n38; Crochet Coral Reef and, 78
syms vs. non-syms, 140, 144, 149–50, 159–62
symthonic hive mind, 187n65
synchronicity, 73, 99; autopoietic, 189n6; evolution and, 176n13; Gaian, 189n4; systems theories and, 6
system change: evolutionary, 61; global, 180n36; invention of multicellularity, 64–67. *See also* synchronicity
system collapse, 100, 102, 221n19; climate change and, 159; of ecosystems, 46; Gaia and, 189n6
systemic urgencies, 102
"systems systematize systems," 101
systems thinking, 52; autopoesis vs. sympoesis, 61–62; complexity in, 61; Extended Evolutionary Synthesis and, 63; Modern Synthesis and, 62
systems theories, 7, 49, 60, 175n12

Tagaq, Tanya, 164, 165
Taimina, Daina, 76
Takahashi, Dean, 199n70
Talen, Rev. Billy, 187n63
TallBear, Kim, 203n82, 207n12, 216n4, 220n16
tangling, tangles, 29, 32, 150, 174n4, 192n28; of animals and people, 16;

chthonic forces and, 52; of Gosiute, 202n80; of histories, 195n57; *khipu* and, 202n79; of practices, 128; of string figures in Chthulucene, 79, 202n80; sympoietic, 97; tentacular ones and, 31, 42, 71; threads and, 3; tracing, 116. *See also* entanglement, entanglements
tar sands oil extraction, 225n31; pollution from, 183–84nn47–48; resistance of Aboriginal peoples to, 184n47, 217n6
Tartarus, 181n38, 186n58
Tauber, Alfred I., 67, 191n21
tavy, 82, 196n58
technology, technologies, 10, 41, 63, 100, 118, 216n4; of coal, 194n37; determinism of, 48; histories of, 148; humans and, 151; imaging, 66; information, 104; pigeons and, 18, 20–23, 23*f*, 27, 132; renewable energy, 46; as rescuer, 3; transformation and, 215–16n3
technoscience, 69, 104, 108. *See also* science; technology
temporality, temporalities, 37, 51, 91, 132, 176n17, 192n28, 199n67; hyphae and, 2; spatialities and, 64, 101
tentacle, tentare, tentaculum: BMWC as, 97; exhibit on, 188n69; Latin roots of, 31; making string figures, 35
tentacularity, tentacular ones, 71, 179n35, 180n38, 183n45; in Chthulucene, 42; definition of, 32; in *Doctor Who*, 187n65; entanglements with, 71; examples of, 32; forces and powers of, 174n4; Gaia and, 181n38; Gorgons as, 53; Hayward on, 174n5, 174n6; kinship with, 176n17; Nagas and, 186n61; names of, 101; Navajo-Churro restoration and, 95; octopi and, 55; speculative fabulation and, 31; string figures and, 31, 32, 34; symchthonic and, 33; thinking of, 5, 30–57
teratogens, 210n2
termites, 62, 190n9
Terra, terra, 60, 101, 174n4, 175n11, 209n18; belief and, 178n34; chthonic powers of, 31; critters of, 97; flourishing on, 10; immiseration of, 46; living and dying well on, 116; other names of, 33; overstressed, 56; staying with the trouble on, 63; stories of, 49; troubling times on, 1; wounded, 105
Terraforming, 117–25; acacias and ants on, 125; bacteria and, 99; companion species and, 11; finding seeds for, 121; recuperation and, 213n8; SF, 120
Terranova, Fabrizio, 134, 136, 215n1
Terrans, 49, 101, 175n11, 184n47
Terrapolis, 10–12, 14–16, 50, 175n11; Despret, da Costa, and Crasset in, 25; Medusa and, 52; PigeonWatch and, 24; worlding and, 29
Terriens. *See* Earthbound
thinker/maker, as art practice, 89, 96, 200n75
thinking: Arendt on, 177nn18–19; art and, 200n75; speculative, 12; string figures as, 14; thoughts and, 199n67
thinking and making practices, 12, 71, 34, 37, 71, 145, 167
thinking from, 131
thinking-with, 177n18, 178n34, 203n80; Despret and, 7, 126; feminist collective, 173n epigraph 2; string figures and, 31
"Think we must!" (Woolf), 34, 36, 40, 47, 130, 174n6, 175n10; meaning of, 173n epigraph 2
Third Carbon Age, 46–48
Thousand Names of Gaia (Os Mil Nomes de Gaia), 52, 90, 101, 119; speculative fabulation and, 101
three-or-more parent practices, 138, 220–21n17
timeplaces, 2, 125, 178n31. *See also* timescapes
timescapes, 5. *See also* Anthropocene; Capitalocene; Chthulucene
timespaces, 125; Chthulucene as, 71, 101; new, 182n44
toxic pollution, 73, 78, 79, 100, 141, 151
trading post system: permanent debt of, 92; price of blankets under, 201n77
transcontextual tangles, and khipu, 202n79

transdisciplinary biologies, 63
Trans-Mexican volcanic belt, 152, 154, 160, 167, 225n31
trials of strength: Latour on, 41, 42, 43, 179n35
trouble, etymology of, 1
Tsing, Anna, 5, 72, 185n53, 206n5, 215n2 (ch. 8), 223n27, 225n32; "arts of living on a damaged planet" and, 37, 87; climate modeling and, 181–82n43; "Feral Biologies," 100; forest refugia and, 72; on Holocene, 100, 178n31, 192n28; human exceptionalism and, 212n1; on response-ability, 38; on resurgence,193n34; on worlds worth fighting for, 97
Turner, Tina, 19

underworld, 173n4; Erinyes as powers of, 54; chthonic serpents of, 174n4
UNICEF, in Madagascar, 81*f*, 84*f*, 195n56
United Nations High Commission for Human Rights, 75
United States: African Americans in, 207n12; Arctic and, 184n48; Black Mesa and, 73; belief in, 178n34; Bureau of Indian Affairs (BIA), 74, 196n58, 283n81; Bureau of Reclamation, 74; Congress, 75; Department of Agriculture, 93; Department of the Interior, 74; DES and, 107, 210n2; Environmental Protection Agency (EPA), 194n37; Fish and Wildlife Service, 222n22; Food and Drug Administration (FDA), 107, 109; Iran nuclear activities monitored by, 172n28; Mine Safety and Health Administration, 74; monarch butterflies in, 8; Navajo in, 170n11; supersizing in, 183n45; War Department, 91–92
units and relations: autopoesis and, 33, 61; holobionts and, 60, 67, 191n21; Modern Synthesis and, 62; organisms as, 33; sympoesis and, 33, 61, 64; unitarian individualism and, 49
University of California, Berkeley, Nicole King and, 65
University of California, Irvine, 20, 23
University of California Santa Cruz Research Cluster of Women of Color in Conflict and Collaboration, 223n25
University of Massachusetts, Amherst, 58
unworlding, 97; of Capitalocene, 56, 57
urban antitoxic coalitions, 205n94
"Urban Penguins: Stories for Lost Places," 39
urgencies, 7, 37; of Anthropocene, 7, 35, 67, 69; apocalypse vs., 35, 37; avoidance of, 6; of Capitalocene, 7, 69; of Chthulucene, 7; climate change and, 6; da Castro and Danowski refiguring, 52; Le Guin and, 40; making alliances and, 207; shared, 88; stories and, 37; systemic, 102; Tsing on, 37; worldly, 7
urinary incontinence, treatment for in dogs and humans, 105, 106, 108, 109, 110, 114, 211n4
urine, 105, 109; of pregnant Canadians, 111, 211n7, DES distilled from, 108; HRT and, 110; Premarin and, 110, 111, 112. *See also* pregnant mare urine
Utah, 31, 74, 170n11, 202–3n80
utilitarian individualism, 49, 57

van Dooren, Thom, 5, 101, 177n24; *Flight Ways*, 38–39, 173n2, 177n25, 223n23; on response-ability, 38
Varley, John: Gaea Trilogy, 175n11
Venezuela: oil reserves of, 184nn47–48
Venus, 188n68
versions: animism of materialism, 88; Communities of Compost and, 134, 136; opening up, 130; to understand, 128
veterinary research, veterinarians, 7, 106, 108, 112–13, 210–11n4
viruses, 189n3: become-with, 65, 189n3; coral biome and, 72; response-ability as, 114
visiting: in Arendt, 127, 177n19; risky but not boring, 128; as subject-and-object making dance, 127; training the mind to go, 127–30, 140, 177n19

Viveiros de Castro, Eduardo, 52, 88, 165, 206n8
volcanoes, 141, 152, 154, 160, 167, 225n31

Washington, D.C.: Boyden and, 75; Pigeon-Watch in, 24, 27
Watanabe, Shigeru, 19
water: in Appalachia, 222n22, 225n31; Black Mesa and, 71–74, 94, 96, 97; chthonic ones in, 2, 71; Communities of Compost and, 8, 137, 140, 141, 144, 151, 152, 154, 155, 157, 160; horses and, 112; pigeons and, 18; reefs in, 72, 79, 193n32; settle troubled, 1; symbiosis in, 45; from tar sands oil extraction, 183–84n47; toxic lakes, 183n47, 184n84; urgency and, 88
water pollution, 4, 8, 45, 71; in Appalachia, 222n22, 225n31; on Black Mesa, 73–74, 94; Communities of Compost and, 140, 144, 152, 154; coral reefs and, 79; from tar sands oil extraction, 183–84n47; toxic lakes, 183n47, 184n84; urgency and, 88
water transfer projects, 157, 226n39; depletion of resources and, 100; relation to indigenous struggles, 73–74
weapons: words and, 39, 42, 46, 118; needing a net, 40, 118
"we are all lichens," 56, 179–80n35; taught by coral reef critters, 72
weaving, weavers: cat's cradles and, 176n17; continuous, 200n74, 201n77, 202n79; of Navajo (Diné), 89–97, 200n74, 200–202nn76–78, 202n79, 204n88; Navajo string figures and, 14; paths and consequences, 16, 31, 33; as relational worldings, 96; in Spitalfields, 192n24
webs of relationships, 216n4
weediness, 182n43
"We Have Never Been Individuals," 67
Werner, Brad, 47, 187n63
Wertheim, Christine, 195n44; Australian birth of, 79; crocheted coral reef and, 76, 78, 80, 194n43; IFF and, 78–79
Wertheim, Margaret, 195n44; Australian birth of, 79; on hyperbolic space, 192n24; crocheted coral reef and, 76, 78, 80, 194n43; IFF and, 78–79; TED talk of, 79
West, 15, 100, 175n12; evolutionary science of, 176n12; gender binarism in, 221n18; medicine of, 108; patrilines of thinking in, 130; people indebted to, 30; philosophy of, 30, 161; politics of, 161; psychology of, 18; thinkers, 51
West Nile Virus, 222n22
West Virginia: Camille's community in, 141, 144, 218–19n11; mountaintop removal in, 175n10; Native Americans in, 153
West Virginia Natural Heritage Program, 222n22
wetlands destruction, 27, 29
Whitehead, Alfred North, 12
Whitesinger, Pauline, 203n81
wiccans, 186n61
Wikipedia, 173n34
Willink, Roseann S., 200–201nn76–77
Wilson, E. O., 213–14n9
winter, as storytelling season, 170n11
Witherspoon, Gary, 201n77
woman, women: angry old, 185n56; "Anthropos" and, 183n45; Crochet Coral Reef and, 76, 78, 89; DES daughters, 105, 106; as elders, 170n11; estrogen and, 6, 105–6, 110–11; feminists and, 6; as gatherer, 177n27, 213n4; as health activists, 7, 210n2; as mothers, 83; murdered and missing aboriginal, 165; Navajo weaving and, 89–93; pregnant Canadian, 111, 115; thinking and, 130; working, 113
Women in Resistance (Black Mesa), 95
Women's Health Initiative, 112
women's health movements, 105, 111, 113, 115; DES and, 107; HorseAid and, 114
Women Who Make a Fuss: The Unfaithful Daughters of Virginia Woolf, 130
Woolf, Virginia: "Think we must!" 130, 173n epigraph 2, 175n10; going visiting and, 130
World-Ecology Research Network, 100

world games, 202n79; as science-art worlding, 86; *Never Alone* as, 86–89; as new genre, 87; sympoietic collaborations with indigenous storytellers, 86, 87
worlding, worldings: activist, 76, 79; alignment in, 42; animal-human, 172n30; art-science, 67, 79; of Burning Man, 182n45; Butler's, 119; of Camille stories, 144, 160; of Chthulucene, 55; companion species and, 110, 118; of coral reef, 56; culture in, 13; curves of, 192n24; cyborg, 115; Despret's, 7, 127; dragon time in, 118; earthly, 97; England's industrial revolution and, 48; evolutionary ecological developmental process of, 212n1; game of living and dying and, 29, 40, 116; Heideggerian, 11; historical, 50; Holocene and, 192n28; human-butterfly, 152, 155; imaginative, 185n53; kin stories and, 114; Le Guin's, 101; linked, 52; matter and, 120; of microbes, 63; multinaturalist, 154; multispecies, 105, 225n32; nature in, 13, 50; of ongoing chthonic powers, 180n36; Oya's power for, 119; pigeons and, 10, 16; place- and travel-based, 182n43; practices of, 86, 88; Premarin in, 115; Protestant, 176n12, 179n35; realm of play and, 23, 24; reimagining, 217n4; of science art, 64, 67, 69, 71–72, 86, 97; SF, 12, 14, 178n32; situated, 87, 91; storying and, 13, 87, 119; as string figure games, 10, 25, 55; sym, 160; sympoietic, 58, 76, 88; tentacular, 42; terran, 61, 105; unfinished, 52; weaving as, 96
worlding, reworlding, 100, 192n28; linked, 52; rendering-capable for flourishing and, 96; in sixteenth and seventeenth centuries, 48
World Wildlife Fund, 193n32
Wright, Patricia, 85, 195–96n58
writing practices: collaborative, 136; in SF, 134, 136; writing without words, 202n79
Wu, Chia-Ling, 216n4
Wurundjeri, 26–27
Wurundjeri Tribe Land Compensation and Cultural Heritage Council, 27
Wyeth-Ayerest, 112

Xena Warrior Princess, 76, 195n44
xkcd, "Bee Orchid," 69, 70

Yarra River, 26, 27
Yellowstone to Yukon Conservation Initiative, 218n9

Zahavi, Amotz, 127–28
Zapatista movement, 155, 156, 173, 225n32, 226n35
zebrafish (*Danio rerio*), as biological model, 63
Zeus, 54
Zolbrod, Paul G., 200–201nn76–77
zooanthellae, symbiosis with cnidarians, 45, 54, 56, 72
zoos, 7, 13, 111
Zubot, Jesse, 165